Coherent Vibrational Dynamics

PRACTICAL SPECTROSCOPY
A SERIES

PRACTICAL SPECTROSCOPY SERIES VOLUME 36

Coherent Vibrational Dynamics

Edited by

Sandro De Silvestri
Giulio Cerullo
Guglielmo Lanzani

CRC Press
Taylor & Francis Group
Boca Raton London New York

CRC Press is an imprint of the
Taylor & Francis Group, an **informa** business

CRC Press
Taylor & Francis Group
6000 Broken Sound Parkway NW, Suite 300
Boca Raton, FL 33487-2742

First issued in paperback 2019

ISBN-13: 978-1-57444-650-0 (hbk)
ISBN-13: 978-0-367-38815-7 (pbk)

Library of Congress Cataloging-in-Publication Data

Coherent vibrational dynamics / editors, Sandro De Silvestri, Giulio Cerullo, and Guglielmo Lanzani.
    p. cm. -- (Practical spectroscopy)
    Includes bibliographical references and index.
    ISBN 978-1-57444-650-0 (alk. paper)
    1. Vibrational spectra. 2. Laser spectroscopy. I. De Silvestri, Sandro. II. Cerullo, G. III. Lanzani, Guglielmo. IV. Title.

QD96.V53C64 2008
539'.6--dc22
                                                2007026621

Visit the Taylor & Francis Web site at
http://www.taylorandfrancis.com

and the CRC Press Web site at
http://www.crcpress.com

Contents

Preface

Vibrational spectroscopy is a powerful investigation tool for a wide class of materials covering diverse areas in physics, chemistry, and biology. Over the years, several techniques have been developed in the stationary regime as well as with temporal resolution. In the stationary regime, Raman and infrared spectroscopy for many years have been providing important information on frequencies and linewidths of vibrational modes coupled to the ground electronic state. Time-resolved techniques such as coherent anti-Stokes Raman scattering have been developed to study excited state vibrations and dephasing. In this technique, two excitation pulses centered at two different frequencies provide the driving force to the vibrational mode with a frequency equal to the difference frequency, and a third pulse is scattered by the induced vibrational population and upshifted in frequency.

The continuous development in the laser field regarding ultrashort pulse generation has led to the possibility of producing light pulses ranging from 50 to 5 fs and frequency tunable from the near infrared to the ultraviolet range. Such pulses enable following vibrational motion coupled to the electronic transitions in molecules and solids in real time. A short light pulse excites in phase many levels of a vibrational mode of the excited state: their coherent superposition forms a vibrational wavepacket. Periodical motion of localized packets formed in this way along displaced bound coordinates can in turn be followed in emission or absorption and allows deducing the vibrational dynamics of the excited state. The short excitation pulse can also induce vibrational coherence in the ground state due to a mechanism known as impulsive stimulated Raman scattering. In this mechanism, the wavepacket generated in the excited state is stimulated back down to the ground state in a nonequilibrium position and begins oscillating. The time domain observation of these vibrational motions can be performed by pump-probe experiments or more sophisticated spectroscopic schemes using multiple pulses. Using this technique, vibrational frequencies as high as 2100 cm^{-1} (corresponding to a period of 16 fs) have been directly observed in the time domain. More recently, due to the development of midinfrared sources of ultrashort pulses, direct excitation of vibrational transitions and measurement of vibrational lifetime have become possible.

This book is addressed to researchers involved in the use of time domain vibrational spectroscopy in a broad range of fields, extending from physics of matter to chemistry and to biology. It provides introductory chapters as well as more advanced contents reporting on recent progress. It is supplemented with a great number of references. These features make this book suitable for both graduate students and senior researchers. It is also a good starting point for those scientists who want to enter this exciting new field with basic knowledge about ultrafast optics and spectroscopic techniques.

The book starts with a chapter on the principles and tools used on time domain vibrational spectroscopy, which involves the coherent optical excitation of an
ensemble of molecules by a short light pulse and the study of the dynamics of their
collective vibrational motion using a suitably delayed pulse (or pulse sequence). A
coherent excitation can be obtained by impulsively exciting a molecular vibration
with light pulses of duration much shorter than the vibrational period. A compre-
hensive overview of the experimental tools and methods commonly used in time
domain vibrational spectroscopy is provided with the basic concepts of ultrafast
optics to facilitate the understanding of the more specialized material presented
in the following chapters. A description of the main properties of ultrashort light
pulses, the concept of dispersion, and the most common pulse characterization
techniques is given. Then, the modern methods for ultrashort pulse generation
from the mode-locked Ti:sapphire oscillator to chirped pulse amplification, spectral broadening by self-phase modulation, and optical parametric amplification
are discussed. The pump-probe technique, which is employed in several of the
following chapters, is explained, focusing in particular on the use of very short
pulses (5–10 fs), which are required for the excitation and the observation of
high-frequency vibrational modes. Finally, a simple impulsive coherent vibrational spectroscopy experiment, providing an explanation of the mechanisms of
excited and ground-state vibrational wavepacket generation, is described in detail.

Hydrogen bonding represents a local interaction of specific functional groups,
which determines the fluctuating structure of liquids and plays a key role for
elementary chemical reactions such as hydrogen and proton transfer. Chapter 2
provides a review of coherent vibrational dynamics of hydrogen bonds in liquids.
Both coherent nuclear motions (vibrational wavepackets) as well as processes of
vibrational dephasing have been studied by ultrafast pump-probe and photo-echo
techniques. The dynamics of hydrogen-bonded systems cover a wide range in time,
from a few tens of femtoseconds up to tens of picoseconds. Techniques for the
generation and characterization of ultrashort pulses in the midinfrared in a wide
range of pulse parameters are fully discussed, followed by a brief introduction to
the methods applied to nonlinear vibrational spectroscopy. A description of recent
results on wavepacket dynamics along intra- and intermolecular hydrogen bonds is
provided, giving a quantitative analysis of the anharmonic coupling underlying the
generation and detection of such wavepackets. The time scale and the mechanisms
of dephasing of the high-frequency O–H stretching excitations and low-frequency
modes in hydrogen bonds are also discussed.

Polyenes and carotenoids are found in living systems, as photoreceptors. Their
role in complex biological processes, the structure–property relationship, and the
detailed electronic structures are subjects of intensive and exciting research. Conjugated polymers are molecular semiconductors that have large technological impact on industrial areas such as lighting, display, automotive, telecommunica-
tion, energy conversion, energy storage, and more. The object of Chapter 3 is
the time-domain investigation of phonon dynamics in carbon-based π-conjugated
materials, including polyenes, carotenoids, and semiconducting polymers. All
have in common π-electron delocalization along the carbon skeleton, which is

responsible for the electronic properties and the coupling to the chain vibrations. Focus is on the specific issue of electron–phonon coupling and vibrational dynamics in carbon-conjugated chains. Electron–phonon coupling governs excited-state relaxation, energy dissipation, thermal properties, and transport. The chapter provides a general introduction to the subject of coherent phonons, which includes quantum mechanics principles and experimental techniques, and some examples regarding their observation in linear conjugated chains. Quantum mechanical calculations are used for reproducing the wavelength-dependent signal due to coherent phonons. By investigating a simple case, the connection between standard Raman and coherent phonon spectra is shown, elucidating the differences and the added values of the latter. The localized nature of the vibrational wavepacket is the key issue in time domain experiments because it allows following the quasi-classical trajectory of motion in the phase space and then the local mapping of the potential energy surfaces. This reveals details on the dynamics of the geometrical evolution that cannot be achieved with other spectroscopy techniques.

The impulsive excitation and phase-sensitive detection of coherent phonons in semiconductors and semiconductor nanostructures provides new insight into light–electron–phonon interaction on a femtosecond time scale. Different excitation mechanisms for the coherent lattice vibrations can be distinguished depending on the peculiarities of the optical excitation. Fingerprints of the excitation mechanisms are the initial phases of coherent phonons and the dependence on excitation fluence, wavelength, and polarization. The simultaneous excitation of charge carriers and coherent phonons gives rise to coupled plasmon–phonon modes that can be studied under nonequilibrium conditions. In low-dimensional systems such as semiconductor quantum wells and superlattices, the electronic density of states as well as the phonon dispersion may be strongly modified, leading to zone-folding effects and new electron–phonon resonances. The phase- and amplitude-resolved detection of coherently excited electronic wavepackets and phonons provides a new access for the study of electron–phonon interaction under resonance conditions. In Chapter 4, the fundamental aspects and recent developments in the field of coherently excited phonons in semiconductors and semiconductor heterostructures are summarized. The physics of coherent phonon states in various condensed matter systems underlies a multitude of excitation and detection mechanisms that have been shown to be distinguishable by specific experimental techniques. It is again worth noting here that, in contrast to continuous wave Raman scattering, combined carrier and phonon dynamics can be studied under nonequilibrium conditions, thus providing profound insight into their interaction processes.

A molecular vibrational wavepacket consists of coherently coupled vibrational eigenfunctions in a molecular electronic potential. *Coherence* means, in this context, that the phase relation among the corresponding eigenfunctions is well-defined and does not change in an arbitrary way. In general, any environment has the potential to destroy the well-defined relative phase among molecular eigenstates, and the wavepacket can decay into an incoherent statistical superposition of eigenstates. These processes are crucial on all levels of complexity: from the simple diatomic molecule in solution to the most complicated biological molecules in their functional environment. Chapter 5 aims at establishing molecule–solvent interactions (energy relaxation, coherence decay) for the conceptually simple system of diatomic halogens in a rare gas solid, which can also be simulated in great detail. Such small molecules as I_2 and Br_2 in the gas phase show long-lasting vibrational and rotational wavepackets, indicating a vibrational and rotational coherence in the several hundred picosecond range. The anharmonicity effects of electronic potentials on vibrational wave packets and a dispersion (anharmonicity-induced broadening) followed by wavepacket revivals are known. This phenomenon is used in this chapter to gain detailed insight into vibrational and electronic coherence decay for such small molecules in a rare gas solid. The rare gas solids have a large ionization energy, allowing use of high-intensity laser pulses. The rare gas atoms are chemically inert; thus, molecules do not react after photoexcitation, and adopted gas phase potentials can be used for the molecular dopants.

The last chapter describes the coherent vibrational dynamics of exciton self-trapping in quasi-one-dimensional systems. The localization of electronic excitations via electron–lattice interactions is an important process in a wide range of condensed matter systems and has a dramatic impact on the optical and transport properties of materials. These localized states not only have clear technological importance, but they also reflect fundamental interactions in the physics of condensed matter systems through the interplay of electron–electron and electron–phonon interactions. Quasi-one-dimensional materials and are ideal systems for studying the localization process since their reduced dimensionality can lead to strong electron–phonon interactions, and the linear structure of the materials simplifies the dynamical configuration space in that the dominant motion is expected to occur along the linear axis. This chapter focuses specifically on transient absorption studies of photoexcitation dynamics in a class of quasi-one-dimensional materials, and the halide-bridged mixed-valence transition metal linear chain (or MX) complexes, which have proven to be excellent model systems for investigating the physics of electron–lattice interactions in low-dimensional systems. The vibrational periods for the metal–halide vibrational modes along the chain axis are relatively long, allowing the motions to be resolved in detail with femtosecond impulsive excitation techniques. These studies of exciton self-trapping in the MX materials provide an unusually clear observation of excited-state vibrational wavepacket dynamics associated with a photoinduced structural change. The dynamics of MX materials have also attracted the attention of theorists and have provided fertile ground for interaction of experimental and theoretical work.

We would like to express our gratitude to all the contributing authors. Without their expertise and work none of this would have been possible.

Giulio Cerullo

Guglielmo Lanzani

Sandro De Silvestri

Editors

Giulio Cerullo was born in 1965 in Italy. He received a degree in electrical engineering from the Politecnico di Milano in 1988. From 1989 to 1991, he was with Baasel Lasertechnik GmbH (Starnberg, Germany), developing laser systems for medical applications. From 1991 to 1999, he was staff researcher with the Department of Physics, Politecnico di Milano. In the years 1995–1996, he visited the Lawrence Berkeley Laboratory (Berkeley, CA) with a NATO Advanced Research Fellowship. In 1999, he became associate professor of general physics with the Department of Physics, Politecnico di Milano. His research activity has focused mainly on the physics and applications of ultrashort pulse lasers, covering a wide range of aspects. His current scientific interests concern (1) generation of broadly tunable few-optical-cycle pulses by optical parametric amplifiers, (2) ultrafast spectroscopy with time resolution down to a few femtoseconds, (3) ultrafast nanooptics, and (4) glass microstructuring by femtosecond pulses. He is the author or coauthor of about 150 papers in international peer-reviewed journals and has given many invited presentations at international conferences. Since 2006, he has been topical editor of the journal *Optics Letters* for the area of ultrafast optical phenomena. He is coordinator of the EU project HIBISCUS, dealing with the integration of femtosecond laser written optical waveguides and microfluidic channels.

Guglielmo Lanzani was born in 1962 in Italy. He graduated in solid-state physics from the University of Milan, Italy, in 1987 and received a doctorate in chemical physics (working on NLO in conjugated polymers) from the University of Genova in 1991. Since 1999, he has been associate professor in physics at Dipartimento di Fisica, Politecnico di Milano (Italy). Starting 1993, he was a researcher (permanent) at Istituto di Matematica e Fisica of the University of Sassari (Italy). His research activity is focused on the experimental investigation of carbon-based π-conjugated materials, with particular interest in elementary excitation dynamics in the ultrafast time domain. He has coauthored about 150 articles and several reviews in journals and books and has given about 50 invited talks at international conferences and institutions. He has three patents on application of organic materials. He is coordinator of the EU project POLYCOM within the IST-FET program. He is involved in two EU networks and several national projects.

Sandro De Silvestri was born in 1951 in Italy. He received a degree in nuclear engineering from the Politecnico of Milan in 1976. From 1977 to 1987, he was a researcher scientist with the National Research Council. In 1987, he become a member of the faculty of Politecnico of Milan as associate professor in physics. Since 1994, he has been a full professor in the same university's Department of Physics. He has made a number of significant contributions to the field of ultrafast phenomena, over a period of about 25 years, on a variety of topics, such as (1) coherent vibrational spectroscopy, (2) development of techniques for the generation of few-optical-cycle pulses either with high energy or tunable from near-infrared to visible, (3) study of ultrafast dynamics in organic and quantum confined systems, (4) carrier envelope phase effects on strong field photoionization and high-order harmonic generation, and (5) generation of attosecond pulses.

His research activity has led to the publication of more than 200 papers in international scientific journals and numerous conference presentations. He has been topical editor of *Optics Letters* in the area of ultrafast phenomena (1998–2003) and is a fellow of the Optical Society of America and member of the European Physical Society (EPS). Since 1998, he has been a member of the board of the Quantum Electronics and Optics Division of EPS, which coordinates activities in Europe in the field of optics (president of the same board 2002–2004). He has belonged to the steering committee of CLEO/Europe-EQEC conference since 1998, sewing as program co-chair of CLEO/Europe 1998 (Glasgow, UK) and general co-chair of CLEO/Europe 2000 (Nice, France). Presently, he is chair of the steering committee. He is director of the European Laser Facility Centre for Ultrafast Science and Biomedical Optics (CUSBO) at the Department of Physics of Politecnico di Milan. This center provides access to European research groups for doing research in a broad area of laser applications. CUSBO is part of LASERLAB-Europe, a network of European facilities. He is director of the National Laboratory for Ultrafast and Ultraintense Optical Science (ULTRAS) and belongs to the National Institute for Physics of Matter (National Research Council), a newly established infrastructure devoted to the developments of ultrashort laser sources, from near-infrared up to the X-ray spectral region, for applications in physics of matter.

Contributors

Giulio Cerullo
National Laboratory for
 Ultrafast and Ultraintense
 Optical Science
Dipartimento di Fisica
Politecnico di Milano
Milan, Italy

Thomas Dekorsy
Universität Konstanz
Fachbereich Physik
Konstanz, Germany

Susan L. Dexheimer
Department of Physics
 and Astronomy
Washington State University
Pullman, Washington

Thomas Elsaesser
Max Born Institut für Nichtlineare
 Optik und Kurzzeitspektroskopie
Berlin, Germany

Michael Först
Institut für Halbleitertechnik
RWTH Aachen University
Aachen, Germany

Markus Gühr
Institut für Experimental-Physik
Freie Universität
Berlin, Germany
 now at
Stanford PULSE Center
Stanford University and SLAC
Stanford, California

Guglielmo Lanzani
Dipartimento di Fisica
Politecnico di Milano
Milan, Italy

Cristian Manzoni
National Laboratory for Ultrafast
 and Ultraintense Optical Science
Dipartimento di Fisica
Politecnico di Milano
Milan, Italy

1 Time Domain Vibrational Spectroscopy: Principle and Experimental Tools

Giulio Cerullo and Cristian Manzoni

CONTENTS

1.1 INTRODUCTION

Time domain vibrational spectroscopy involves the *coherent* optical excitation of an ensemble of molecules by a short light pulse and the study of the dynamics of their collective vibrational motion using a suitably delayed pulse (or pulse sequence) to visualize it. To impulsively excite a molecular vibration and to visualize the ensuing dynamics, light pulses with duration much shorter than its period are required.

In this chapter, we give a brief overview of the experimental tools and methods commonly used in time domain vibrational spectroscopy. This review is not meant to be exhaustive but should provide the basic concepts of ultrafast optics and facilitate understanding of the more specialized material presented in the following chapters. In Section 1.2, we discuss the main properties of ultrashort light pulses (Section 1.3.1), introduce the concept of dispersion, present methods for its control (Section 1.2.2), and describe the most common pulse characterization techniques (Section 1.2.3). In Section 1.3, we outline the modern methods for ultra-short pulse generation, from the mode-locked Ti:sapphire oscillator (Section 1.3.1) to chirped pulse amplification (Section 1.3.2), spectral broadening by self-phase modulation (Sections 1.3.3 and 1.3.4), and optical parametric amplification (Sections 1.3.5 and 1.3.6). In Section 1.4, we describe the pump-probe technique, which is employed in several of the following chapters, focusing in particular on the use of very short pulses (5–10 fs), which are required for the excitation and the observation of high-frequency vibrational modes. Finally, in Section 1.5 we introduce a simple impulsive coherent vibrational spectroscopy experiment, pro-viding an explanation of the mechanisms of excited- and ground-state vibrational wavepacket generation.

1.2 ULTRASHORT LIGHT PULSES: MANIPULATION AND CHARACTERIZATION

In time domain vibrational spectroscopy, the system under study interacts with a sequence of ultrashort light pulses. For full understanding of the details of the experimental techniques, it is therefore important to recall the main properties of femtosecond light pulses, the effects of their propagation in a linear optical system, and the methods for their characterization.

1.2.1 ULTRASHORT LIGHT PULSES

The electric field of an ultrashort light pulse can be written as [1]

$$\tilde{E}(t) = \Re\{E(t)\} = \Re\{A(t) \exp[j\omega_0 t + j\varphi(t)]\} \tag{1.1}$$

where $A(t)$ is the slowly varying pulse amplitude, ω_0 is the carrier frequency, and $\varphi(t)$ is the temporal phase; use of the complex notation simplifies the mathematical treatment and allows focusing only on $E(t)$. As an example and to ease calculations, one can consider a pulse with a Gaussian profile:

$$E(t) = A_0 \exp\left(-t^2/2\tau_p^2\right) \exp(j\omega_0 t) \tag{1.2}$$

where τ_p is the 1/e half-width of the pulse intensity. The full width at half maximum (FWHM) intensity pulsewidth can then be calculated as

$$\Delta\tau_{FWHM} = 2\sqrt{\ln 2}\,\tau_p \tag{1.3}$$

By taking a Fourier transform of $E(t)$, one obtains the pulse spectrum:

$$E(\omega) = \int\limits_{-\infty}^{+\infty} E(t) \exp(-j\omega t)\, dt = A(\omega) \exp[-j\phi(\omega)] \tag{1.4}$$

where $A(\omega)$ is the spectral amplitude, peaked at the carrier frequency ω_0, and $\phi(\omega)$ is the spectral phase. For the case of the Gaussian pulse, one gets

$$E(\omega) = \sqrt{2\pi} A_0 \cdot \exp\left[-\frac{\tau_p^2}{2}(\omega - \omega_0)^2\right] \tag{1.5}$$

To describe a pulse, it is useful to expand the spectral phase in a Taylor series around the carrier frequency ω_0:

$$\phi(\omega) = \phi(\omega_0) + \phi'(\omega_0)(\omega - \omega_0) + \frac{1}{2}\phi''(\omega_0)(\omega - \omega_0)^2 + \frac{1}{6}\phi'''(\omega_0)(\omega - \omega_0)^3 + \cdots \tag{1.6}$$

Usually, an expansion up to the third order is sufficient unless very broadband pulses are considered. The derivative of the spectral phase with respect to frequency is called the *group delay*: $\tau_g(\omega) = d\phi/d\omega$. The group delay has a straightforward physical interpretation: it represents the relative arrival time within the pulse of a quasi-monochromatic wavepacket at frequency ω. $D_2 = \phi''(\omega_0)$ is called *second-order dispersion* or *group delay dispersion* (GDD), and $D_3 = \phi'''(\omega_0)$ is called *third-order dispersion* (TOD). Using this expansion, one can express the group delay as

$$\tau_g(\omega) = \phi'(\omega_0) + D_2(\omega - \omega_0) + \frac{1}{2}D_3(\omega - \omega_0)^2 + \cdots \tag{1.7}$$

When $D_2 = D_3 = 0$, one has $\tau_g(\omega) = \text{const.}$; that is, all the frequency components of the pulse arrive simultaneously. In this case, the pulse is said to be *transform limited* (TL), and it has the shortest possible duration compatible with its spectrum. If the dispersion terms are nonzero, then the frequency components of the pulse arrive at different times, and the pulse is said to be temporally *chirped*. In particular, if $D_2 > 0$, the blue components are delayed with respect to the red ones, and the pulse is said to have *positive chirp* (PC) or *blue chirp*; if $D_2 < 0$, the red components are delayed with respect to the blue ones, and the pulse is said to have *negative chirp* (NC) or *red chirp*.

Linear pulse propagation in a block of material of length L changes its spectral phase, according to the expression

$$\phi_{out}(\omega) = \phi_{in}(\omega) + L\frac{n(\omega)}{c}\omega \tag{1.8}$$

introducing a second-order dispersion, which can be written as

$$D_2 = \phi''(\omega_0) = \frac{\lambda^3}{2\pi c^2}\frac{d^2n}{d\lambda^2}L \tag{1.9}$$

Second-order dispersion is positive in all materials in the visible wavelength range, while it becomes negative in the near infrared (IR; for example, in fused silica $D_2 < 0$ for $\lambda > 1.2\ \mu m$). Let us now study the simple case of propagation of a Gaussian pulse in a dispersive medium, considering dispersion only up to the second order. Assuming a flat spectral phase at the input $[\phi_{in}(\omega) = 0]$ and substituting (1.8) into (1.4), one obtains [2, 3]

$$E_{out}(\omega) = E_{in}(\omega) \exp[-j\phi_{out}(\omega)]$$
$$= \exp\left[-j\phi(\omega_0) - j\tau_g(\omega_0)(\omega - \omega_0) - \frac{\tau_p^2}{2}(\omega - \omega_0)^2 \left(1 + \frac{jD_2}{\tau_p^2}\right)\right]$$

$$(1.10)$$

which can be easily Fourier transformed back to the time domain as

$$E_{out}(t) = \frac{2\pi A_0 \tau_p}{\left(\tau_p^2 + jD_2\right)^{1/2}} \exp j[\omega_0 t - \phi(\omega_0)] \exp\left[-\frac{(t - \tau_g)^2}{2(\tau_p^2 + jD_2)}\right] \quad (1.11)$$

Equation (1.11) shows that the pulse envelope is temporally translated by the group delay $\tau_g(\omega_0)$. By moving to a new time frame $t' = t - \tau_g(\omega_0)$ and by some straightforward manipulations, one can rewrite (1.11) as

$$E_{out}(t') = A_{out}(t') \exp[+j\phi_{out}(t')] \quad (1.12)$$

The pulse amplitude can be written as

$$A_{out}(t') = \frac{2\pi A_0 \tau_p}{\left(\tau_p^4 + D_2^2\right)^{1/2}} \exp\left(-\frac{t'^2}{\tau_{out}^2}\right) \quad (1.13)$$

where

$$\tau_{out} = \tau_p \left[1 + \left(\frac{D_2}{\tau_p^2}\right)^2\right]^{1/2} \quad (1.14)$$

while the pulse phase becomes

$$\phi_{out}(t') = \omega_0 t' - \phi(\omega_0) + \frac{D_2 t'^2}{2(D_2^2 + \tau_p^4)} \quad (1.15)$$

Equations (1.14) and (1.15) show the main effects of pulse propagation in a dispersive medium. First, the pulse experiences a temporal broadening (irrespective of the sign of dispersion). For large dispersion or short pulsewidths, when $D_2 \gg \tau_p^2$, Eq. (1.14) can be approximated as $\tau_{out} \approx \frac{D_2}{\tau_p}$, indicating that a short pulse is more susceptible to dispersive broadening. This can be intuitively understood by considering that a short pulse contains many frequencies, which travel with different group velocities in the medium. To illustrate the effects of dispersive pulse broadening, Figure 1.1 plots the output pulsewidth, as a function of input pulsewidth, after propagation in a 1-cm block of fused silica ($D_2 = 362$ fs^2) or a 1-cm block of SF18 ($D_2 = 1543$ fs^2) at 800-nm carrier wavelength. One can see that relatively long pulses ($\Delta\tau_{FWHM} > 100$ fs) are robust with respect to dispersion, while a 10-fs pulse is dramatically broadened.

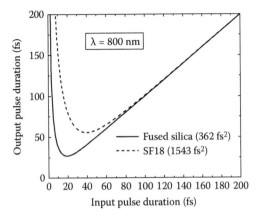

FIGURE 1.1 Output pulsewidth, as a function of input pulsewidth, after propagation in a 1-cm block of fused silica (solid line) or a 1-cm block of SF18 (dashed line) at 800-nm carrier wavelength.

In addition to temporal broadening, dispersion introduces a time-varying temporal phase resulting in a frequency sweep of the pulse. From Eq. (1.15), the instantaneous frequency can be calculated as

$$\omega_i(t') = \frac{d\phi_{out}(t')}{dt'} = \omega_0 + \frac{D_2 t'}{D_2^2 + \tau_p^4} \tag{1.16}$$

and contains a term that varies linearly with time. In particular, one can see that for $D_2 > 0$ the frequency increases with time, so the blue components are delayed with respect to the red ones (PC), while the red components of the spectrum are delayed with respect to the blue ones for $D_2 < 0$ (NC).

1.2.2 DISPERSION MANAGEMENT

Dispersion control is a key issue in the generation and exploitation of ultrashort pulses. In particular, since materials provide a positive dispersion ($D_2 > 0$, $D_3 > 0$) in the visible frequency range, it is important to have optical systems introducing negative dispersion, the so-called pulse compressors. A few such systems are shown in Figure 1.2:

1. A pair of prisms at distance L, adjusted to be at minimum deviation for the incident beam [4, 5] (Figure 1.2a). The angular dispersion of the first prism creates a longer path through the second prism material for the red wavelengths with respect to the blue ones, thus providing negative dispersion. To achieve a negative GDD, the prism distance L must be sufficiently large to compensate for the positive dispersion induced by the prism material. GDD can be coarsely adjusted by setting the distance L and finely tuned by translating one of the prisms along

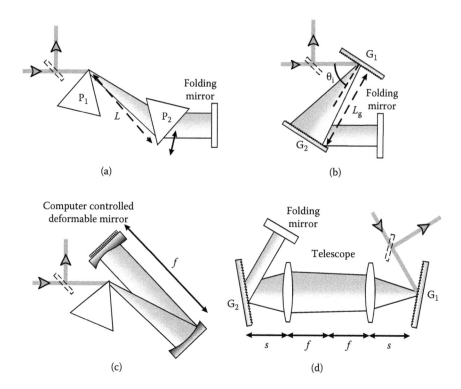

FIGURE 1.2 Optical systems used to control dispersion of ultrashort pulses: (a) prism pair, (b) grating pair, (c) deformable mirror compressor, (d) pulse stretcher.

a direction perpendicular to its base, thus varying the positive material dispersion. To remove the spatial dispersion introduced by the prism pair, the beam is reflected by a mirror and passed again through the prism sequence (if needed, a small vertical offset can be introduced in the second pass to allow extraction of the beam). The prism pair has the advantage of very high throughput: if the prisms have apex angle cut such that at minimum deviation the angle of incidence is Brewster's angle, then there is virtually no reflection for the correct linear polarization, and the system is essentially loss free. On the other hand, this compressor introduces not only a GDD, but also a large negative TOD, which cannot be independently controlled; if the prism pair GDD is set to compensate for that of the material, then the overall dispersion is dominated by TOD. Since both D_2 and D_3 are proportional to the prism separation L, the ratio D_3/D_2 is a characteristic of the prism's material and the wavelength and can be minimized by choosing materials with low dispersion (such as fused silica, MgF_2, or CaF_2). In this case, however, the prism distance required to achieve a given value of D_2 increases. In summary, the prism pair is a simple compressor but cannot be used with very broadband pulses due to the residual TOD.

2. A pair of parallel and identical gratings [6] (Figure 1.2b). For an incidence angle θ_i on the grating, the reflected beam direction is given by the grating equation: $\sin \theta_r = \sin \theta_i + m\frac{\lambda}{d}$, where d is the grating period, and m is the diffraction order. For $m = -1$, the red frequencies travel a longer path with respect to the blue ones (see Figure 1.2b), resulting in negative dispersion. The GDD can be expressed as

$$D_2 = -\frac{\lambda^3 L_g}{\pi^2 c^2 d^2} \frac{1}{\left[1 - \left(\sin \theta_i - \lambda/d\right)^2\right]^{3/2}} \qquad (1.17)$$

where λ is the carrier wavelength and L_g is the gratings distance. Also for the grating pair, spatial dispersion can be removed by a double pass after reflection from a mirror. Grating pairs provide, for a given separation, much greater dispersion with respect to prism pairs and are thus able to introduce large-frequency chirps, as required in the chirped pulse amplification systems. In a grating pair, the sign of TOD is opposite to that of the prisms, so a combined prism-grating compressor can cancel both GDD and TOD [7, 8].

3. The so-called chirped mirrors [9, 10]. These are multilayer dielectric mirrors consisting of a large number (~ 50) of alternating low and high refractive index layers, suitably designed so that the layer thickness increases going toward the substrate. In this way, the high-frequency components of the laser spectrum are reflected first, while the low-frequency components penetrate more into the multilayer, thus acquiring an additional group delay. Suitable computer optimization allows avoiding spurious resonances and designing a custom-tailored, frequency-dependent spectral phase. Chirped mirrors have several advantages: the dispersion can be arbitrarily controlled over broad bandwidths (approaching an octave [11]); they have high energy throughput and are particularly insensitive to misalignment. On the other hand, their fabrication is not trivial since tight tolerances on each layer thickness are required, achievable only by sophisticated techniques such as ion beam sputtering. In addition, the negative GDD obtainable from a single bounce is typically rather small (≈ -50 fs^2), so that many bounces are generally required.

4. Adaptive compressors, in which the spectral phase is actively controlled and optimized for the application. Figure 1.2c shows a compressor using a deformable mirror [12], which is an electrostatically actuated, metal-coated, silicon nitride membrane. The mirror is used in the so-called 4-f arrangement [13]: after a dispersive element (grating or prism), a lens (or a spherical mirror) located at a distance f from the dispersive element performs a Fourier transform that converts the angular dispersion from the grating to a spatial separation in its back focal plane. The deformable mirror, located in the focal plane, allows controlling the delay of each

spectral component of the spatially dispersed broadband beam. Finer control of both amplitude and phase of each spectral component in the Fourier plane can be achieved by using a liquid crystal spatial light modulator [14].

Figure 1.2d also shows an optical system that is important for the amplification of short pulses; it is the so-called pulse stretcher, which consists of a pair of gratings in antiparallel configuration with a 1:1 inverting telescope between them [15]. Calling s the grating lens distance and f the focal length of the telescope lenses, the expression for the GDD of this arrangement is similar to Eq. (1.17), valid for a grating pair, but with

$$L = 2(s - f) \cos \theta_i \qquad (1.18)$$

The stretcher can thus provide a positive GDD for $f > s$. The dispersion of the stretcher is matched to all orders, but with opposite sign, with respect to that of a grating compressor, making this combination useful in the so-called chirped pulse amplification systems (see Section 1.3.2). In practical implementations, to avoid chromatic aberrations and dispersion, the lenses are often replaced by spherical or cylindrical mirrors.

1.2.3 MEASUREMENT TECHNIQUES

The bandwidth of electronic devices, such as photodetectors and oscilloscopes, is limited in the best cases to a few tens of gigahertz; therefore, any optical signal shorter than a few tens of picoseconds cannot be detected by electronic means, but requires optical measurement techniques.

The simplest and most widely used technique for ultrashort pulse characterization is autocorrelation, which uses the light pulse to measure itself. In an autocorrelator, a pulse at the fundamental frequency (FF) is divided into two replicas; one is imparted a variable-delay τ, and both are spatially overlapped in an instantaneously responding nonlinear medium, such as a second harmonic (SH) generation crystal. Being $I_{SH} \propto I_{FF}^2$, the SH signal is higher when the two pulses are temporally overlapped than when they are separated. Measurement of I_{SH} as a function of time delay gives information on the pulsewidth. In the noncollinear autocorrelator shown in Figure 1.3, the two beams form an angle, and the SH crystal is oriented so that a signal is produced only when the phase-matching condition

$$k_1 + k_2 = k_{SH} \qquad (1.19)$$

is satisfied. In this case, the SH signal is proportional to the product of the two signals:

$$I_{SH}(t, \tau) = c \, I(t) I(t + \tau) \qquad (1.20)$$

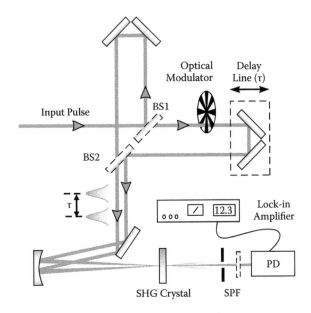

FIGURE 1.3 Schematic of a noncollinear autocorrelator. BS, beam splitters; PD, photodetector; SPF, short-pass filter for rejecting the FF beams.

and the SH energy for a given time delay is

$$E_{SH}(\tau) = c \int\limits_{-\infty}^{+\infty} I_{SH}(t, \tau) \, dt = c \int\limits_{-\infty}^{+\infty} I(t) \, I(t + \tau) \, dt \qquad (1.21)$$

where c is a proportionality factor depending on the crystal nonlinearity. Measurement of the SH energy, selected spatially by an aperture and spectrally by a short-pass filter, as a function of τ gives the autocorrelation function of the pulse intensity. Note that this method does not require a fast detector since only the SH pulse energy is measured (and often averaged over several consecutive pulses for a given time delay) and is background free since the signal goes to zero for well-separated pulses. Modulation of one of the beams by an optical modulator and synchronous detection by a lock-in amplifier are often used to improve the signal-to-noise ratio. Typically, modulation is provided by a mechanical chopper for laser frequencies up to 3 kHz, and by an acousto-optic modulator in the MHz regime.

To extract the pulsewidth from the autocorrelation measurement, one has to make an assumption on the pulse shape: the FWHM pulsewidth is obtained by dividing the FWHM of the autocorrelation by a suitable constant. For the case, often encountered with femtosecond pulses, of a pulse with square hyperbolic secant shape,

$$I(t) = I_0 \, \mathrm{sech}^2 \left(\frac{t}{\tau_p} \right) \qquad (1.22)$$

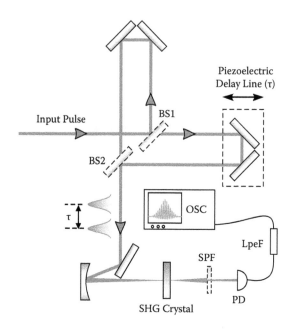

FIGURE 1.4 Schematic of a collinear autocorrelator. BS, beam splitters; SPF, short-pass filter for rejecting the FF beams; PD, photodetector; LPeF, low-pass electronic filter; OSC, oscilloscope.

the constant is 1.55, while for a Gaussian pulse [Eq. (1.2)] it is 1.44 [16]. Note that by definition an autocorrelation is always symmetric, even for an asymmetric pulse shape (although it can become asymmetric if the autocorrelator is misaligned). In addition, the noncollinear autocorrelation does not provide any information on pulse chirp; the noncollinear autocorrelation of a chirped pulse just appears to be longer than expected from the Fourier transform of its spectrum.

Figure 1.4 shows a schematic of the collinear autocorrelator; it is basically a Michelson interferometer, with two collinear and delayed replicas of the pulse. Their interference on the nonlinear crystal gives rise to the SH intensity:

$$I_{SH}(t, \tau) = k| [E(t) + E(t + \tau)]^2 |^2 \tag{1.23}$$

The energy of the SH pulse, spectrally selected from the FF by a spectral filter, is then measured as a function of time delay τ. Due to the interference between the two delayed collinear beams, the autocorrelation signal will rapidly oscillate in time according to the expression

$$E_{SH}(\tau) = \int_{-\infty}^{+\infty} I_{SH}(t, \tau) \, dt = k \, [1 + (2 + 2 \cos 2\omega_0\tau)A(\tau) + 2 \cos \omega_0\tau \, g(\tau)] \tag{1.24}$$

where

$$A(\tau) = \frac{\int\limits_{-\infty}^{+\infty} I(t) I(t + \tau) \, dt}{\int\limits_{-\infty}^{+\infty} I^2(t) \, dt} \qquad (1.25)$$

$$g(t) = \frac{\int\limits_{-\infty}^{+\infty} [I(t) + I(t + \tau)] \sqrt{I(t) I(t + \tau)} \, dt}{\int\limits_{-\infty}^{+\infty} I^2(t) \, dt} \qquad (1.26)$$

and ω_0 is the carrier frequency. The expression inside the square brackets rapidly oscillates in time, at frequencies ω_0 and $2\omega_0$. By choosing the speed of the delay line so that the detection system can resolve those oscillations and rejecting the frequency components at the laser repetition rate by a low-pass electronic filter, one has a fringe-resolved autocorrelation (FRAC). In this case, the term in square brackets in Eq. (1.24) has a maximum value of 8 for $\tau = 0$ [in fact $g(0) = 2$ and $A(0) = 1$] and a background of 1 when the pulses are well separated [$g(\infty) = A(\infty) = 0$].

Advantages of the interferometric autocorrelator are (i) direct and accurate self-calibration provided by the interferometric fringes; (ii) the possibility to check proper alignment by verifying the contrast ratio of the trace; (iii) sensitivity to pulse chirp, which results in a loss of modulation contrast on the tails of the pulse; (iv) absence of pulse smearing due to the noncollinear geometry (which becomes relevant for pulsewidths on the order of 10 fs). Note that it is important that the two pulse replicas created by the Michelson interferometer are exactly equal; to this purpose, the balanced autocorrelator displayed in Figure 1.4 can be used [17], in which both pulse replicas see the same amount of glass and undergo the same number of reflections from the beam splitter.

Autocorrelation techniques provide only limited information on the pulse characteristics. In particular, they require the a priori assumption of a given pulsewidth and do not give direct information on the spectral phase. Other, more sophisticated techniques allow a full reconstruction of the pulse in both amplitude and phase, that is, to access both $A(\omega)$ and $\phi(\omega)$ in Eq. (1.4). The first of such techniques is called frequency-resolved optical gating (FROG) [1, 18, 19]; it consists of measuring an autocorrelation (typically noncollinear) as a function of both time delay and frequency, obtaining a sequence of SH spectra at different time delays:

$$I_{FROG}(\omega, \tau) = \left| \int\limits_{-\infty}^{+\infty} E(t) E(t - \tau) \exp(-j\omega t) \, dt \right|^2 \qquad (1.27)$$

This function is known as a spectrogram because it consists of a series of gated spectra of the pulse, where the variable-delay gate function is given by the pulse

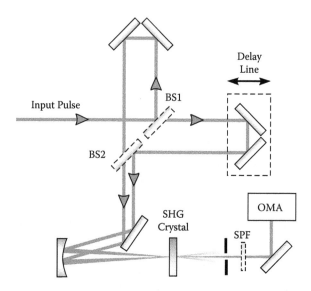

FIGURE 1.5 Schematic of the experimental setup used for FROG measurements. BS, beam splitters; SPF, short-pass filter for rejecting the FF beams; OMA, optical multichannel analyzer.

itself. Recovery of the electric field profile requires inversion of the spectrogram, which can be performed using suitable algorithms.

A typical experimental setup for SH-FROG is shown in Figure 1.5; it is similar to that for a noncollinear autocorrelator except that the photodetector is replaced by a spectrometer. The sequence of SH spectra for different values of the delay τ yields the spectrogram, from which the pulse intensity and phase are determined using a phase-retrieval algorithm. The SH-FROG has the advantages of simplicity and sensitivity, but it has an ambiguity on the sign of the spectral phase; for high-energy pulses, other versions of FROG employing third-order nonlinear processes (third harmonic generation, self-diffraction, polarization gating, etc.) are available that unambiguously provide the sign of the phase.

An alternative technique allowing complete pulse reconstruction in amplitude and phase is called spectral phase interferometry for direct electric field reconstruction (SPIDER) [20,21]. This technique extracts the spectral phase of the pulse from an interferogram obtained from the interference of two spectrally sheared replicas of the pulse to be characterized. Measurement of the spectral phase $\phi(\omega)$ can be combined with measurement of the power spectrum of the pulse by an ordinary spectrometer to obtain the spectral amplitude $A(\omega)$. The combined spectral amplitude and phase information then gives access to the complex electric field of the optical pulse in the frequency domain, from which an inverse Fourier transform returns the temporal amplitude and phase of the pulse.

To retrieve the spectral phase, two replicas of the pulse to be characterized are shifted in frequency by a spectral shear Ω and delayed in time by τ. The resulting spectral interference pattern they generate is

$$S(\omega) = |E(\omega)|^2 + |E(\omega + \Omega)|^2 + 2|E(\omega)E(\omega + \Omega)| \cdot \cos[\phi(\omega + \Omega) - \phi(\omega) + \omega\tau]$$

$$(1.28)$$

where $\phi(\omega)$ is the spectral phase of the input pulse. This interferogram can be easily recorded by a spectrometer. The cosine term of Eq. (1.28) contains all the spectral phase information needed, and it is responsible for the fringes of the interferogram. The argument of the cosine can be extracted from Eq. (1.28) by straightforward Fourier transform and filtering operations. Note that information on the amplitudes of the interferogram is at no point required for analysis of the data. The constant delay τ is determined by separate spectral interferometry of the two pulse replicas. After subtraction of the linear phase term $\omega\tau$, one is left with the phase difference $\phi(\omega + \Omega) - \phi(\omega)$, from which the spectral phase is extracted by integration.

The spectral shear Ω is generated by up-converting two replicas of the pulse with a strongly stretched pulse using sum-frequency generation (SFG) in a non-linear optical crystal. The up-converter pulse has to be temporally stretched such that its instantaneous frequency can be considered constant for the duration of the pulse to be measured. The interferogram is then obtained between the two up-converted pulses, which are exact but frequency-shifted replicas of the pulse to be characterized.

SPIDER is a self-referencing interferometric technique; that is, there is no need for a well-characterized reference. When applied to ultrabroadband pulses, SPIDER has several advantages with respect to other pulse characterization techniques. In particular, the accuracy of spectral phase reconstruction is widely insensitive to the phase-matching bandwidth of the up-conversion crystal and the spectral responsivity of the detector since it depends only on the fringe spacing. In addition, there are a direct, noniterative phase retrieval algorithm and no moving parts in the apparatus. On the other hand, a very accurate measurement of the delay τ between the interfering pulses is required.

A typical experimental setup for a SPIDER capable of measuring very short pulses is shown in Figure 1.6 [22]. A pulse entering the apparatus is divided by reflection onto a thick glass block, which serves as the stretcher for the transmitted pulse. The reflected pulse is sent to a balanced Michelson interferometer, forming two time-delayed replicas. The beams are combined noncollinearly in a nonlinear crystal for frequency up-conversion; the chirp introduced by the glass block is sufficiently high to ensure that each pulse replica is up-converted by a quasi-CW (continuous wave) field. The sum frequency is detected by a high spectral resolution spectrometer, yielding the spectral interferogram corresponding to Eq. (1.28). The time delay τ is determined by measuring the interferogram of the SH of the two pulse replicas generated by the Michelson interferometer.

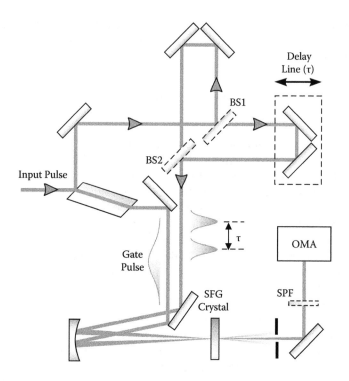

FIGURE 1.6 Schematic of the experimental setup of a SPIDER capable of measuring very short pulses. BS, beam splitters; SPF, short-pass filter for rejecting the FF beams; OMA, optical multichannel analyzer.

1.3 ULTRASHORT PULSE GENERATION METHODS

Early sources of tunable femtosecond optical pulses were based on dye laser technology [23]; by use of spectral broadening in an optical fiber and subsequent amplification, 10-fs pulses with center wavelength tunable in the visible and near-IR range could be obtained [24–26], enabling the first time domain detection of coherent vibrational motion in molecules and solids [27–31]. However, the system complication and the painstaking requirement of changing the laser dye to achieve frequency tunability greatly limited the application of such instruments.

The 1990s witnessed a revolution in ultrafast laser technology thanks to the advent of solid-state active materials, such as Ti:sapphire [32], and powerful mode locking techniques, such as Kerr lens mode-locking (KLM) [33]. With these advances, femtosecond lasers have gained tremendously in reliability and user friendliness, becoming "turnkey" devices available to a wide community of nonspecialists. Another landmark of femtosecond technology has been the chirped pulse amplification (CPA) technique [34, 35], which enables increasing the energy of a femtosecond laser by five to six orders of magnitude, from the nanojoule to the millijoule level. Now, almost all ultrafast spectroscopy systems are based on the Ti:sapphire technology.

KLM Ti:sapphire lasers amplified by the CPA technique are now widely used sources of stable, energetic femtosecond pulses; however, their frequency tunability is limited to a narrow range around the FF of 0.8 μm or around the SH of 0.4 μm. Their tuning range can be greatly extended by optical parametric amplifiers (OPAs) [36], which have become the workhorses for generation of tunable, high-energy femtosecond pulses; they can cover, with the help of frequency doubling and sum and difference-frequency generation stages, a broad spectral range from the ultraviolet to the infrared. A single amplified Ti:sapphire system can drive two synchronized OPAs, providing two independently tunable pulses necessary in pump-probe experiments. Another unique characteristic of the OPAs is their capability of significantly shortening the pulse duration with respect to that of the driving pulses since the minimum pulsewidth achievable from an OPA is limited by the phase-matching bandwidth of the parametric interaction rather than by the pump pulse duration. This feature is exploited especially in the noncollinear OPA (NOPA) pumped by the SH of a Ti:sapphire laser, which can generate sub-10-fs pulses broadly tunable in the visible range [37–42]. OPAs are now the most widely used sources for coherent vibrational spectroscopy [43–50].

In this section, we provide a brief outline of current methods for femtosecond pulse generation, with the aim of giving a basic guideline for understanding the most commonly used experimental tools in coherent vibrational spectroscopy.

1.3.1 Ti:Sapphire Laser Oscillator

Figure 1.7 shows the scheme of a typical Ti:sapphire laser oscillator passively mode-locked by the KLM technique. The Ti:sapphire rod, with 5-mm length and Brewster-cut faces, is enclosed between two curved folding mirrors (100-mm radius of curvature), which tightly focus the laser mode inside the rod; the two end mirrors complete the laser cavity, which has a length of about 1.5 m. One of the two cavity long arms contains a pair of Brewster-cut fused-silica prisms, used to provide a negative net intracavity second-order dispersion necessary for

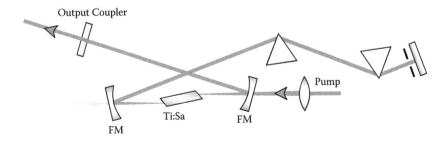

FIGURE 1.7 Schematic of a KLM Ti:sapphire laser oscillator. FM, folding mirror.

the laser to operate in the solitonlike regime. The laser is pumped in the green ($\lambda = 532$ nm) by a CW frequency-doubled Nd:YVO$_4$ laser (typically Verdi from Coherent or Millennia from Spectra Physics) with an average power around 5 W; the pump light is delivered to the rod through one of the folding mirrors. When operated in the CW regime in the middle of the optical stability range, the laser provides output power up to 1 W.

To achieve KLM operation, the laser is brought close to a border of the optical stability range, and the rod is offset with respect to the center of the folding [51]; KLM is initiated by a slight perturbation of the resonator, usually the rapid insertion of one of the intracavity prisms. KLM is a passive mode-locking technique employing a fast saturable absorber. Its basic principle can be explained as follows: during pulsed operation, the peak intensity in the cavity is so high that the refractive index of the Ti:sapphire rod becomes intensity dependent (Kerr effect; see Section 1.3.3). Since the laser mode has a spatial Gaussian intensity profile, the rod acquires a radially varying refactive index, making it equivalent to a lens (Kerr lens). This lens, which is present only during pulsed operation, modifies the laser mode decreasing the resonator losses, either by higher transmission through an intracavity aperture (hard-aperture KLM) or through a better overlap between the laser and the pump mode inside the Ti:sapphire rod (soft-aperture KLM).

It is important that the laser operates in a regime of negative net intracavity dispersion, where solitonlike pulse shaping takes place. In this case, the nonlinear self-phase modulation (SPM) in the Ti:sapphire rod and dispersion are balanced, while the pulsewidth can be calculated as [52, 53]

$$\Delta\tau_p = k \frac{|D_2|}{\delta E_p} \tag{1.29}$$

where k is a numerical factor, D_2 is the GDD for a round-trip in the laser cavity, δ is the nonlinear round-trip phase shift per unit power in the Kerr medium, and E_p is the intracavity pulse energy. In this regime, solitonlike pulse shaping determines the pulse duration, while KLM provides the self-amplitude modulation needed to stabilize the cavity against perturbations.

A typical KLM Ti:sapphire oscillator as the one described can deliver output powers of 500 mW (corresponding to a 5 nJ pulse energy for a 100-MHz repetition rate) with pulsewidths down to 10 fs (in this case, the pulse bandwidth, centered at 800 nm, ranges from 750 to 850 nm) [54, 55]. More sophisticated systems, using chirped dielectric mirrors for intracavity dispersion control, can generate much shorter pulses, with duration down to 5 fs [56, 57], exploiting the full gain bandwidth of the Ti:sapphire active material. The output pulse energy can be increased to the 100-nJ level using either cavity dumping [58, 59] or multipass stretched cavities [60, 61].

1.3.2 CHIRPED PULSE AMPLIFICATION

The energy of the pulses produced by a Ti:sapphire oscillator, on the order of few nanojoules, is not sufficient for many experiments. Straightforward amplification

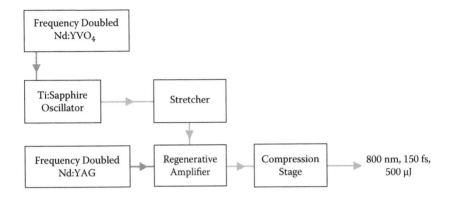

FIGURE 1.8 Principle scheme of a femtosecond laser system with chirped pulse amplification.

of these pulses runs into the problem that their peak power becomes so high that it induces nonlinear optical effects, causing optical damage to the amplifier components; this problem can be solved using the CPA technique. A principle scheme of a CPA system is shown in Figure 1.8. The pulses from the oscillator are first lengthened in a dispersive delay line (the so-called stretcher) by three to four orders of magnitude, achieving a corresponding decrease in their peak power; note that the pulses acquire a large frequency chirp while their bandwidth is preserved. The stretched pulses are then sent to an amplifier, which increases their energy by many orders of magnitude, from the nanojoule to the millijoule level, while their repetition rate is reduced to the kilohertz range. After amplification, the pulses are sent to a *compressor*, introducing a dispersion of opposite sign with respect to the stretcher, which brings the pulsewidth back to the original duration. It is important that the stretcher and the compressor are matched, introducing dispersion of opposite signs but equal to all orders. Standard CPA systems use a positive dispersion stretcher, consisting of a grating pair in antiparallel configuration with a 1:1 inverting telescope between the gratings (see Section 1.2.1), and a negative dispersion compressor, consisting of a grating pair.

There are two basic schemes for the amplifier: regenerative [62, 63] and multipass [64, 65]. In both cases, the pump beam is provided by a CW-pumped, Q-switched frequency doubled Nd:YAG (or Nd:YLF) laser delivering 10- to 20-mJ, 200-ns green pulses at 1-kHz repetition rate.

A scheme of a regenerative amplifier setup is shown in Figure 1.9; it consists of a Ti:sapphire amplifier and an electro-optic Pockels cell, oriented to produce a static $\lambda/4$ retardation, located in a three-mirror folded resonator. Its operation can be understood as follows: a vertically polarized pulse coming from the stretcher is reflected into the cavity by polarizer P and, after a double pass through the Pockels cell, becomes horizontally polarized and is thus transmitted by polarizer P toward the Ti:sapphire amplifier. After returning from the amplifier, the pulse double passes the Pockels cell again, becomes vertically polarized, and is reflected

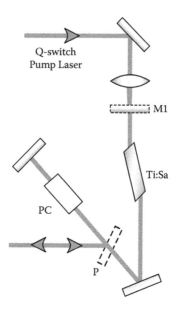

FIGURE 1.9 Setup of a femtosecond Ti:sapphire regenerative amplifier. PC, Pockels cell; P, polarizing beam splitter.

by the polarizer out of the cavity. In this regime of operation, each pulse passes only twice through the Ti:sapphire rod so its amplification is negligible. If, however, a $\lambda/4$ voltage is applied to the Pockels cell while the pulse is between the polarizer and mirror M_1, the cell becomes equivalent to a $\lambda/2$ plate and does not change the polarization state of the pulse after a double pass. In this case, one pulse gets trapped in the regenerative amplifier and is amplified on each pass through the Ti:sapphire crystal, while the following ones are not coupled into the cavity. After a suitable number of round-trips in the cavity (typically 15 to 20), the pulse reaches its maximum energy value and can be extracted from the cavity by applying an additional $\lambda/4$ voltage to the PC. In this case, in fact, after a double pass through the cell, the pulse becomes vertically polarized and is ejected out of the cavity by the polarizer.

The multipass amplifier is in principle much simpler: a pulse, temporally gated by a Pockels cell, makes several passes (typically 8 to 10) in a Ti:sapphire amplifying crystal, with a slight spatial offset at each one, before being picked off and ejected. Both the multipass and regenerative amplifier concepts have their merits. The multipass allows the extraction of higher pulse energies due to the significantly lower losses and the generation of shorter pulses due to the reduced material dispersion during the amplification process. On the other hand, the regenerative amplifier has generally better mode quality, which is defined by the regenerative amplifier cavity, and higher energy and pointing stability. In general, the regenerative amplifier is preferred in time-resolved spectroscopy applications, in which stability is of paramount importance; in this case, pulses generated by CPA systems are often shortened by other means, such as the use of OPAs (see Section 1.3.5).

1.3.3 SUPERCONTINUUM GENERATION

Pulses produced by KLM Ti:sapphire oscillators are tunable in a broad spectral range, from 700 to 1000 nm; however, CPA systems can operate only in a narrow spectral range around 800 nm due to gain narrowing and spectral gain-shifting effects. The bandwidth of an ultrashort light pulse, however, can be easily increased by an effect known as *supercontinuum* (white light) *generation* [66], occurring when an intense light pulse is focused inside a transparent material, such as fused silica or sapphire. White light generation can be understood as a result of SPM of the pulse. When an intense laser beam impinges on a material of thickness L, its refractive index becomes a linear function of the intensity (Kerr effect):

$$n(t) = n_0 + n_2 I(t) \tag{1.30}$$

so that the phase shift experienced by a quasi-monochromatic pulse of carrier frequency ω_0 is

$$\phi(t) = \omega_0 t - \frac{\omega_0}{c_0} n(t) L = \omega_0 t - \frac{\omega_0}{c_0} n_0 L - \frac{\omega_0}{c_0} n_2 L\, I(t) \tag{1.31}$$

and the instantaneous frequency becomes

$$\omega(t) = \frac{d\phi(t)}{dt} = \omega_0 - \frac{\omega_0 n_2 L}{c_0}\frac{dI}{dt} = \omega_0 \left(1 - \frac{n_2 L}{c_0}\frac{dI}{dt} \right) \tag{1.32}$$

Equation (1.32) suggests that the instantaneous carrier frequency of the pulse changes in time; in particular, during the leading edge of the pulse $(dI/dt > 0)$ it is shifted to the red, while during the trailing edge $(dI/dt < 0)$ it is shifted to the blue, leading to a broadening of the spectrum. Note that this is an oversimplified description of the white light generation phenomenon; in fact, other effects in addition to SPM, such as spatial self-focusing, temporal self-steepening, and space-time focusing, play a role (see [67] and [68] for a detailed description).

Practically, white light generation is typically achieved by focusing 800-nm, 100-fs pulses with energy from 1 to 3 μJ into a sapphire plate, with thickness ranging between 1 and 3 mm. Sapphire is chosen because of excellent thermal conductivity and low UV absorption, preventing long-term degradation. The white light extends throughout the visible (down to $\approx 0.4\ \mu$m) and the near-IR (up to $\approx 1.5\ \mu$m) range, with an energy of approximately 10 pJ per nanometer of bandwidth. Under the correct conditions (i.e., a single self-focused filament), the white light has an excellent spatial quality, with a circular Gaussian beam, and a very high pulse-to-pulse stability.

The white light supercontinuum generated in a sapphire plate can be used as a probe in pump-probe experiments; alternatively, it can be used as the seed pulse for an OPA (see Section 1.3.5).

1.3.4 PULSE COMPRESSION

Although KLM Ti:sapphire oscillators can generate very short pulses, with du-ration down to 5 fs, the bandwidth of conventional CPA amplifiers, limited by gain-narrowing effects, gives high-energy amplified pulses in the 20 to 30-fs range. The duration of both low- and high-energy pulses can be dramatically shortened by so-called pulse compression techniques. In a pulse compressor, first the pulse bandwidth is increased by SPM in a nonlinear medium; then, propagation in a suit-able dispersive delay line rephases all the new frequency components generated by the phase modulation to produce a pulse with TL duration.

SPM can in principle take place in any medium (gas, liquid, or solid) when the peak intensity is sufficiently high. In a bulk medium, however, the spatial intensity profile of the propagating light pulse leads to a spatially nonuniform SPM, so different frequency chirps are generated across the transverse beam distribution [69]. This problem can be solved by propagating the pulse in a guiding medium, such as an optical fiber [70], which homogenizes the beam profile, removing any spatial chirp effects. This concept has been initially implemented with low pulse energies (a few tens of nanojoules). Using a short (≈ 1 cm long) single-mode optical fiber for SPM followed by a prism-grating compressor, pulses as short as 6 fs at 620 nm were obtained from 50-fs pulses generated by an amplified dye laser. More recently, 13-fs pulses at 800 nm from a cavity-dumped Ti:sapphire laser were compressed to 4.5 fs with the same technique using a compressor consisting of a quartz prism pair combined with broadband chirped mirrors [71]. However, the use of a single-mode optical fiber limits the pulse energy to a few tens of nanojoules before the onset of optical damage to the fiber facets.

An alternative technique capable of handling much higher pulse energies uses as phase modulator a hollow fiber filled with noble gases [72]. This approach presents the advantages of a guiding element with a large-diameter mode and of a fast nonlinear medium with high damage threshold (tunnel ionization), enabling the propagation of energies as high as a few hundred microjoules. The implementation of the hollow-fiber technique using 20-fs seed pulses from a Ti:sapphire system and a high-throughput broadband dispersive delay line has led to the generation of pulses with duration down to 4.5 fs [73] and energy up to 0.5 mJ [74].

In a typical hollow-fiber pulse compressor, input pulses with 20-fs duration and energy up to 1 mJ, generated by an amplified Ti:sapphire laser, are coupled into a 60-cm-long, 500-μm diameter hollow fiber, filled with argon at pressures ranging between 0.4 and 1 bar. Wave propagation along hollow fibers can be thought of as occurring by grazing incidence reflections at the dielectric inner sur-face [75]. Since the losses caused by these reflections greatly discriminate against higher-order modes, only the fundamental mode, with large and scalable size, will be transmitted through a sufficiently long fiber. By proper mode matching, transmission efficiencies up to 80% can be achieved. The output pulse spectrum, broadened by SPM in the noble gas, spans the 650- to 950-nm wavelength range. Pulse compression is achieved by multiple bounces on ultrabroadband chirped di-electric mirrors, introducing a nearly constant negative GDD. Typical compressed pulsewidths, characterized by FRAC [73], FROG [76], and SPIDER [77], are around 5 fs; the maximum output energies, considering the overall throughput of the system, are around 0.5 mJ, corresponding to a peak intensity of 0.1 TW.

An alternative high-energy pulse compression technique has been proposed that involves filament formation in a gas-filled cell [78–80]. If the high-energy pulse is loosely focused into the cell, then the filament forms through a dynamic balance between Kerr-lensing and plasma-induced defocusing. The filament has length ranging from a few centimeters to a few tens of centimeters and provides the guiding medium necessary to obtain a homogeneous beam profile, removing the difficulty of coupling into the hollow fiber. By a combination of SPM in the filament and compression using a suitable delay line, high-energy few-optical-cycle pulses can be generated. The filamentation technique can in principle be scaled to higher energies with respect to the hollow fiber, but this gives worse beam quality and requires spatial beam clipping by a suitable aperture to avoid a spatial variation of the temporal properties of the pulse.

1.3.5 OPTICAL PARAMETRIC AMPLIFIERS

The OPAs are powerful devices for extending the tuning range of virtually fixed-wavelength sources, such as amplified Ti:sapphire lasers. The principle of an OPA is quite simple [81, 82]: in a suitable second-order nonlinear crystal, a high-frequency and high-intensity beam (the pump beam, at frequency ω_3) amplifies a lower-frequency, lower-intensity beam (the signal beam, at frequency ω_2); in addition, a third beam (the idler beam, at frequency ω_1) is generated. The OPA process can be given a simple corpuscular interpretation: a photon at frequency ω_3 is absorbed by a virtual level of the material, and two photons at frequencies ω_2 and ω_1 are emitted. In this interaction, both energy conservation

$$\hbar\omega_3 = \hbar\omega_1 + \hbar\omega_2 \qquad (1.33)$$

and momentum conservation

$$\hbar k_3 = \hbar k_1 + \hbar k_2 \qquad (1.34)$$

must be fulfilled. The last condition is also known as *phase matching*, and it guarantees that the signal/idler components originating from different positions in the nonlinear crystal all add up in phase. The signal frequency can in principle vary from $\omega_3/2$ (the so-called degeneracy condition) to ω_3, and correspondingly the idler varies from $\omega_3/2$ to 0 (at degeneracy, signal and idler coincide).

A general scheme of an ultrafast OPA is presented in Figure 1.10. The system is powered by energetic femtosecond pulses, typically coming from a CPA Ti:sapphire laser. Pumping can take place either at the FF or at the SH (i.e., 800 or 400 nm). A fraction of the beam is split and used to generate the seed beam. Since the seed beam is at a different frequency from the pump beam, a nonlinear optical process is required for its production. Two different techniques are

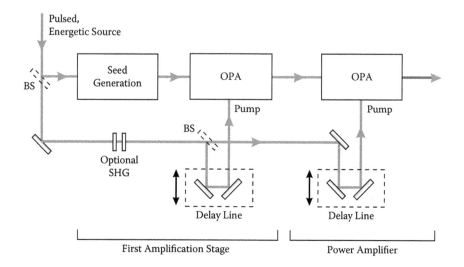

FIGURE 1.10 Principle scheme of an ultrafast optical parametric amplifier. BS, beam splitter; OPA, optical parametric amplification stage.

used for the seed generation: parametric superfluorescence and white light continuum generation. Parametric superfluorescence [83] is parametric amplification of vacuum noise and can also be thought of as two-photon emission from a virtual level excited by the pump field. In practice, it is achieved by pumping a suitable second-order nonlinear crystal (often of the same kind as those employed in the later OPA stages); amplification of vacuum noise occurs at those wavelengths for which the phase-matching condition is satisfied. Disadvantages of this technique are the poor spatial quality of the generated seed beam and its large fluctuations (inherent in a process that starts from noise); for these reasons, it is now seldom used in OPAs. Supercontinuum generation, described in Section 1.3.3, provides very high pulse-to-pulse stability and excellent spatial beam quality and is thus the preferred seed generation technique.

Following generation of the seed pulse, the pump and seed pulses are combined in a suitable nonlinear crystal in a first parametric amplification stage (preamplifier). To achieve temporal overlap, their relative timing must be adjusted by a delay line. Often, the pump spot size in the nonlinear crystal is set by a telescope and is chosen to achieve the highest possible gain without causing optical damage to the crystal or inducing third-order nonlinear effects (self-focusing, SPM, or white light generation) that would cause beam distortion or breakup. In case of parametric superfluorescence seed, the preamplifier is also used as a spatial filter to improve the spatial coherence of the signal beam by amplifying only those spatial components of the superfluorescence that overlap the pump beam in the crystal.

After the first amplification stage, the signal beam can be further amplified in a second stage, the power amplifier. Usually, this stage is driven into saturation (i.e., with significant pump depletion and conversion efficiency above 30%). In this regime, the amplified energy is less sensitive to seed fluctuations, and high pulse stability can be achieved.

The purpose of using two amplification stages instead of one long crystal is twofold: (i) the group velocity mismatch between pump and signal pulses in the first stage can be compensated by a delay line; (ii) this scheme gives the flexibility of separately adjusting the pump intensity, and thus the parametric gain, in the two stages. After the power amplifier, signal and idler beams are separated from the pump and from each other using dichroic filters or mirrors. Finally, in case of broadband amplification, a pulse compressor is used to obtain TL pulse duration.

OPAs pumped by the FF of Ti:sapphire and working in the near-IR range are the most straightforward to operate [84–87]. They have the following advantages: (i) high available pump energies (up to the millijoule level) and (ii) low pump-signal and pump-idler group velocity mismatch values, allowing the use of long nonlinear crystals and the obtainment of high gains.

A typical setup for a white light-seeded, near-IR OPA is shown in Figure 1.11 [88]; the OPA is pumped by an amplified Ti:sapphire laser generating 500-μJ, 50-fs pulses at a 1-kHz repetition rate. A small fraction of the pump light ($\approx$2 μJ) is used to generate the white light seed in a 2-mm-thick sapphire plate; the near-IR fraction of the continuum is amplified in a first stage consisting of a 3-mm-long β-barium borate (BBO) crystal cut for type II phase matching ($\theta = 28°$). Pumping with $\approx$50 μJ of energy, up to 6 μJ energy at the signal wavelength is obtained. Wavelength tuning is achieved by changing the phase-matching angle of the crystal and simultaneously optimizing the pump-seed delay. The second stage consists of an identical BBO crystal pumped by $\approx$450 μJ of energy and generates up to 200 μJ of amplified signal + idler light, corresponding to a conversion efficiency of 45%; the pulsewidth ranges from 30 to 50 fs across the tuning range.

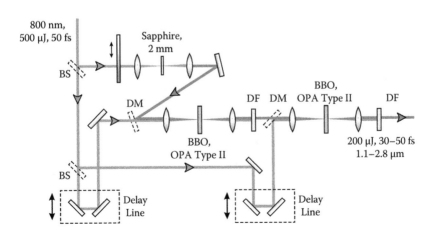

FIGURE 1.11 Scheme of a near-IR OPA. BS, beam splitter; DF, dichroic filter; DM, dichroic mirror.

A straightforward way of generating tunable visible pulses consists of frequency doubling the output of an 800-nm pumped near-IR OPA [87]; however, since infrared absorption of the idler in the nonlinear crystal prevents signal amplification for wavelengths shorter than ≈1100 nm, the SH would be tunable down to only 550 nm, leaving a substantial part of the visible range uncovered. Pumping with the SH of an amplified Ti:sapphire laser around 400 nm [89–91], the signal can be tuned through most of the visible range, from ≈ 450 nm to degeneracy (800 nm). Correspondingly, the idler tunes from 800 nm to 3 μm; this fills the gap in the tuning range left by near-IR OPAs.

Visible OPAs in general produce lower energies than near-IR ones because of the lower energy available from a frequency-doubled pump. Furthermore, the differences in group velocities of the interacting pulses are much larger in the visible range, preventing the use of long nonlinear crystals. This disadvantage is partially compensated for by the larger gain for parametric interaction in the visible. Also in visible OPAs the most popular nonlinear material is BBO. Type II phase matching provides gain bandwidths that are narrower and stay essentially constant over the tuning range, which may be beneficial for some spectroscopic applications [90]. Using type I phase matching, the amplified pulse bandwidth strongly depends on signal wavelength, increasing in the red as degeneracy is approached. The collinear interaction geometry limits the available phase-matching bandwidth. A solution to this problem, which consists of using a noncollinear interaction geometry, is discussed in Section 1.3.6.

The generation of ultrashort pulses tunable in the mid-IR spectral region (3–10 μm) is spectroscopically interesting because it covers the vibrational transitions in molecules. The most widely used approach is based on difference frequency generation (DFG) between signal and idler pulses produced by a near-IR OPA pumped at 800 nm [92–94]. A typical experimental setup is shown in Figure 1.12; the signal and idler pulses are generated by a type II near-IR OPA and thus have

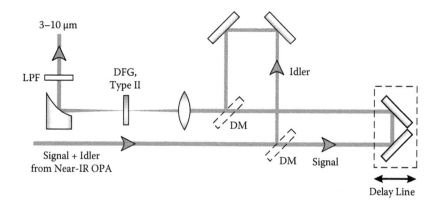

FIGURE 1.12 Scheme of a mid-IR pulse generation stage. DM, dichroic mirrors; LPF, long-pass filter.

perpendicular polarization, as required in the DFG process. The two pulses, with a combined energy of $\approx 200 \ \mu J$, are separated by a dichroic mirror, reflecting the idler and transmitting the signal, and recombined by an identical mirror; in this way, their relative delay can be adjusted. The two collinear and temporally overlapped pulses are then focused on a 1-mm-thick $AgGaS_2$ crystal, cut for type II phase matching for the DFG process; the mid-IR pulses are separated from the residual near-IR ones by a long-pass filter and collimated by an off-axis paraboloid. By tuning the OPA and simultaneously readjusting the phase-matching angle of the DFG crystal, the mid-IR pulses can be tuned from 3 to 10 μm with energies in excess of 1 μJ and pulsewidths down to 70 fs, corresponding to just few cycles of the mid-IR electric field.

1.3.6 THE NONCOLLINEAR OPTICAL PARAMETRIC AMPLIFIER

The OPA acts as a broadband optical amplifier with continuously tunable center frequency, which can amplify by many orders of magnitude a suitably generated weak signal beam (the so-called seed beam). Ideally, one would like to have a broadband amplifier, that is, an amplifier that, for a fixed pump frequency ω_3, provides a more or less constant gain over a broad range of signal frequencies. Practically, however, the phase-matching condition can be satisfied only for a given set of frequencies $(\omega_1, \omega_2, \omega_3)$. If the pump frequency is fixed at ω_3 and the signal frequency changes to $\omega_2 + \Delta\omega$, then by energy conservation the idler frequency changes to $\omega_1 - \Delta\omega$. The ensuing wave vector mismatch can be approximated to the first order as

$$\Delta k \cong -\frac{\partial k_2}{\partial \omega_2}\Delta\omega + \frac{\partial k_1}{\partial \omega_1}\Delta\omega = \left(\frac{1}{v_{g1}} - \frac{1}{v_{g2}}\right)\Delta\omega \qquad (1.35)$$

where $v_g = d\omega/dk$ is the *group velocity*. A broad gain bandwidth therefore requires the group velocity matching between signal and idler waves. For an OPA pumped by the SH of Ti:sapphire and with signal in the visible range, the idler falls in the near-IR range and moves at a significantly larger group velocity with respect to the signal due to the lower refractive index, resulting in narrow gain bandwidths. In this configuration, the signal and idler group velocities are fixed and so is the phase-matching bandwidth of the process.

An additional degree of freedom can be introduced using noncollinear geometry, such as that shown in Figure 1.13a; pump and signal wave-vectors form an angle α (independent of signal wavelength), and the idler is emitted at an angle Ω with respect to the signal. In this case, the phase-matching condition is a vector equation, which, projected on directions parallel and perpendicular to the signal wave-vector, becomes

$$\Delta k_{par} = k_3 \cos\alpha - k_2 - k_1 \cos\Omega = 0 \qquad (1.36a)$$

$$\Delta k_{perp} = k_3 \sin\alpha - k_1 \sin\Omega = 0 \qquad (1.36b)$$

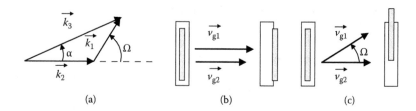

FIGURE 1.13 (a) Schematic of a noncollinear interaction geometry; (b) representation of signal and idler pulses in the case of collinear interaction; and (c) same as (b) for noncollinear interaction.

Note that the angle Ω is not fixed but depends on the signal frequency. If the signal frequency increases by $\Delta\omega$, then the idler frequency decreases by $\Delta\omega$, and the wave-vector mismatches along the two directions can be approximated, to the first order, as

$$\Delta k_{par} \cong -\frac{\partial k_2}{\partial \omega_2}\Delta\omega + \frac{\partial k_1}{\partial \omega_1}\cos\Omega\,\Delta\omega - k_1\sin\Omega\frac{\partial\Omega}{\partial\omega_1}\Delta\omega \qquad (1.37\text{a})$$

$$\Delta k_{perp} \cong \frac{\partial k_1}{\partial \omega_1}\sin\Omega\,\Delta\omega + k_1\cos\Omega\frac{\partial\Omega}{\partial\omega_1}\Delta\omega \qquad (1.37\text{b})$$

To achieve broadband phase matching, both Δk_{par} and Δk_{perp} must vanish. On multiplying Eq. (1.37a) by $\cos\Omega$ and Eq. (1.37b) by $\sin\Omega$ and adding the results, we get

$$\frac{\partial k_1}{\partial \omega_1} - \cos\Omega\frac{\partial k_2}{\partial \omega_2} = 0 \qquad (1.38)$$

which is equivalent to

$$v_{g2} = v_{g1}\cos\Omega. \qquad (1.39)$$

Equation (1.39) shows that broadband phase matching can be achieved for a signal-idler angle Ω such that the signal group velocity equals the projection of the idler group velocity along the signal direction. This effect is shown pictorially in Figure 1.13; for a collinear geometry (Figure 1.13b), signal and idler moving with different group velocities are quickly separated, giving rise to pulse lengthening and bandwidth reduction, while in the noncollinear case (Figure 1.13c), the two pulses manage to stay effectively overlapped. Note that Eq. (1.39) can be satisfied only if $v_{g1} > v_{g2}$; this is always true in the visible range in the commonly used type I phase matching in negative uniaxial crystals, where both signal and idler see the ordinary refractive index. Equation (1.39) allows determination of the signal-idler angle Ω required for broadband phase matching; from a practical point of view, it is more useful to know the pump-signal angle α, which is given by

$$\alpha = \arcsin\left(\frac{1 - v_{g2}^2/v_{g1}^2}{1 + 2v_{g2}n_2\omega_2/v_{g1}n_1\omega_1 + n_2^2\omega_2^2/n_1^2\omega_1^2}\right)^{1/2} \qquad (1.40)$$

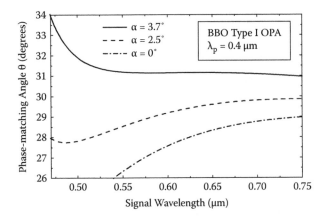

FIGURE 1.14 Phase-matching angle as a function of signal wavelength for a noncollinear BBO Type I OPA pumped at 0.4 μm as a function of pump-signal angle α.

Given the pump-signal angle α, determined by the propagation direction of the seed beam, the signal-idler angle Ω adjusts itself, according to Eq. (1.36b), to satisfy the phase-matching condition; so, the idler is emitted at a different angle for each wavelength; that is, it is angularly dispersed.

As an example, let us consider a NOPA pumped at 400 nm (the SH of a Ti:sapphire laser) using a BBO nonlinear crystal. In Figure 1.14, we show the phase-matching angle θ (the angle between the propagation direction of the pump and the optical axis) as a function of signal wavelength for a type I interaction (pump extraordinary, signal and idler ordinary). For a collinear configuration ($\alpha = 0°$), the angle rapidly varies with the signal wavelength so that, for a fixed crystal orientation, phase matching can be achieved only over a narrow signal frequency range. By going to a noncollinear configuration and increasing α, the wavelength dependence of θ becomes progressively weaker until, for the optimum value $\alpha = 3.7°$, a single crystal orientation ($\theta = 31.5°$) allows simultaneous phase matching over an ultrabroad bandwidth, extending from 500 to 750 nm.

A schematic of an ultrabroadband visible NOPA is shown in Figure 1.15 [37, 39, 95, 96]. The system is driven by a CPA Ti:sapphire laser, generating 500-μJ, 150-fs pulses at 1 kHz and 800-nm wavelength. A fraction of this light is frequency doubled in a 1-mm-thick BBO crystal, yielding pulses at 400 nm with an energy of 20 μJ, which are used to pump the NOPA. Another small fraction of the light, with energy of approximately 2 μJ, is focused into a 1-mm-thick sapphire plate to produce the seed pulses. By carefully controlling the energy incident on the plate (using a variable-optical-density attenuator) and the position of the plate around the focus, a highly stable single-filament white light continuum is obtained. Pump and seed pulses are then spatially and temporally overlapped on the NOPA crystal. Use of reflective optics for the white light avoids the introduction of additional chirp. A thin short-pass glass filter is inserted in the seed beam path to suppress the

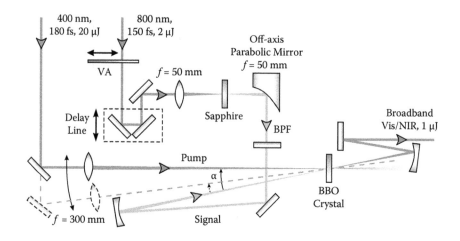

FIGURE 1.15 Scheme of a NOPA pumped at 0.4 μm. VA, variable attenuator; BPF, color filter (short-pass filter for the visible and long-pass filter for the near-IR).

strong residual peak at 800 nm. Parametric gain is achieved in a 1-mm-thick BBO crystal cut at $\theta = 31.5°$ to obtain the broadest phase-matching bandwidth (see Figure 1.14); a single-pass configuration is used to maximize the gain bandwidth. To avoid self-focusing, the BBO crystal is placed beyond the focus of the pump beam, and the pump intensity is kept below 120 GW/cm². The amplified pulses have energy of approximately 2 μJ, peak-to-peak fluctuations lower than 5%, and good TEM$_{00}$ beam quality. The energy can be increased by adding a second amplification stage up to a few hundred microjoules [97], but the microjule range is sufficient for most spectroscopic experiments.

For the optimum pump-signal angle ($\alpha = 3.7°$ inside the crystal, corresponding to 6.2° outside) and with optimized pump-seed delay, the amplified spectrum displays an ultrabroad bandwidth that extends over most of the visible range. Such ultrabroadband pulses are ideally suited as probe pulses; on the other hand, many experiments require pump pulses that are sufficiently narrowband to be able to resonantly excite just one electronic transition. To this purpose, it is sometimes useful to reduce the NOPA bandwidth, generating pulses that are still broadband but have center frequency tunable across the visible range. This can be achieved by detuning the pump-signal angle α from the optimum value (e.g., using $\alpha = 2.5°$) [98]; in this case, 10- to 15-fs pulses are generated, and tunability is achieved by tilting the crystal toward the desired θ.

The NOPA pulses need to be temporally compressed to obtain the minimum pulsewidth compatible with their bandwidth (TL duration). Several compressor schemes have been proposed for the visible NOPA, including prism-gratings [40] and prism-chirped mirrors combinations [41,99], as well as adaptive compressors using deformable mirrors [42,100,101]. Deformable-mirror compression enables spectral phase correction over ultrabroad bandwidths and the generation of the shortest pulses, with duration down to 4 fs [42, 102]. On the other hand, a compressor design employing exclusively chirped mirrors [95, 96, 103], besides having the advantages of high-energy throughput and broadband phase correction, greatly simplifies the system design, allowing for compactness, insensitivity to misalignment, and high day-to-day reproducibility, which are of great importance in spectroscopic applications.

The NOPA can also be tuned to the near-IR range [98, 104], from 0.8 to 1.7 μm, by simply replacing the short-pass filter with a long-pass one and using a 2-mm-thick sapphire plate to increase the white light spectral energy density in the near-IR range. For the near-IR NOPA, the idler is at shorter wavelengths with respect to the signal, so that $v_{g1} < v_{g2}$; as a consequence, the broadband noncollinear amplification scheme cannot be used, and the largest amplification bandwidth is achieved in a collinear configuration. However, a small noncollinear angle does not significantly decrease the gain bandwidth and is preferred because it makes it easier to spatially separate the signal beam from the pump and idler. For center wavelengths shorter than 1200 nm, the pulses can be compressed by a Brewster-cut fused-silica prism pair to durations between 12 and 20 fs; for longer wavelengths, no compression is necessary because of the lower dispersion and the narrower pulse bandwidths, and pulsewidths range from 25 to 40 fs. The opportunity of easily reconfiguring the NOPA to the near-IR range gives rise to a very broad spectral tuning range with a single source and very short pulses.

1.4 FEMTOSECOND PUMP-PROBE SPECTROSCOPY

1.4.1 THE PRINCIPLE

Figure 1.16 shows the conceptual scheme of a pump-probe experiment. A first energetic pulse (the "pump" pulse), resonant with an electronic or vibrational transition of the system under study, perturbs its absorption; the subsequent system evolution is monitored by measuring the transmission change of a delayed, weak "probe" pulse. The probe can be either an attenuated replica of the pump pulse, obtained by a beam splitter (*degenerate* pump-probe) or a pulse with a different color (*nondegenerate* or *two-color* pump-probe). One usually detects the pump-induced

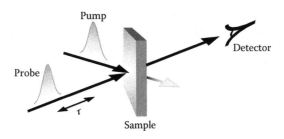

FIGURE 1.16 Principle conceptual scheme of a pump-probe experiment.

variation of the probe energy, measured by a slow detector, as a function of the time delay τ between pump and probe pulses.

To describe a pump-probe experiment, let us consider a situation in which the sample is illuminated by two pulses: a pump pulse centered at time zero, $I_{pu}(t) = I_1(t)$, and a probe pulse delayed by a time τ, $I_{pr}(t) = I_2(t - \tau)$. Calling α_0 the unperturbed optical absorption coefficient of the sample, the pump-pulse induces a variation given by [105, 106]

$$\Delta\alpha(t) = \int\limits_{-\infty}^{+\infty} I_1(t')\, A(t - t')\, dt' = I_1(t) * A(t) \tag{1.41}$$

where the symbol * stands for convolution, and $A(t)$ is the impulse response of the medium, the determination of which is usually the object of the experiment. The probe pulse intensity changes by

$$\Delta I_{pr}(t) = I_{pr}(t)\exp(-\alpha_0 d)\,(\exp[-\Delta\alpha\,(t)d] - 1) \approx -I_{pr}(t)\exp(-\alpha_0 d)\Delta\alpha\,(t)d \tag{1.42}$$

where d is the sample thickness, and the perturbation induced by the pump pulse is assumed to be small ($\Delta\alpha\, d \ll 1$). The pump-induced variation of probe energy is

$$\Delta E_{pr}(\tau) = \int\limits_{-\infty}^{+\infty} \Delta I_{pr}(t')dt' = k \int\limits_{-\infty}^{+\infty} I_2\,(t' - \tau)\, dt' \int\limits_{-\infty}^{t'} I_1\,(t'')\, A\,(t' - t'')\, dt''$$

$$= kI_2(t) \otimes [I_1(t) * A(t)] \tag{1.43}$$

where $k = -d\,\exp(-\alpha_0 d)$, and the symbol $\otimes$ stands for cross-correlation. By a change of variables, Eq. (1.43) can be cast in the form

$$\Delta E_{pr}(\tau) = A(t) * [I_1(t) \otimes I_2(t)] = A(t) * C(t) \tag{1.44}$$

where

$$C(t) = \int\limits_{-\infty}^{+\infty} I_1(t')I_2(t + t')\, dt' \tag{1.45}$$

is the cross-correlation of the pump and probe pulse intensity profiles. The pump-probe signal is thus given by the convolution of the system response with the cross-correlation of pump and probe pulses. Equation (1.44) highlights the need to use very short pump and probe pulses to resolve fast temporal dynamics. All dynamical processes taking place on a timescale much shorter than the pump-probe cross-correlation are averaged out by the experiment.

For time delays τ much longer than the duration of the pump-probe cross-correlation, when pump and probe pulses are not temporally overlapped, one can simplify Eq. (1.43) to

$$\Delta E_{pr} = -\Delta\alpha(\tau)d\,\exp(-\alpha_0 d) \int\limits_{-\infty}^{+\infty} I_{pr}(t)dt \tag{1.46}$$

so that the normalized probe energy variation becomes

$$\frac{\Delta T}{T} = \frac{\Delta E_{pr}}{E_{pr}} = -\Delta \alpha(\tau) d \qquad (1.47)$$

To calculate the absorption change $\Delta\alpha$, let us assume that the system under study consists of n electronic states, each with a population N that is changed by the pump pulse by an amount ΔN. One can then write

$$\Delta \alpha(\nu, \tau) = \sum_{i,i=1}^{n} \sigma_{ij}(\nu)[\Delta N_i(\tau) - \Delta N_j(\tau)] \qquad (1.48)$$

where $\sigma_{ij}(\nu)$ is the frequency-dependent cross-section for the transition between state N_i and state N_j. The resulting differential transmission change is

$$\frac{\Delta T}{T}(\nu, \tau) = -\Delta \alpha \, d = -\sum_{i,i=1}^{n} \sigma_{ij}(\nu)[\Delta N_i(\tau) - \Delta N_j(\tau)] \, d \qquad (1.49)$$

The $\Delta T/T$ signal depends both on pump-probe delay and on the probe frequency ν; at each frequency, the signal can be the result of the overlap of several transitions, each weighed with its cross-section.

Figure 1.17 shows an example of an energy-level scheme of a molecular system under study, indicating the possible signals that can be observed by the pump-probe experiment. The pump pulse reduces the number of absorbing molecules in the ground state, inducing, at probe frequencies equal to or higher than the ground-state absorption, an absorption decrease; this is the ground-state *photobleaching* (PB), giving rise to a transmission increase ($\Delta T/T > 0$). At the same time, the pump pulse populates the excited state, so a probe photon can stimulate it to emit back to the ground state; this *stimulated emission* (SE) signal, also causing a

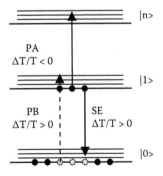

FIGURE 1.17 Energy-level scheme of a molecular system and possible signals in a pump-probe experiment. PB, photobleaching; SE, stimulated emission; PA, photoinduced absorption.

transmission increase ($\Delta T/T > 0$), occurs at probe frequencies equal to or lower than the ground-state absorption. For some probe frequencies, PB and SE overlap, while for others a pure SE signal is observed. Finally, the excited state populated by the pump pulse can absorb to some other higher-lying level; this *photoinduced absorption* (PA) signal causes a transmission decrease ($\Delta T/T < 0$). PA can occur at any probe frequency, depending on the energy-level structure of the molecule under study; in particular, it can sometimes spectrally overlap the PB and SE signals and even overwhelm them.

1.4.2 PUMP-PROBE EXPERIMENTAL SETUPS

Figure 1.18 shows a typical experimental setup used for degenerate pump-probe: a beam splitter creates two replicas of the incoming pulse, one of which is sent to a variable delay line with a computer-controlled stepping motor. The two pulse replicas are focused on the sample by a lens (or a spherical mirror), and the transmitted probe pulse is spatially selected with an iris and sent to a slow detector, usually a photodiode. In case of broadband pulses, a single wavelength of the probe can be selected by an interference filter or a monochromator. To increase signal-to-noise ratio, the pump pulse is modulated by a mechanical chopper or an acousto-optic modulator, and the pump-induced transmission change of the probe is detected by a lock-in amplifier. The sensitivity of this system depends on the pulse repetition rate, and it ranges from $\Delta T/T = 10^{-4}$ for amplified systems

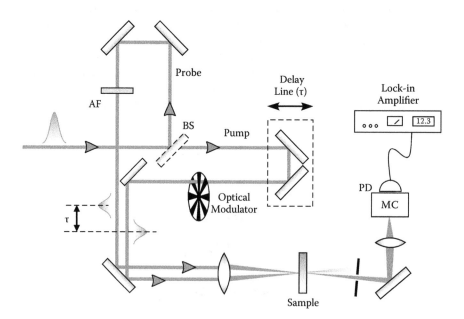

FIGURE 1.18 Setup for a degenerate pump-probe experiment. BS, beam splitter; AF, attenuating filter; MC, monochromator; PD, photodetector.

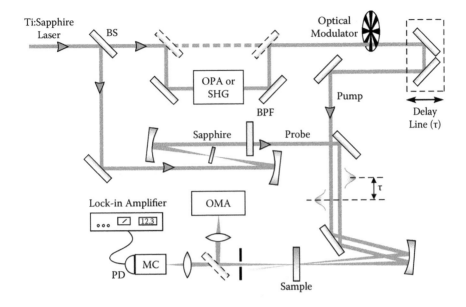

FIGURE 1.19 Femtosecond pump-probe setup with broadband detection using white light generation. BS, beam splitter; OMA, optical multichannel analyser; MC, monochromator; PD, photodetector; BPF, color filter (short-pass filter for the visible and long-pass filter for the near-IR).

at 1 kHz to $\Delta T/T = 10^{-6}$ or lower for oscillators at 100 MHz. The temporal resolution ranges from 100 to 10 fs depending on the input pulsewidth.

Figure 1.19 shows an experimental setup for a two-color pump-probe in which the broadband probe light is obtained by white light generation. This system is normally powered by a fixed-wavelength femtosecond laser system, such as a CPA Ti:sapphire laser. A fraction of the pulse energy (around 1–2 μJ) is focused in a thin sapphire plate (1–2 mm thickness) to generate a white light supercontinuum (see Section 1.3.3). A typical white light spectrum extends from 400 to 1600 nm (use of a CaF$_2$ plate enables extending the white light spectrum to the blue down to 320 nm). The white light is collected and focused on the sample using only reflective optics to prevent pulse-chirping effects. The pump pulse is provided either directly by the driver laser (through its FF or its SH) or by an OPA. After the probe pulse has traversed the sample, two types of measurement are possible: (i) measurement of the whole probe spectrum by an optical multichannel analyzer (in this case $\Delta T/T$ spectra at a fixed pump-probe delay can be obtained by subtracting pump-on and pump-off spectra) and (ii) measurement of the $\Delta T/T$ signal as a function of pump-probe delay at a fixed pump wavelength, selected by a monochromator or an interference filter. In both cases, the typical sensitivity ranges between 10^{-3} and 10^{-4}, depending on the probe wavelength. The temporal resolution, limited by the chirp of the white light probe, is usually around 100 fs; such resolution may not be sufficient for the time domain observation of many high-frequency vibrations.

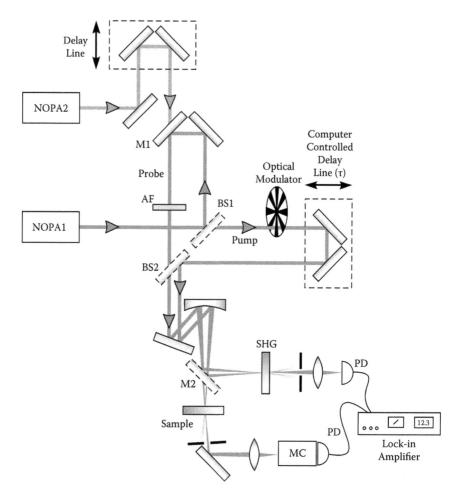

FIGURE 1.20 Scheme of the two-color pump-probe system based on NOPAs. BS, beam splitters; AF, attenuating filter; M1, M2, metal mirrors; MC, monochromator; PD, photodetector; SHG, second-harmonic generation crystal for auto-/cross-correlation.

Figure 1.20 shows the schematic of a two-color pump-probe system based on NOPAs, providing sub-20-fs temporal resolution [98, 106] over a broad spectral range. The laser source is a CPA Ti:sapphire. The system is used to pump two independent NOPAs, each of which can be configured to deliver ultrashort pulses from the visible (500–720 nm) to the near-IR (900–1500 nm). The pump-probe setup is designed to accommodate a broad range of pump and probe wavelengths, maintaining a very high temporal resolution. It is based on a balanced Michelson interferometer: beam splitters BS1 and BS2 consist of 120-μm-thick glass slides coated with Inconel. This coating guarantees a nearly uniform 40% reflection and 20% transmittance over a very broad bandwidth, from 300 to 2000 nm. The two pulse replicas are combined onto beam splitter BS2, which is flipped by 180° with respect to BS1 in such way that both pulse replicas cross the glass substrate only once [17]. The two beams are focused on the sample by a spherical mirror. After the sample, the probe beam is selected by an iris and focused on the entrance slit of a monochromator; the spectrally filtered probe light is detected by a photodiode, either silicon (in the visible) or InGaAs (in the near-IR). The pump beam is modulated at ≈500 Hz by a mechanical chopper, and synchronous detection with a lock-in amplifier is applied to the probe signal, enabling the detection of $\Delta T / T$ signals down to 10^{-4}.

This setup can be employed for degenerate pump-probe experiments in which pump and probe pulses are derived from the same NOPA; by removing mirror M1 and aligning a second input pulse derived from the second NOPA, one can rapidly switch to a nondegenerate configuration, enabling two-color experiments. In this case, the pulse timing must be properly adjusted by acting on the second delay line shown in Figure 1.20. By inserting a 45° mirror (M2) before the sample, it is possible to deviate the beams to a nonlinear crystal, thus obtaining an auto-/cross-correlation setup. Since both pulses follow exactly the same path in reaching the sample and the nonlinear crystal, the auto-/cross-correlation provides both the instrumental response function, to be deconvolved from the excited-state dynamics [see Eq. (1.44)], and an accurate determination of the zero time delay between pump and probe pulses.

1.5 IMPULSIVE COHERENT VIBRATIONAL SPECTROSCOPY

Impulsive coherent vibrational spectroscopy can be understood using either a time domain or an eigenstate picture [107–109]. In the time domain description, the molecule is excited essentially without nuclear motion due to the short pump pulse-width. Therefore, it is left in a nonequilibrium position in the excited state and starts to oscillate; the modulations of the absorption spectrum induced by these oscillations can be experimentally detected by various techniques. In the eigenstate description, nuclear and electronic wavefunctions can be decoupled applying the Born–Oppenheimer approximation. Nuclear motion occurs on potential energy surfaces (PESs), which are defined by the electronic energy levels and can be approximated by harmonic potentials. The short pulse has a bandwidth that is broad enough to excite simultaneously, according to the Franck–Condon (FC) principle, several different vibrational levels of the electronic excited state, separated by the oscillator frequency ω (see Figure 1.21a). These states, which are excited in phase, form an oscillating wavepacket [110, 111] of the form

$$\psi_1 = \sum_{n=0}^{\infty} c_n \phi_n \exp[-i\omega_n t] \tag{1.50}$$

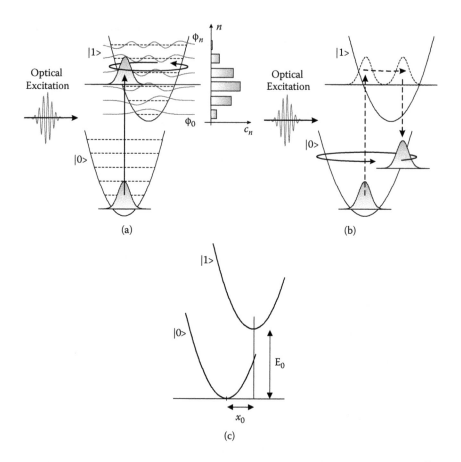

FIGURE 1.21 (a) Scheme of vibrational wavepacket generation in the excited state. The wavepacket can be seen as a linear combination of the oscillator eigenfunctions through the population coefficients c_n. (b) Scheme of vibrational wavepacket generation in the ground state. (c) Ground- and excited-state potential energy surfaces used for the calculation of vibrational dynamics.

where $\omega_n = \omega \left(\frac{1}{2} + n\right)$ and c_n are the population coefficients. The vibrational wavepacket is therefore a coherent superposition of vibrational eigenstates of the electronic excited state. The evolution of the wavepacket following its excitation by a short pulse depends on the shape of the excited-state PES. In particular, there can be different cases, as shown in Figure 1.22:

1. The excited state PES is dissociative; in this case, the wavepacket moves away from the FC region and spreads in space (Figure 1.22a) [112,113].
2. The excited-state PES is harmonic with the same curvature as the ground state; in this case, the wavepacket oscillates back and forth from the FC region, keeping the same shape, forming the so-called coherent state (Figure 1.22b).

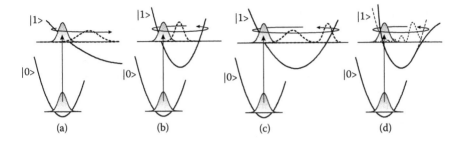

(a)   (b)   (c)   (d)

FIGURE 1.22 Excited-state wavepacket evolution following its excitation by a short pulse for various shapes of the excited-state PES: (a) dissociative, (b) harmonic with the same curvature as the ground state, (c) harmonic with a different curvature from the ground state, (d) anharmonic.

3. The excited-state PES is harmonic but with different curvature with respect to the ground state; in this case, the wavepacket remains Gaussian but its width changes during propagation (Figure 1.22c).
4. The excited-state PES is anharmonic; in this case, the wavepacket breaks up after several oscillations. However, it can experience periodic "revivals" and re-form after a while [114, 115] (Figure 1.22d).

The short pulse excitation can create a vibrational wavepacket not only on the excited-state, but also on the ground-state PES [116, 117]. Pump-probe spectroscopy can be understood as a three-field interaction with the sample [118–120]; two-field interactions are with the pump, creating a population that is then interrogated by the probe field. For short pump pulses, excited-state oscillations are observed when the two fields in the pump pulse excite a population wavepacket on the excited-state PES, which then subsequently oscillates back and forth, leaving and returning to the FC region. Ground-state oscillations are generated when the first field induces a polarization wavepacket on the excited-state PES, which is then allowed to propagate for some time so that the second field brings the wavepacket back down to the ground state, displaced from the hole left behind (see Figure 1.21b). This mechanism is known as impulsive stimulated Raman scattering (ISRS) [29]. In this case, the two sequential field interactions result in an impulsive resonant Raman process that transfers momentum from the light pulse to the ground-state wavefunction using the excited state as an intermediary. To significantly excite ground-state oscillations, the time between the first and the second interaction must be long enough to allow the polarization wavepacket to propagate on the excited-state PES so as to be displaced from the hole in the ground state. The relative contributions of excited and ground state to the oscillatory pattern are strongly dependent on the ratio between pulse length and oscillation period. For very short pump pulses, the pump-probe signal is dominated by excited-state dynamics because the polarization wavepacket created by the first field does not move significantly from the FC region on arrival of the second field. As the pump pulse length increases, the polarization wavepacket has time to propagate further out of the FC region within the pump pulse, thus increasing the amplitude of ground-state oscillations. The previous analysis holds for a harmonic excited-state PES. For a dissociative PES (leading to a photochemical reaction on photoexcitation), excitation of ground-state coherence has been interpreted in terms of the pump pulse drilling a localized coherent "hole" in the ground-state wavefunction [121, 122], which then evolves in time, giving rise to the observed vibrational dynamics

This analysis can be made more quantitative with the help of a numerical solution of the Schrödinger equation for two dipole-coupled electronic levels with displaced harmonic potentials [111]. In the following, we study a system consisting of two PES S_0 and S_1 of equal frequency ω and offsets x_0 and E_0 between ground- and excited-state minima (see Figure 1.21c), subject to an electric field $\tilde{E}(t) = \Re[A(t) \exp(-i\omega_0 t)]$. We make the electric dipole and the rotating wave approximation, which consists in neglecting rapidly oscillating terms of the form $\exp[\pm j(\omega + \omega_0)t]$; One can write for the S_0 (ψ_0) and S_1 (ψ_1) wavefunctions the set of coupled differential equations [123, 124]:

$$i\frac{\partial \psi_0}{\partial \tau} = -\frac{1}{2}\frac{\partial^2 \psi_0}{\partial \xi^2} + \frac{1}{2}\xi^2 \psi_0 - \frac{\mu_{21} A^*}{\hbar\omega}\tilde{\psi}_1 \qquad (1.51a)$$

$$i\frac{\partial \tilde{\psi}_1}{\partial \tau} = -\frac{1}{2}\frac{\partial^2 \tilde{\psi}_1}{\partial \xi^2} + \frac{1}{2}\left[(\xi - \Delta)^2 + \delta\right]\tilde{\psi}_1 - \frac{\mu_{21} A^*}{\hbar\omega}\tilde{\psi}_0 \qquad (1.51b)$$

where $\tilde{\psi}_1 = \psi_1 \exp(i\omega_0 t)$ and the following dimensionless coordinates have been used:

$$\tau = \omega t \quad \xi = x\sqrt{\frac{m\omega}{\hbar}} \quad \Delta = x_0\sqrt{\frac{m\omega}{\hbar}} \quad \delta = \frac{E_0 - \hbar\omega_0}{\hbar\omega} \qquad (1.52)$$

Equations (1.51a) and (1.51b) can be solved numerically using the split-step Fourier method [125] and assuming that $A(t)$ is a Gaussian pulse:

$$A(t) = A_0 \exp\left(-t^2/2\tau_p^2\right) \qquad (1.53)$$

According to our previous discussion, if the excitation pulse is much shorter than the vibrational period, only an excited-state wavepacket is formed. This is confirmed by Figures 1.23a and 1.23b, showing the temporal evolution of ground- and excited-state wavefunctions for an excitation pulse with 5-fs FWHM pulsewidth and a mode frequency $\omega = 150$ cm^{-1} (corresponding to a 220-fs oscillation period). On the other hand, for pulsewidths comparable to the vibrational period, the wavepacket is formed also in the ground state by the above-described ISRS mechanism; this is illustrated in Figs. 1.23c and 1.23d, showing the results for the same pulsewidth and for a mode frequency $\omega = 1500$ cm^{-1} (corresponding to a 22-fs oscillation period).

The above discussion holds for a TL pulsewidth, in which all frequency components of the pulse arrive simultaneously. An additional knob to control the ratio between ground- and excited-state wavepackets is the chirp of the excitation pulse [126–128]. Since the wavepacket on S_1 moves from higher optical frequencies to lower, the ISRS process will be enhanced when the frequency components of the pulse are ordered in time so that red follows blue. Thus, an NC pulse

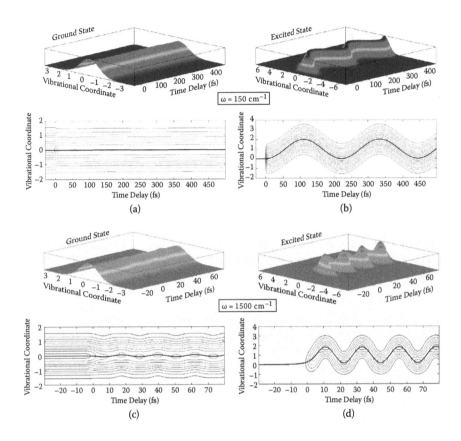

FIGURE 1.23 Probability densities ($|\psi|^2$) for the ground- and excited-state wavefunctions following excitation by a TL 5-fs pulse: (a) $\omega = 150\,\text{cm}^{-1}$, ground state; (b) $\omega = 150\,\text{cm}^{-1}$, excited state; (c) $\omega = 1500\,\text{cm}^{-1}$, ground state; (d) $\omega = 1500\,\text{cm}^{-1}$, excited state.

(see Section 1.2.1) favors the creation of a ground-state wavepacket, while a PC pulse will discriminate against the formation of the oscillating ground-state component, leaving only the excited-state coherence. Chirping the pump pulse provides a way to experimentally enhance wavepacket motion on either the ground or excited state and allows one to selectively probe dynamics on those states. The effect of pulse chirp is demonstrated in Figure 1.24, in which we consider again a 1500-cm^{-1} mode frequency and a pulse with a bandwidth corresponding to a 5-fs TL pulsewidth; in this figure, we plot the temporal evolutions of the mean values of the (a) excited- and (b) ground-state wavepacket positions, calculated as

$$\langle \xi_{0,1} \rangle = \int\limits_{-\infty}^{+\infty} |\psi_{0,1}|^2 \xi \, d\xi \tag{1.54}$$

We consider different cases: TL pulses (solid lines), PC pulses ($D_2 = +12.6\,\text{fs}^2$, dotted lines), and NC pulses ($D_2 = -12.6\,\text{fs}^2$, dashed lines). We can see that

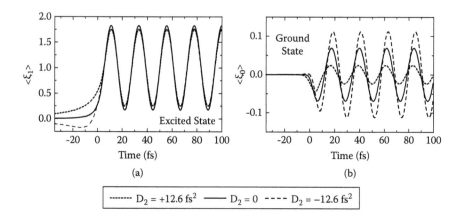

FIGURE 1.24 Temporal evolution of the mean values of the (a) excited- and (b) ground-state wavepacket positions following excitation by TL (solid lines), PC (dotted lines), and NC (dashed lines) pulses.

excited-state dynamics is not significantly altered by the pulse chirp, while the ground-state oscillation is reduced by a PC and enhanced by an NC. Application of a linear frequency chirp is just the simplest way of manipulating vibrational dynamics by acting on the excitation pulse shape; more sophisticated pulse-shaping techniques [13] allow full control of the spectral and temporal characteristics of the pulse and exploit vibrational coherence to steer the excited-state evolution toward the desired target (*coherent control*) [130].

REFERENCES

1. R. Trebino, K. W. DeLong, D. N. Fittinghoff, J. N. Sweetser, M. A. Krumbügel, B. A. Richman, and D. J. Kane. Measuring ultrashort laser pulses in the time-frequency domain using frequency-resolved optical gating. *Rev. Sci. Instrum.*, 68:3277–3295, 1997.

2. A. E. Siegman. *Lasers*, University Science Books, Mill Valley, CA, 1986.

3. H. A. Haus. *Waves and Fields in Optoelectronics*, Prentice-Hall, Englewood Cliffs, NJ, 1984.

4. O. E. Martinez, J. P. Gordon, and R. L. Fork. Negative group-velocity dispersion using refraction. *J. Opt. Soc. Am. A*, 1:1003–1006, 1984.

5. R. L. Fork, O. E. Martinez, and J. P. Gordon. Negative dispersion using pairs of prisms. *Opt. Lett.*, 9:150–152, 1984.

6. E. B. Treacy. Optical pulse compression with diffraction gratings. *IEEE J. Quantum Electron.*, QE-5:454–458, 1969.

7. R. L. Fork, C. H. Brito Cruz, P. C. Becker, and C. V. Shank. Compression of optical pulses to 6-fs by using cubic phase compensation. *Opt. Lett.*, 12:483–485, 1987.

8. A. Baltuska, Z. Wei, M. S. Pshenichnikov, and D. A. Wiersma. Optical pulse compression to 5 fs at a 1-MHz repetition rate. *Opt. Lett.*, 22:102–104, 1997.

9. R. Szipocs, K. Ferencz, C. Spielmann, and F. Krausz. Chirped multilayer coatings for broadband dispersion control in femtosecond lasers. *Opt. Lett.*, 19:201–203, 1994.

10. F. X. Kärtner, N. Matuschek, T. Schibli, U. Keller, H. A. Haus, C. Heine, R. Morf, V. Scheuer, M. Tilsch, and T. Tschudi. Design and fabrication of double-chirped mirrors. *Opt. Lett.*, 22:831–833, 1997.

11. G. Steinmeyer. Femtosecond dispersion compensation with multilayer coatings: toward the optical octave. *Appl. Opt.*, 45:1484–1490, 2006.

12. E. Zeek, K. Maginnis, S. Backus, U. Russek, M. Murnane, G. Mourou, H. Kapteyn, and G. Vdovin. Pulse compression by use of deformable mirrors. *Opt. Lett.*, 24:493–495, 1999.

13. A. M. Weiner. Femtosecond pulse shaping using spatial light modulators. *Rev. Sci. Instrum.*, 71:1930–1960, 2000.

14. T. Binhammer, E. Rittweger, U. Morgner, R. Ell, and F. X. Kärtner. Spectral phase control and temporal superresolution toward the single-cycle pulse. *Opt. Lett.*, 31:1552–1154, 2006.

15. O. E. Martinez. 3000 times grating compressor with positive group velocity dispersion: application to fiber compensation in 1.3–1.6 μm region. *IEEE J. Quantum Electron.*, 23:59–64, 1987.

16. K. Sala, G. Kenney-Wallace, and G. Hall. CW autocorrelation measurements of picosecond laser pulses. *IEEE J. Quantum Electron.*, 16:990–996, 1980.

17. C. Spielmann, L. Xu, and F. Krausz. Measurement of interferometric autocorrelations: comment. *Appl. Opt.*, 36:2523–2525, 1997.

18. D. J. Kane and R. Trebino. Characterization of arbitrary femtosecond pulses using frequency-resolved optical gating. *IEEE J. Quantum Electron.*, 29:571–579, 1993.

19. R. Trebino. *Frequency-Resolved Optical Gating: The Measurement of Ultrashort Laser Pulses*, Kluwer Academic Publishers, Boston, 2002.

20. C. Iaconis and A. Walmsley. Spectral phase interferometry for direct electric-field reconstruction of ultrashort optical pulses. *Opt. Lett.*, 23:792–794, 1998.

21. C. Iaconis and A. Walmsley. Self-referencing spectral interferometry for measuring ultrashort optical pulses. *IEEE J. Quantum Electron.*, 35:501–509, 1999.

22. L. Gallmann, D. H. Sutter, N. Matuschek, G. Steinmeyer, U. Keller, C. Iaconis, and I. A. Walmsley. Characterization of sub-6-fs optical pulses with spectral phase interferometry for direct electric-field reconstruction. *Opt. Lett.*, 24:1314–1316, 1999.

23. F. J. Duarte (Ed.). *Tunable Lasers Handbook*, Academic, New York, 1995.

24. R. L. Fork, B. I. Greene, and C. V. Shank. Generation of optical pulses shorter than 0.1 psec by colliding pulse mode locking. *Appl. Phys. Lett.*, 38:671–672, 1981.

25. W. H. Knox, M. C. Downer, R. L. Fork, and C. V. Shank. Amplified femtosecond optical pulses and continuum generation at 5-kHz repetition rate. *Opt. Lett.*, 9:552–554, 1984.

26. C. H. Brito Cruz, P. C. Becker, R. L. Fork, and C. V. Shank. Phase correction of femtosecond optical pulses using a combination of prisms and gratings. *Opt. Lett.*, 13:123–125, 1988.

27. M. J. Rosker, F. W. Wise, and C. L. Tang. Femtosecond relaxation dynamics of large molecules. *Phys. Rev. Lett.*, 57:321–324, 1986.

28. S. De Silvestri, J. G. Fujimoto, E. P. Ippen, E. B. Gamble, Jr., L. R. Williams, and K. A. Nelson. Femtosecond time-resolved measurements of optic phonon dephasing by impulsive stimulated Raman scattering in a-perylene crystal from 20–300K. *Chem. Phys. Lett.*, 116:146–152, 1985.

29. S. Ruhman, A. G. Joly, and K. A. Nelson. Time resolved observations of coherent molecular vibrational motion and the general occurrence of impulsive stimulated scattering. *J. Chem. Phys.*, 86:6563–6565, 1987.

30. J. Chesnoy and A. Mokhtari. Resonant impulsive-stimulated Raman scattering on malachite green. *Phys. Rev. A*, 38:3566–3576, 1988.

31. H. L. Fragnito, J. Y. Bigot, P. C. Becker, and C. V. Shank. Evolution of the vibronic absorption spectrum in a molecule following impulsive excitation with a 6 fs optical pulse. *Chem. Phys. Lett.*, 160:101–104, 1989.

32. P. F. Moulton. Spectroscopic and laser characteristics of $Ti:Al_2O_3$. *J. Opt. Soc. Am. B*, 3:125–133, 1986.

33. D. E. Spence, P. N. Kean, and W. Sibbett. 60-fs pulse generation from a self-mode-locked Ti:sapphire laser. *Opt. Lett.*, 16:42–44, 1991

34. D. Strickland and G. Mourou. Compression of amplified chirped optical pulses. *Opt. Commun.*, 56:219–221,1985.

35. S. Backus, C. G. Durfee III, M. M. Murnane, and H. C. Kapteyn. High power ultrafast lasers. *Rev. Sci. Instrum.*, 69:1207–1223, 1998.

36. G. Cerullo and S. De Silvestri. Ultrafast optical parametric amplifiers. *Rev. Sci. Instrum.*, 74:1–18, 2003.

37. G. Cerullo, M. Nisoli, and S. De Silvestri. Generation of 11 fs pulses tunable across the visible by optical parametric amplification. *Appl. Phys. Lett.*, 71:3616–3618, 1997.

38. T. Wilhelm, J. Piel, and E. Riedle. Sub-20-fs pulses tunable across the visible from a blue pumped single pass noncollinear parametric converter. *Opt. Lett.*, 22:1494–1496, 1997.

39. G. Cerullo, M. Nisoli, S. Stagira, and S. De Silvestri. Sub-8-fs pulses from an ultrabroadband optical parametric amplifier in the visible. *Opt. Lett.*, 23:1283–1285, 1998.

40. A. Shirakawa, I. Sakane, and T. Kobayashi. Pulse-front-matched optical parametric amplification for sub-10-fs pulse generation tunable in the visible and near infrared. *Opt. Lett.*, 23:1292–1294, 1998.

41. A. Shirakawa, I. Sakane, M. Takasaka, and T. Kobayashi. Sub-5-fs visible pulse generation by pulse-front-matched noncollinear optical parametric amplification. *Appl. Phys. Lett.*, 74:2668–2670, 1999.

42. A. Baltuska, T. Fuji, and T. Kobayashi. Visible pulse compression to 4 fs by optical parametric amplification and programmable dispersion control. *Opt. Lett.*, 27:306–308, 2002.

43. G. Cerullo, G. Lanzani, M. Muccini, C. Taliani, and S. De Silvestri. Real time vibronic coupling dynamics in a prototypical conjugated oligomer. *Phys. Rev. Lett.*, 83:231–234, 1999.

44. A. Sugita, T. Saito, H. Kano, M. Yamashita, and T. Kobayashi. Wave packet dynamics in a quasi-one-dimensional metal-halogen complex studied by ultrafast time-resolved spectroscopy. *Phys. Rev. Lett.*, 86:2158–2161, 2001.

45. T. Kobayashi, T. Saito, and H. Ohtani. Real-time spectroscopy of transition states in bacteriorhodopsin during retinal isomerization. *Nature*, 414:531–534, 2001.

46. S. Adachi, V. M. Kobryanskii, and T. Kobayashi. Excitation of a breather mode of bound soliton pairs in *trans*-polyacetylene by sub-5-fs optical pulses. *Phys. Rev. Lett.*, 89:027401, 2001.

47. S. Lochbrunner, A. J. Wurzer, and E. Riedle. Microscopic mechanism of ultrafast excited-state intramolecular proton transfer: a 30-fs study of 2-(2′-hydroxyphenyl)benzothiazole. *J. Phys. Chem. A*, 107:10580–10590, 2003.

48. G. Lanzani, G. Cerullo, Ch. Brabec, and N. S. Sariciftci. Time domain investigation of the intrachain vibrational dynamics of a prototypical light-emitting conjugated polymer. *Phys. Rev. Lett.*, 90:047402, 2003.

49. M. Ikuta, Y. Yuasa, T. Kimura, H. Matsuda, and T. Kobayashi. Phase analysis of vibrational wave packets in the ground and excited states in polydiacetylene. *Phys. Rev. B*, 70:214301, 2004.

50. G. Lanzani, M. Zavelani-Rossi, G. Cerullo, D. Comoretto, and G. Dellepiane. Real time observation of coherent nuclear motion in polydiacetylene isolated chains. *Phys. Rev. B*, 69:134302, 2004.

51. G. Cerullo, S. De Silvestri, V. Magni, and L. Pallaro. Resonators for Kerr-lens mode-locked femtosecond Ti:sapphire lasers. *Opt. Lett.*, 19:807–809, 1994.

52. T. Brabec, C. Spielmann, and F. Krausz. Mode locking in solitary lasers. *Opt. Lett.*, 16:1961–1963, 1991.

53. F. X. Kärtner, J. A der Au, and U. Keller. Mode-locking with slow and fast saturable absorbers—what's the difference? *IEEE J.S.T.Q.E.*, 4:159–168, 1998.

54. M. T. Asaki, C.-P. Huang, D. Garvey, J. Zhou, H. C. Kapteyn, and M. M. Murnane. Generation of 11-fs pulses from a self-mode-locked Ti:sapphire laser. *Opt. Lett.*, 18:977–979, 1993.

55. A. Stingl, C. Spielmann, F. Krausz, and R. Szipocs. Generation of 11-fs pulses from a Ti:sapphire laser without the use of prisms. *Opt. Lett.*, 19:204–206, 1994.

56. D. H. Sutter, G. Steinmeyer, L. Gallmann, N. Matuschek, F. Morier-Genoud, U. Keller, V. Scheuer, G. Angelow, and T. Tschudi. Semiconductor saturable-absorber mirror-assisted Kerr-lens mode-locked Tisapphire laser producing pulses in the two-cycle regime. *Opt. Lett.*, 24:631–633, 1999.

57. R. Ell, U. Morgner, F. X. Kärtner, J. G. Fujimoto, E. P. Ippen, V. Scheuer, G. Angelow, T. Tschudi, M. J. Lederer, A. Boiko, and B. Luther-Davies. Generation of 5-fs pulses and octave-spanning spectra directly from a Ti:sapphire laser. *Opt. Lett.*, 26:373–375, 2001.

58. M. S. Pshenichnikov, W. P. de Boeij, and D. A. Wiersma. Generation of 13-fs, 5-MW pulses from a cavity-dumped Ti:sapphire laser. *Opt. Lett.*, 19:572–574, 1994.

59. M. Ramaswamy, M. Ulman, J. Paye, and J. G. Fujimoto. Cavity-dumped femtosecond Kerr-lens mode-locked Ti:Al$_2$O$_3$ laser. *Opt. Lett.*, 18:1822–1824, 1993.

60. S. H. Cho, F. X. Kärtner, U. Morgner, E. P. Ippen, J. G. Fujimoto, J. E. Cunningham, and W. H. Knox. Generation of 90-nJ pulses with a 4-MHz repetition-rate Kerr-lens mode-locked TiAl$_2$O$_3$ laser operating with net positive and negative intracavity dispersion. *Opt. Lett.*, 26:560–562, 2001.

61. A. Fernandez, T. Fuji, A. Poppe, A. Fürbach, F. Krausz, and A. Apolonski. Chirped-pulse oscillators: a route to high-power femtosecond pulses without external amplification. *Opt. Lett.*, 29:1366–1368, 2004.

62. J. V. Rudd, G. Korn, S. Kane, J. Squier, G. A. Mourou, and P. Bado. Chirped-pulse amplification of 55-fs pulses at a 1-kHz repetition rate in a Ti:Al$_2$O$_3$ regenerative amplifier. *Opt. Lett.*, 18:2044–2046, 1993.

63. C. P. J. Barty, G. Korn, F. Raksi, C. Rose-Petruck, J. Squier, A.-C. Tien, K. R. Wilson, V. V. Yakovlev, and K. Yamakawa. Regenerative pulse shaping and amplification of ultrabroadband optical pulses. *Opt. Lett.*, 21:219–221, 1996.

64. S. Backus, J. Peatross, C. P. Huang, M. M. Murnane, and H. C. Kapteyn. Ti:sapphire amplifier producing millijoule-level, 21-fs pulses at 1 kHz. *Opt. Lett.*, 20:2000–2002, 1995.

65. M. Lenzner, C. Spielmann, E. Wintner, F. Krausz, and A. J. Schmidt. Sub-20-fs, kilohertz-repetition-rate Ti:sapphire amplifier. *Opt. Lett.*, 20:1397–1399, 1995.

66. R. R. Alfano (Ed.). *The Supercontinuum Laser Source*, Springer-Verlag, New York, 1989.

67. J. K. Ranka and A. L. Gaeta. Breakdown of the slowly varying envelope approximation in the self-focusing of ultrashort pulses. *Opt. Lett.*, 23:534–536, 1998.

68. A. L. Gaeta. Catastrophic collapse of ultrashort pulses. *Phys. Rev. Lett.*, 84:3582–3585, 2000.

69. C. Rolland and P. B. Corkum. Compression of high-power optical pulses. *J. Opt. Soc. Am. B*, 5:641–647, 1988.

70. H. Nakatsuka, D. Grischkowsky, and A. C. Balant. Nonlinear picosecond-pulse propagation through optical fibers with positive group velocity dispersion. *Phys. Rev. Lett.*, 47:910–913, 1981.

71. A. Baltuška, Z. Wei, R. Szipöcs, M. S. Pshenichnikov, and D. A. Wiersma. All solid-state cavity-dumped sub-5-fs laser. *Appl. Phys. B*, 65:175–188, 1997.

72. M. Nisoli, S. De Silvestri, and O. Svelto. Generation of high energy 10-fs pulses by a new pulse compression technique. *Appl. Phys. Lett.*, 68:2793–2795, 1996.

73. M. Nisoli, S. De Silvestri, O. Svelto, R. Szipöcs, K. Ferencz, Ch. Spielmann, S. Sartania, and F. Krausz. Compression of high-energy laser pulses below 5 fs. *Opt. Lett.*, 22:522–524, 1997.

74. S. Sartania, Z. Cheng, M. Lenzner, G. Tempea, Ch. Spielmann, F. Krausz, and K. Ferencz. Generation of 0.1-TW 5-fs optical pulses at a 1-kHz repetition rate. *Opt. Lett.*, 22:1526–1528, 1997.

75. E. A. J. Marcatili and R. A. Schmeltzer. Hollow metallic and dielectric waveguides for long distance optical transmission and lasers. *Bell Syst. Tech. J.*, 43:1783–1809, 1964.

76. Z. Cheng, A. Fürbach, S. Sartania, M. Lenzner, C. Spielmann, and F. Krausz. Amplitude and chirp characterization of high-power laser pulses in the 5-fs regime. *Opt. Lett.*, 24:247–249, 1999.

77. W. Kornelis, J. Biegert, J. W. G. Tisch, M. Nisoli, G. Sansone, C. Vozzi, S. De Silvestri, and U. Keller. Single-shot kilohertz characterization of ultrashort pulses by spectral phase interferometry for direct electric-field reconstruction. *Opt. Lett.*, 28:281–283, 2003.

78. C. P. Hauri, W. Kornelis, F. W. Helbing, A. Heinrich, A. Couairon, A. Mysyrowicz, J. Biegert, and U. Keller. Generation of intense, carrier-envelope phase-locked few-cycle laser pulses through filamentation. *Appl. Phys. B*, 79:673–677, 2004.

79. C. Hauri, A. Guandalini, P. Eckle, W. Kornelis, J. Biegert, and U. Keller. Generation of intense few-cycle laser pulses through filamentation—parameter dependence. *Opt. Express*, 13:7541–7547, 2005.

80. G. Stibenz, N. Zhavoronkov, and G. Steinmeyer. Self-compression of millijoule pulses to 7.8 fs duration in a white-light filament. *Opt. Lett.*, 31:274–276, 2006.

81. Y. R. Shen. *The Principles of Nonlinear Optics*, John Wiley & Sons, New York, 1984.

82. R. W. Boyd. *Nonlinear Optics*, 2nd ed., Academic Press, Boston, 2003.

83. S. E. Harris, M. K. Oshman, and R. L. Byer. Observation of tunable optical parametric fluorescence. *Phys. Rev. Lett.*, 18:732–734, 1967.

84. G. P. Banfi, P. Di Trapani, R. Danielius, A. Piskarkas, R. Righini, and I. Sa'nta. Tunable femtosecond pulses close to the transform limit from traveling-wave parametric conversion. *Opt. Lett.*, 18:1547–1549, 1993.

85. F. Seifert, V. Petrov, and F. Noack. Sub-100-fs optical parametric generator pumped by a high-repetition-rate Ti:sapphire regenerative amplifier system. *Opt Lett.*, 19:837–839, 1994.
86. M. Nisoli, S. De Silvestri, V. Magni, O. Svelto, R. Danielius, A. Piskarkas, G. Valiulis, and A. Varavinicius. Highly efficient parametric conversion of femtosecond Ti:sapphire laser pulses at 1 kHz. *Opt. Lett.*, 19:1973–1975, 1994.
87. V. V. Yakovlev, B. Kohler, and K. R. Wilson. Broadly tunable 30-fs pulses produced by optical parametric amplification. *Opt. Lett.*, 19:2000–2002, 1994.
88. K. R. Wilson and V. V. Yakovlev. Ultrafast rainbow: tunable ultrashort pulses from a solid-state kilohertz system. *J. Opt. Soc. Am. B*, 14:444–448, 1997.
89. M. K. Reed, M. S. Armas, M. K. Steiner-Shepard, and D. K. Negus. 30-fs pulses tunable across the visible with a 100-kHz Ti:sapphire regenerative amplifier. *Opt. Lett.*, 20:605–607, 1995.
90. S. R. Greenfield and M. R. Wasielewski. Near-transform-limited visible and near-IR femtosecond pulses from optical parametric amplification using type II β-barium borate. *Opt. Lett.*, 20:1394–1396, 1995.
91. P. Di Trapani, A. Andreoni, C. Solcia, G. P. Banfi, R. Danielius, A. Piskarskas, and P. Foggi. Powerful sub-100-fs pulses broadly tunable in the visible from a blue-pumped parametric generator and amplifier. *J. Opt. Soc. Am. B*, 14:1245–1248, 1997.
92. F. Seifert, V. Petrov, and M. Woerner. Solid-state laser system for the generation of midinfrared femtosecond pulses tunable from 3.3 to 10 μm. *Opt. Lett.*, 19:2009–2011, 1994.
93. M. K. Reed and M. K. Steiner Shepard. Tunable infrared generation using a femtosecond 250 kHz Ti:sapphire regenerative amplifier. *IEEE J. Quantum Electron.*, 32:1273–1277, 1996.
94. B. Golubovic and M. K. Reed. All-solid-state generation of 100-kHz tunable mid-infrared 50-fs pulses in type I and type II AgGaS$_2$. *Opt. Lett.*, 23:1760–1762, 1998.
95. G. Cerullo, M. Nisoli, S. Stagira, S. De Silvestri, G. Tempea, F. Krausz, and K. Ferencz. Mirror-dispersion-controlled sub-10-fs optical parametric amplifier tunable in the visible. *Opt. Lett.*, 24:1529–1531, 1999.
96. M. Zavelani-Rossi, G. Cerullo, S. De Silvestri, L. Gallmann, N. Matuschek, G. Steinmeyer, U. Keller, G. Angelow, V. Scheuer, and T. Tschudi. Pulse compression over 170-THz bandwidth in the visible using only chirped mirrors. *Opt. Lett.*, 26:1155–1157, 2001.
97. P. Tzankov, J. Zheng, M. Mero, D. Polli, C. Manzoni, and G. Cerullo. 300 μJ noncollinear optical parametric amplifier in the visible at 1 kHz repetition rate. *Opt. Lett.*, 31:3629–3631, 2006.
98. C. Manzoni, D. Polli, and G. Cerullo, Two-colour pump-probe system broadly tunable over the visible and the near infrared with sub-30-fs temporal resolution. *Rev. Sci. Instrum.*, 77:023103, 2006.
99. P. Baum, M. Breuer, E. Riedle, and G. Steinmeyer. Brewster-angled chirped mirrors for broadband pulse compression without dispersion oscillations. *Opt. Lett.*, 31:2220–2222, 2006.
100. M. R. Armstrong, P. Plachta, E. A. Ponomarev, and R. J. D. Miller. Versatile 7-fs optical parametric pulse generation and compression by use of adaptive optics. *Opt. Lett.*, 26:1152–1154, 2001.
101. P. Baum, S. Lochbrunner, L. Gallmann, G. Steinmeyer, U. Keller, and E. Riedle. Real-time characterization and optimal phase control of tunable visible pulses with a flexible compressor. *Appl. Phys. B*, 74:s219–s224, 2004.

102. A. Baltuška and T. Kobayashi. Adaptive shaping of two-cycle visible pulses using a flexible mirror. *Appl. Phys. B*, 75:427–443, 2004.

103. M. Zavelani-Rossi, D. Polli, G. Cerullo, S. De Silvestri, L. Gallmann, G. Steinmeyer, and U. Keller. Few-optical-cycle laser pulses by OPA: broadband chirped mirror compression and SPIDER characterization. *Appl. Phys. B*, 74:S245–S251, 2002.

104. J. Piel, M. Beutter, and E. Riedle. 20–50-fs pulses tunable across the near infrared from a blue-pumped noncollinear parametric amplifier. *Opt. Lett.*, 25:180–182, 2000.

105. Z. Vardeny and J. Tauc. Picosecond coherence coupling in the pump and probe technique. *Opt. Commun.*, 39:396–400, 1981.

106. G. Cerullo, C. Manzoni, L. Lüer, and D. Polli. Broadband pump-probe spectroscopy system with sub-20-fs temporal resolution for the study of energy transfer processes in photosynthesis. *Photochem. Photobiol. Sci.*, 6:135–144, 2007.

107. B. M. Garraway and K.-A. Suominen. Wave-packet dynamics: new physics and chemistry in femto-time. *Rep. Progr. Phys.*, 58:365–419, 1995.

108. L. Dhar, J. A. Rogers, and K. A. Nelson. Time-resolved vibrational spectroscopy in the impulsive limit. *Chem. Rev.*, 94:157–193, 1994.

109. D. J. Tannor. *Introduction to Quantum Mechanics: a Time-Dependent Perspective*, University Science Books, Sausalito, CA, 2006.

110. H. L. Fragnito, J. Y. Bigot, P. C. Becker, and C. V. Shank. Evolution of the vibronic absorption spectrum in a molecule following impulsive excitation with a 6 fs optical pulse. *Chem. Phys. Lett.*, 160:101–105, 1989.

111. W. T. Pollard, H. L. Fragnito, J.-Y. Bigot, C. V. Shank, and R. A. Mathies. Quantum-mechanical theory for 6 fs dynamic absorption spectroscopy and its application to nile blue. *Chem. Phys. Lett.*, 168:239–245, 1990.

112. A. H. Zewail. Femtochemistry: recent progress in studies of dynamics and control of reactions and their transition states. *J. Phys. Chem.*, 100:12701–12724, 1996.

113. A. H. Zewail. Femtochemistry: atomic-scale dynamics of the chemical bond. *J. Phys. Chem. A*, 104:5660–5694, 2000.

114. M. J. J. Vrakking, D. M. Villeneuve and A. Stolow. Observation of fractional revivals in molecular wavepackets. *Phys. Rev. A*, 54:R37–R40, 1996.

115. J. Heufelder, H. Ruppe, S. Rutz, E. Schreiber, and L. Wöste. Fractional revivals of vibrational wave packets in the NaK $A_1 \Sigma^+$ state. *Chem. Phys. Lett.*, 269:1–8, 1997.

116. S. L. Dexheimer, Q. Wang, L. A. Peteanu, W. T. Pollard, R. A. Mathies, and C. V. Shank. Femtosecond impulsive excitation of nonstationary vibrational states in bacteriorhodopsin. *Chem. Phys. Lett.*, 188:61–66, 1992.

117. A. T. N. Kumar, F. Rosca, A. Widom, and P. M. Champion. Investigations of ultrafast nuclear response induced by resonant and nonresonant laser pulses. *J. Chem. Phys.*, 114:6795–6815, 2001.

118. S. Mukamel. *Principles of Nonlinear Optical Spectroscopy*, Oxford University Press, New York, 1995.

119. W. T. Pollard, and R. A. Mathies. Analysis of femtosecond dynamic absorption spectra of nonstationary states. *Annu. Rev. Phys. Chem.*, 43:497–523, 1992.

120. W. T. Pollard, S. L. Dexheimer, Q. Wang, L. A. Peteanu, C. V. Shank, and R. A. Mathies. Theory of dynamic absorption spectroscopy: 4. application to 12-fs resonant impulsive raman spectroscopy of bacteriorhodopsin. *J. Phys. Chem.*, 96:6147–6158, 1992.

121. U. Banin, A. Bartana, S. Ruhman, and R. Kosloff. Impulsive excitation of coherent vibrational motion ground surface dynamics induced by intense short pulses. *J. Chem. Phys.*, 101:8461–8481, 1994.
122. E. Gershgoren, J. Vala, R. Kosloff, and S. Ruhman. Impulsive control of ground surface dynamics of I_3^- in solution. *J. Phys. Chem. A*, 105:5081–5095, 2001.
123. D. J. Tannor, R. Kosloff and S. Rice. Coherent pulse sequence induced control of selectivity of reactions: exact quantum mechanical calculations. *J. Chem. Phys.*, 85:5805–5820, 1986.
124. R. Kosloff. Time dependent methods in molecular dynamics. *J. Phys. Chem.*, 92:2087–2100, 1988.
125. G. P. Agrawal. *Nonlinear Fiber Optics*, 3rd ed., Academic Press, Boston, 2001.
126. S. Ruhman and R. Kosloff. Application of chirped ultrafast pulses for generating large-amplitude ground-state vibrational coherence: a computer simulation. *J. Opt. Soc. Am. B*, 7:1748–1752, 1990.
127. C. J. Bardeen, Q. Wang, and C. V. Shank. Selective excitation of vibrational wave packet motion using chirped pulses. *Phys. Rev. Lett.*, 75:3410–3413, 1995.
128. C. J. Bardeen, Q. Wang and C. V. Shank. Femtosecond chirped pulse excitation of vibrational wave packets in LD690 and bacteriorhodopsin. *J. Phys. Chem. A*, 102:2759–2766, 1998.
129. A. Kahan, O. Nahmias, N. Friedman, M. Sheves, and S. Ruhman. Following photoinduced dynamics in bacteriorhodopsin with 7-fs impulsive vibrational spectroscopy. *J. Am. Chem. Soc.*, 129:537–546, 2007.
130. J. Hauer, T. Buckup, and M. Motzkus. Enhancement of molecular modes by electronically resonant multipulse excitation: Further progress towards mode selective chemistry. *J. Chem. Phys.*, 125:061101, 2006.

2 Coherent Dynamics of Hydrogen Bonds in Liquids Studied by Femtosecond Vibrational Spectroscopy

Thomas Elsaesser

CONTENTS

2.1 INTRODUCTION

Numerous basic processes in condensed-phase physics and chemistry involve nuclear motions, among them lattice dynamics and structural phase transitions in solids, structural fluctuations in liquids, and chemical reactions leading to new molecular structures. In most cases, vibrational degrees of freedom are coupled to or populated through other (e.g., electronic) excitations and thus play a central role for nonequilibrium phenomena in the condensed phase. Nuclear configurations as well as nuclear motions and their couplings are reflected in the vibrational absorption or Raman spectra of the system, providing insight into both structural properties and microscopic dynamics.

Vibrational spectra of molecules in solution and molecular liquids have provided detailed insight into molecular structure. In particular, local interactions of specific functional groups that are difficult to grasp by other techniques have been identified and analyzed to a large extent by steady-state vibrational spectroscopy. Hydrogen bonding represents such an interaction that determines the (fluctuating) structure of liquids such as water and alcohols and plays a key role for elementary chemical reactions such as hydrogen and proton transfer [1]. Moreover, the supramolecular structure of biological systems (e.g., proteins) is determined to a large extent by hydrogen bonds. In a hydrogen bond, there is an attractive interaction between a so-called hydrogen donor group (e.g., an O–H or N–H group), and a neighboring electronegative acceptor atom (or group). The binding energy of such a bond ranges between about 4 and 40 kJ/mol (i.e., the interaction is weak compared with a covalent bond). This limited interaction strength allows for structural flexibility and for "making and breaking" of bonds that strongly influence the vibrational spectra.

The dynamics of hydrogen bonded systems cover a wide range in time, from about 50 fs up to tens of picoseconds [2]. Steady-state linear vibrational spectroscopy gives only limited insight into the underlying processes. In most cases, there is no quantitative understanding of vibrational line shapes and the different broadening mechanisms in spite of extensive theoretical work on molecular potential energy surfaces and vibrational couplings. Methods of nonlinear ultrafast spectroscopy allow for observing such phenomena in real time and for separating different microscopic couplings in the nonlinear response [3, 4]. Coherent vibrational dynamics of hydrogen bonds in liquids is a topic of substantial current interest [5], and both coherent nuclear motions (i.e., vibrational wavepackets) and processes of vibrational dephasing have been studied recently by ultrafast pump-probe and photon echo techniques. In addition, vibrational population relaxation and energy redistribution in hydrogen-bonded systems have been investigated by both infrared and Raman techniques.

In this chapter, recent results for wavepacket dynamics along intra- and intermolecular hydrogen bonds are reviewed. The anharmonic couplings underlying the generation and detection of such wavepackets are identified and analyzed in a quantitative way. The time scale and the mechanisms of dephasing of high-frequency O–H stretching excitations and low-frequency modes in hydrogen bonds represent the second topic addressed here. The chapter is organized as follows. In Section 2.2, techniques for generating and characterizing ultrashort midinfrared pulses in a wide range of pulse parameters are discussed. This is followed by a brief introduction to methods applied in nonlinear vibrational spectroscopy with ultrashort pulses (Section 2.3). Recent studies of coherent nuclear motions in hydrogen bonds are presented in Section 2.4. Vibrational dephasing and relaxation of stretching excitations of hydrogen-bonded O–H and O–D groups are addressed in Section 2.5. Conclusions are presented in Section 2.6.

2.2   GENERATION OF FEMTOSECOND INFRARED PULSES

Femtosecond vibrational spectroscopy monitors the nonlinear vibrational response of the molecular system in the time domain. To induce a nonlinear response via a resonant dipole excitation, femtosecond pulses in the frequency range between several tens and several thousands of wavenumbers (cm^{-1}) are required. The corresponding mid- to far-infrared wavelength range extends from approximately 2 μm up to several hundreds of micrometers. In comparison to dipole-allowed electronic or vibronic transitions, vibrational transition dipoles are small, corresponding in molar extinction coefficients to $\epsilon \approx 10 - 500$ M^{-1} cm^{-1}. Such weak absorption results in high fluences of several millijoules per square centimeter for saturating a homogeneously broadened vibrational transition by a pulse short compared to the vibrational lifetime [6]. Using infrared excitation pulses of approximately 100 fs duration, pulse energies in the microjoule range and spot diameters in the sample of 100 μm or less are required.

The vibrational oscillation periods cover a wide range from approximately 10 fs for O–H stretching vibrations up to 1 ps for low-frequency modes. Thus, the oscillation period of the vibration under study may be similar to or even longer than the duration of the femtosecond excitation pulse, allowing for an impulsive excitation of vibrations (i.e., the generation of a nonstationary superposition of vibrational quantum states). Such vibrational wavepackets, which can be followed in time by techniques of nonlinear ultrafast spectroscopy, reflect nuclear motions directly, as discussed in more detail in Section 2.4.

There is a very limited number of active laser media which allow a direct generation of femtosecond pulses in the mid- to far-infrared spectral range. Instead, femtosecond mid-infrared generation relies mainly on nonlinear frequency conversion of pulses at shorter wavelength in the near-infrared or—to a lesser extent—visible range. Free-electron lasers represent another type of infrared source presently undergoing rapid development, which has not found much application in nonlinear vibrational spectroscopy so far.

The second-order nonlinearity $\chi^{(2)}$ of transparent bulk crystals has mainly been used to generate femtosecond mid-infrared pulses by parametric amplification, difference frequency mixing, or optical rectification [7,8]. In such mixing schemes, pulses at center frequencies ω_p, ω_s, and ω_i with $\omega_p = \omega_s + \omega_i$ and $\omega_i < \omega_s < \omega_p$ (p, pump; s, signal; i, idler) interact with each other in the nonlinear medium [7] under nonresonant conditions (i.e., absorption of the nonlinear material should be negligible at all three frequencies). In parametric generation and amplification, pulses at ω_s and ω_i are generated/amplified by transferring energy from an intense pump pulse at ω_p (Figure 2.1a). In difference frequency mixing, pulses at $\omega_i = \omega_p - \omega_s$ are derived from two input pulses at ω_p and ω_s (Figure 2.1b). Optical rectification represents a special case of difference frequency mixing where $\omega_p \simeq \omega_s$, and ω_i is on the order of $\Delta\omega$, the spectral bandwidth of the input pulses (Figure 2.1c).

The phase relationship between the electric fields of the three interacting pulses is set by the phase-matching condition $\mathbf{k}_p = \mathbf{k}_s + \mathbf{k}_i$, where $|\mathbf{k}_{p,s,i}| = n_{p,s,i}(\omega_{p,s,i}/c)$ are the respective wave vectors. This phase-matching condition

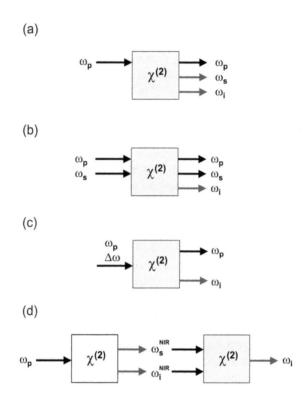

FIGURE 2.1 Parametric mixing schemes for the generation of femtosecond mid-infrared pulses via second-order nonlinearities $\chi^{(2)}$ in crystals. The center frequencies of the three interacting pulses are ω_p (pump), ω_s (signal), and ω_i (idler). (a) Parametric amplification generating pulses at ω_s and ω_i. (b) Difference frequency mixing generating pulses at $\omega_i = \omega_p - \omega_s$. (c) Optical rectification: two spectral components within the bandwidth $\Delta\omega$ of a pulse centered at ω_p generate the difference frequency ω_i. (d) Schematic of a two-step mixing scheme for generating midinfrared pulses. In a first stage, tunable near-infrared pulses at ω_s^{NIR} and ω_i^{NIR} are generated by parametric amplification. In the second stage, such pulses generate the difference frequency $\omega_i = \omega_s^{NIR} - \omega_i^{NIR}$ in the mid-infrared range.

can be fulfilled by adjusting the refractive indices $n_{p,s,i}$ in birefringent nonlinear media (c, vacuum velocity of light) [7]. Changing the phase-matching angle through rotation or variation of the temperature of the nonlinear crystal allows for frequency tuning of the generated pulses. Even for phase-matched parametric frequency conversion, there is a substantial mismatch of the group velocities of the three interacting pulses. The different group velocities limit the effective interaction length in the nonlinear medium and determine the minimum pulse duration achieved [9].

Phase-matched parametric frequency conversion to the midinfrared range requires nonlinear birefringent materials with a sufficiently broad range of infrared transparency and high optical quality of the crystal [8]. For idler wavelengths of up to 5 μm, bulk and periodically poled $LiNbO_3$, $LiIO_3$, $KNbO_3$, β-barium borate (BBO), and $KTiOPO_4$ (KTP) have been used. $AgGaS_2$ represents a standard material for wavelengths up to 12 μm; $AgGaSe_2$ and GaSe allow parametric mixing at even longer wavelengths up to approximately 20 μm. Data on nonlinear materials and their application for frequency conversion has been collected in Ref. [8]. In most sources for femtosecond infrared pulses, the peak intensities of the input pulses are between 1 GW/cm^2 and 1 TW/cm^2, resulting in an energy conversion efficiency into the mid-infrared between several 10^{-5} and 10^{-2}.

There are essentially two classes of femtosecond parametric sources for the mid-infrared: (i) generation schemes for pulses at megahertz repetition rate and comparably low peak intensities and energies per pulse and (ii) sources for intense pulses working at a much lower kilohertz repetition rate. For generating pulses of 30 to 300 fs duration at high repetition rate, optical rectification of the broadband output of mode-locked Ti:sapphire oscillators [10–14], optical parametric oscillators synchronously pumped by femtosecond pulse trains [15–17], and a combination of both techniques [18–20] have been applied. Such sources have found widespread applications (e.g., in nonlinear spectroscopy of solids). The generated pulse energies, however, are too small to generate nonlinear vibrational excitations, and thus sources of megahertz repetition rate play a minor role for femtosecond vibrational spectroscopy. Instead, most experiments have been performed with midinfrared pulses derived from amplified Ti:sapphire or—in the early days of femtosecond spectroscopy—dye lasers [21, 22].

Intense near-infrared pulses generated in regenerative Ti:sapphire and—in some cases—Cr:forsterite amplifiers at kilohertz repetition rates have been used for pumping a variety of parametric conversion schemes. For a recent overview, refer to Refs. [23] and [24]. In parametric amplification, two-photon absorption of the intense pump pulses in the nonlinear material can represent a major limitation when driving the mixing process with 800-nm pulses from amplified Ti:sapphire lasers. For this reason, one frequently applies two-step conversion schemes (Figure 2.1d): the 800-nm output of the Ti:sapphire laser is converted first into pulses tunable in the near-infrared range between about 1 and 2.5 μm. In a second stage, the midinfrared difference frequency of the near-infrared signal and idler pulses is generated. In the wavelength range from 2.5 to approximately 10 μm, such sources provide pulses of 50 to 150 fs duration and energies up to 20 μJ [25–27]. Very recently, tunable infrared pulses with an electric field amplitude of up to megavolts per centimeter have been generated between 10 and 20 μm by difference frequency mixing of spectral components of an amplified 25-fs pulse in a thin GaSe crystal [28]. This technique allows for the generation of well-defined few-cycle pulses of very high peak intensity.

A complete characterization of femtosecond pulses requires measurement of their time-dependent electric field in amplitude and (absolute) carrier phase. In contrast to the visible and near-infrared spectral range, such measurements are possible for midinfrared electric field transients by electrooptic sampling, a technique that originally developed for the far infrared around 300 μm [29]. The infrared pulse induces a change of the refractive index in a birefringent electrooptic crystal (e.g., ZnTe) that is proportional to its momentary electric field and monitored through the polarization rotation of an ultrashort probe pulse. Changing the delay between infrared and probe pulse, the time-dependent electric field of the infrared pulse is determined. For a quantitative measurement, the probe pulse has to be short compared with the period of the infrared field, and group velocity dispersion in the electrooptic crystal has to be limited. Using 10-fs pulses at 800 nm and ZnTe crystals typically 10 μm thick, infrared pulses up to a wavelength of about 7 μm have been characterized [29]. This technique has been extended to characterize intense midinfrared pulses generated at a 1-kHz repetition rate [28]. In such recent experiments, 12-fs pulses from the 75-MHz repetition rate oscillator of an amplified Ti:sapphire laser system sample the electric field of the midinfrared pulses at a 1-kHz repetition rate. An electronic gating technique is applied to measure the polarization rotation of the probe pulse overlapping in time with the mid-infrared pulse. As an example, the time-dependent electric field of a pulse centered at 15 μm is plotted in Figure 2.2 together with its intensity spectrum. The electric field transient gives a width of the temporal intensity envelope (FWHM [full width at half maximum] pulse duration) of approximately 50 fs.

In addition to electrooptic sampling, other standard techniques for characterizing ultrashort pulses have been applied in the midinfrared. Measurements of interferometric autocorrelations and frequency-resolved optical gating (FROG) have allowed for an analysis of the relative optical phase of the pulses and thus the frequency chirp [30–32]. Intensity auto- and cross-correlation measurements represent the most conventional approach giving information on the intensity envelope and the timing jitter.

Different techniques for generating phase-shaped midinfrared pulses have been proposed and demonstrated [33–36]. Direct pulse shaping in a reflective grating stretcher/compressor setup using amplitude masks has been applied [26]. Other schemes are based on shaping the near-infrared input pulses for the subsequent difference frequency mixing stage [34–36] or on shaping a broadband driving pulse in optical rectification [33]. There are first applications of shaped mid-infrared pulses in nonlinear vibrational spectroscopy, for example, for studying vibrational ladder climbing [37].

In recent years, generation of ultrashort infrared pulses in free-electron lasers (FELs) has made substantial progress [38]. At wavelengths between 5 and 10 μm, macropulses consisting of a sequence of 0.5-ps micropulses and microjoule energies/micropulse have been produced using short electron bunches from accelerators. The wide tuning range and the microjoule energies per micropulse make FELs attractive for spectroscopic studies. For studies of liquids, however, averaging of the signals over the macropulse and heating of the samples due to repetitive excitation are experimental limitations. Synchronization of FEL pulses with pulses from a mode-locked Ti:sapphire laser has been demonstrated with a timing jitter of about 400 fs [39].

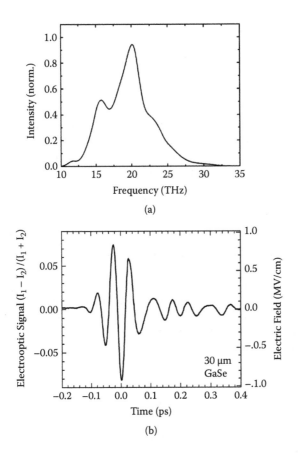

FIGURE 2.2 (a) Intensity spectrum of a femtosecond mid-infrared pulse generated by optical rectification in a 30-μm-thick GaSe crystal. The center frequency of 20 THz corresponds to a wavelength of 15 μm. (b) Time-dependent electric field of the mid-infrared pulse as measured by electrooptic sampling. The absolute field amplitude reaches very high values around 1 MV/cm.

2.3 TECHNIQUES OF ULTRAFAST NONLINEAR VIBRATIONAL SPECTROSCOPY

In this section, techniques of ultrafast nonlinear infrared spectroscopy are outlined briefly. For a more extensive discussion and theoretical treatment of the nonlinear response underlying the measured signals, refer to Refs. [3–5].

 The pump-probe technique is based on a two-pulse interaction scheme in which a first pump pulse generates a polarization on the vibrational transition and a population change of the optically coupled vibrational levels. The resulting change of transmission/absorption is probed by a weak second pulse at the same or a different spectral position and measured as a function of the delay time between pump and probe. In most cases, the transmitted probe intensity is detected in a time-integrating way, either by detecting the full probe spectrum (Figure 2.3a) or by dispersing the probe light (Figure 2.3b) to monitor the time evolution of vibrational spectra. Instead of probing changes of vibrational absorption, Raman scattering of the probe pulse from the excited sample can be applied to measure vibrational coherences or populations. In hydrogen-bonded systems, changes of vibrational population have been monitored by spontaneous anti-Stokes Raman scattering of a visible or near-infrared pulse [40]. To get insight into time-dependent vibrational polarizations, pump-probe propagation experiments have been performed in which the time structure of the transmitted probe pulse has been analyzed [30, 41]. For this, the pulse characterization techniques outlined in Section 2.3 can be applied.

In the perturbative limit, pump-probe studies of nonlinear absorption are third order in the electric field. A third-order polarization is generated in the sample by two interactions with the pump and one interaction with the probe field. This polarization is detected through the field it radiates into the propagation direction of the probe and the two electric fields interfer on the time-integrating detector. The measured signal reflects transient polarizations and population changes (i.e., both real and imaginary parts of the nonlinear susceptibility contribute). For a pulse envelope varying slowly in time compared with the period of the optical field, the spectrally integrated absorbance change $\Delta A(\omega_{pr})$ measured by a probe pulse with center frequency ω_{pr} signal is given by [3, 42]

$$\Delta\alpha(\omega_{pr}) = -\log\left[\frac{T(\omega_{pr})}{T_0(\omega_{pr})}\right] = \frac{4\pi\omega_{pr}}{cn(\omega_{pr})}\text{Im}\int_{-\infty}^{\infty} dt\, E_{pr}^*(t)P^{(3)}(t) / \int_{-\infty}^{\infty} dt\, |E_{pr}(t)|^2$$

$$(2.1)$$

Here, $T(\omega_{pr})$ and $T_0(\omega_{pr})$ represent the transmission of the sample with and without excitation, respectively; c is the velocity of light; $n(\omega_{pr})$ the refractive index of the sample; $E_{pr}(t)$ and $P^{(3)}(t)$ are the time-dependent electric field of the probe pulse and the third-order polarization in the sample, respectively. The third-order polarization contains the nonlinear response of the sample induced by the pump pulse [3]. For a spectrally resolved detection of the probe pulse, the measured signal is given by

$$\Delta\alpha(\omega) = \frac{4\pi\omega_{pr}}{cn(\omega)}\frac{|E_{pr}(\omega)|^2\text{Im}(P^{(3)}(\omega)/E_{pr}^*(\omega))}{\int_0^{\infty} d\omega\, |E_{pr}(\omega)|^2}$$

$$(2.2)$$

Here, $E_{pr}(\omega) = \int_{-\infty}^{\infty} dt\exp(i\omega t)E_{pr}(t)$ is the Fourier transform of the time-dependent probe field, and $P^{(3)}(\omega)$ is the nonlinear polarization component at the frequency ω.

Two-dimensional pump-probe spectroscopy of infrared active vibrations represents a recently developed method that provides insight into couplings between different vibrational oscillators [43, 44]. One of the oscillators is excited by a sub-picosecond pulse of a bandwidth comparable to the vibrational linewidth, and the frequency-resolved transmission change is probed over the full spectral range in

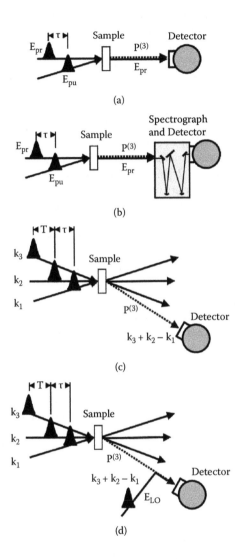

FIGURE 2.3 Experimental techniques of third-order nonlinear vibrational spectroscopy. (a) Pump-probe configuration with a spectrally integrated detection of the probe pulse. (b) Spectrally resolved pump-probe experiment. (c) Three-pulse photon echo scheme with time-integrating (homodyne) detection of the signal diffracted from pulse 3. The time intervals τ and T represent the coherence and the population time, respectively. (d) Three-pulse photon echo with field-resolved (heterodyne) detection of the diffracted signal. The diffracted electric field is interferometrically convoluted with the field E_{LO} of a fourth pulse, the local oscillator.

which the vibrational transitions of the different oscillators occur. The pump-probe signal consists of a transmission change originating from the oscillator excited by the pump, the so-called diagonal signals in a plot of transmission changes as a function of pump and probe frequencies, and from oscillators that are not resonant to the pump but are coupled to the oscillator excited. Such "off-diagonal" signals, which may be observed on the $v = 0 \rightarrow 1$ or $v = 1 \rightarrow 2$ transitions of the coupled oscillator, reflect the microscopic coupling strength. Information on the orientation of the different transition dipoles can be derived from polarization-resolved measurements. The couplings between different oscillators contain information on the underlying molecular structure (e.g., the spatial arrangement of the coupled functional groups) [45]. Using independently tunable pump and probe pulses, couplings among a large variety of vibrational modes can be determined [46].

The dynamics of the macroscopic vibrational polarization in the molecular ensemble is accessible in a direct way through femtosecond photon echo techniques [47–49]. Two- and three-pulse photon echo methods have been applied to study the coherent response of resonant vibrational excitations in hydrogen bonds, in particular, coherent O–H stretching excitations [50–56]. In the most frequently applied photon echo scheme. Three femtosecond pulses of a center frequency ω and propagation directions $\mathbf{k}_1$, $\mathbf{k}_2$, and $\mathbf{k}_3$ interact with the vibrational transition under study (Figure 2.3c,d). There is a time delay τ, the so-called coherence time, between pulses 1 and 2, and a delay T, the population time, between pulses 2 and 3. Interference of pulse 2 with the polarization generated by pulse 1 generates a transient frequency grating in the sample from which pulse 3 is scattered into the phase-matched directions $-\mathbf{k}_1+\mathbf{k}_2+\mathbf{k}_3$ and $+\mathbf{k}_1-\mathbf{k}_2+\mathbf{k}_3$. The diffracted signal is either homodyne detected in a time-integrated way or heterodyne detected by interferometric superposition with the field from a fourth phase-locked pulse that serves as a local oscillator. For an optically thin sample, small diffraction efficiency, perfect phase matching, and the slowly varying amplitude approximation, the homodyne detected intensity I_{HOM} is given by:

$$I_{HOM} = \frac{\pi\omega^2}{2n(\omega)c}l^2 \int_{-\infty}^{\infty} dt\,|P^{(3)}(t, \tau, T)|^2 \qquad (2.3)$$

Here, l represents the interaction length in the sample, and $P^{(3)}(t, \tau, T)$ is the third-order polarization depending on the real time t and time delays τ and T. For heterodyne detection with a local oscillator of identical frequency ω, the signal I_{HET} is given by

$$I_{HET} = \frac{n(\omega)c}{4\pi} \int_{-\infty}^{\infty} dt \mathrm{Re}[E_{LO}^*(t) \cdot E_s(t)]$$

$$\propto -(\omega l) \int_{-\infty}^{\infty} dt \mathrm{Im}[E_{LO}^*(t) \cdot P^{(3)}(t, \tau, T)] \qquad (2.4)$$

where $E_{LO}(t)$ and $E_s(t)$ is the electric field originating from the local oscillator and the nonlinear polarization, respectively [3].

It is important to note that photon echo spectroscopy is intrinsically multidimensional in the time domain, with the coherence time τ, the population time T, and the "real" time t related to the local oscillator [49]. This translates into two frequency dimensions in Fourier space, which may be distinctly different when using pulses of different center frequency [57]. Additional frequency dimensions can be added by using pulses of different center frequency [57]. Depending on the particular choice of the two time delays τ, the coherence time, and T, the population time, the following four techniques have been distinguished: First, in transient grating scattering, the pulses 1 and 2 of zero mutual delay ($\tau = 0$) generate a population grating, and the third pulse of delay T is diffracted from this grating. The diffracted intensity is measured as a function of the population time T, giving insight into population relaxation, diffusion, and reorientation processes.

Second, in a two-pulse photon echo (2PE), the population time $T = 0$, that is, the second and third pulse interact with the sample simultaneously. The signal is measured as a function of the coherence time τ and gives insight into the overall decay of the macroscopic coherent polarization in the sample. For a homogeneously broadened ensemble of oscillators, the coherent response follows a free-induction decay. The homodyne-detected 2PE signal peaks at $\tau = 0$ and decays with $T_2/2$ (T_2, dephasing time). In contrast, inhomogeneous broadening results in a photon-echo-like behavior, that is, a finite shift τ_{max} of the maximum signal to positive τ values and a decay with $T_2/4$.

Third, in a three-pulse photon echo (3PE), both time delays τ and T are varied to monitor the polarization dynamics in a more complete way. Three-pulse photon echoes allow for a distinction of different dephasing processes [58]. Heterodyne detection of the 3PE is a particularly powerful technique for bringing out the different vibrational couplings of the system and has led to substantial new insight [59].

Fourth, in a three-pulse echo peak shift (3PEPS) measurement, one determines for each value of T the coherence time $\tau_{max}(T)$ at which the photon echo signal is maximum. The decrease of $\tau_{max}(T)$ with T indicates the decay of inhomogeneity in the optical response by spectral diffusion processes. It has been shown for two-level systems that the echo peak shift mimics the frequency fluctuation correlation function for time delays T at which the temporal overlap of the three optical pulses is negligible [60, 61].

2.4 COHERENT NUCLEAR MOTIONS IN HYDROGEN BONDS

2.4.1 VIBRATIONAL EXCITATIONS OF HYDROGEN-BONDED SYSTEMS

Linear vibrational spectroscopy of hydrogen-bonded systems has concentrated on (i) analyzing vibrational bands of the functional groups that are part of the hydrogen bond and (ii) identifying new modes occurring on formation of a hydrogen bond [1, 62]. Among the vibrations of hydrogen-bonded groups, the X–H stretching mode has received by far the most attention. The stretching band of the hydrogen donor group undergoes a red shift and, in most cases, substantial spectral broadening when a hydrogen bond is formed. The red shift reflects the reduced force constant or enhanced (diagonal) anharmonicity of the oscillator and has been used to characterize the strength of hydrogen bonds [63]. Spectral broadening can arise from a number of mechanisms, among them anharmonic coupling to low-frequency modes, Fermi resonances with overtone and combination tone levels of fingerprint modes, vibrational dephasing, and inhomogeneous broadening due to different hydrogen-bonding geometries in the molecular ensemble.

On formation of a hydrogen bond, other modes of the hydrogen-bonded groups (e.g., in the fingerprint range between 1000 and 2000 cm^{-1}) undergo spectral shifts and reshaping of their vibrational line shapes. Such effects are less pronounced than the changes of the X–H stretching band and differ substantially for different molecular systems. For a review, refer to Ref. [1].

The so-called hydrogen bond modes represent new modes that occur due to the attractive interaction between the hydrogen donor and acceptor groups. Such modes are connected with motions of the heavy atoms in the hydrogen bond, thus affecting its geometry (e.g., the length or relative orientation of the groups linked). The weak interaction between the hydrogen-bonded groups (i.e., the small force constants) and the large reduced mass of the oscillator result in low frequencies of the hydrogen bond modes, which lie in the range between 50 and 300 cm^{-1}. In many cases, the frequencies of hydrogen bond modes are close to those of low-frequency excitations of the surrounding liquid, resulting in congested vibrational spectra. It is important to note that the vibrational period of hydrogen bond modes of 100 to 700 fs is much longer than that of the X–H stretching mode, which is on the order of 10 fs. Thus, there is a clear separation of the time scales of hydrogen bond and X–H stretching motions.

Couplings between the different vibrational degrees of freedom of the hydrogen bond and with the surroundings play a central role for the ultrafast nonequilibrium dynamics of hydrogen-bonded systems. The most relevant couplings are [64–68] (i) (off-diagonal) anharmonicities of the vibrational potential energy surface, (ii) excitonic couplings between hydrogen-bonded oscillators of similar geometry, (iii) Fermi resonances of vibrational levels, and (iv) fluctuating forces exerted on the hydrogen bond by the surroundings.

Third- and higher-order terms of the vibrational potential containing mixed products of vibrational coordinates are substantially enhanced on hydrogen bond formation and lead to coupling of different modes. Anharmonic coupling of the high-frequency X–H stretching mode to low-frequency hydrogen bond modes has been regarded as a broadening mechanism of the X–H, such as the O–H stretching band [64,65]. The separation of time scales of the low- and high-frequency degrees of freedom allows for a description of such coupling in a picture in which the different states of the O–H stretching oscillator define adiabatic potential energy surfaces for the low-frequency modes (Figure 2.4a). This scheme is equivalent to the separation of electronic and nuclear degrees of freedom in the Born–Oppenheimer picture of vibronic transitions. Vibrational transitions from different levels of the low-frequency oscillator in the $v_{OH} = 0$ state to different low-frequency levels in

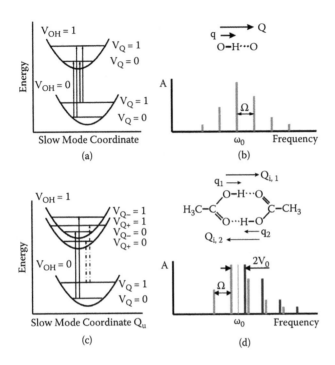

FIGURE 2.4 Anharmonic coupling of the O–H stretching mode q and a low-frequency hydrogen bond (O . . . O) mode Q. (a) Potential energy diagram for the low-frequency mode in a single hydrogen bond. The potential energy surfaces as defined by the stretching mode and the quantum levels of the low-frequency mode are plotted for the $v_{OH} = 0$ and 1 states as a function of the coordinate Q. (b) Progression of vibrational lines centered at the pure O–H stretching transition ω_0 with a line separation ω, the frequency of the mode Q. (c) Potential energy diagram for two excitonically coupled O–H stretching oscillators. The two $v_{OH} = 1$ potentials are separated by $2V_0$ (V_0, excitonic coupling). (d) Progressions of vibrational lines resulting from the coupling scheme in (c).

the $v_{OH} = 1$ state with a shifted origin of the potential result in a progression of lines that is centered at the pure X–H stretching transition and displays a mutual line separation by one quantum of the low-frequency mode (Figure 2.4b). The absorption strength is determined by the dipole moment of the $v_{OH} = 0 \rightarrow 1$ transition of the O–H stretching mode and the Franck–Condon factors between the optically coupled levels of the low-frequency mode. With increasing difference in quantum number of the low-frequency mode in the $v_{OH} = 0$ and 1 states, the Franck–Condon factors decrease, and the progression lines become weaker with increasing frequency separation from the center of the progression. The thermal population of the low-frequency levels in the $v_{OH} = 0$ state decreases with quantum number, resulting in an additional reduction of the progression strength on the low-frequency side of the pure O–H stretching transition. For each low-frequency mode coupling to an O–H stretching oscillator, an independent progression of lines occurs. In addition, combination tones with low-frequency modes contribute to the overall line shape [62]. Such mechanisms result in a strong broadening or spectral substructure of the overall O–H stretching band, even for a small number of absorption lines with large Franck–Condon factors. Third-order coupling strengths of the O–H stretching mode and hydrogen bond modes in acetic acid dimers of up to 100 cm^{-1} have been calculated recently [69].

Excitonic coupling occurs between different resonant oscillators (e.g., in systems with several identical O–H groups or hydrogen bonds). Such interaction can result in vibrational excitations delocalized over the different oscillators and influences both the ultrafast coherent response and incoherent population and energy transfer processes. Cyclic carboxylic acid dimers represent important model systems with coupled O–H stretching oscillators that have been analyzed theoretically by Marechal and Witkowski [65]. In this approach, the $v_{OH} = 0$ states of the two O–H oscillators are considered degenerate, and the coupling comes into play whenever one of the oscillators is excited. The excitonic coupling leads to a splitting of the $v_{OH} = 1$ states (Figure 2.4c). Considering both anharmonic and excitonic coupling, the coupled system has been described by taking into account the C_2 symmetry of the cyclic dimer and introducing symmetrized vibrational coordinates $q_{g,u} = (1/\sqrt{2})(q_1 \pm q_2)$ and $Q_{i,g,u} = (1/\sqrt{2})(Q_{i,1} \pm Q_{i,2})$ for the stretching and the low-frequency modes i, respectively. The symmetry operator C_2 (in-plane rotation of the dimer by 180°) has the following properties: $C_2|v_g(q_g)\rangle = +|v_g(q_g)\rangle$ and $C_2|v_u(q_u)\rangle = (-1)^{v_u}|v_u(q_u)\rangle$ for the high-frequency O–H stretching mode, $C_2|n_{i,g}(Q_{i,g})\rangle = +|n_{i,g}(Q_{i,g})\rangle$ and $C_2|n_{i,u}(Q_{i,u})\rangle = (-1)^{n_{i,u}}|n_{i,u}(Q_{i,u})\rangle$ for the low-frequency hydrogen bond mode [with $v_{g,u}$ and $n_{i,g,u}$ the quantum numbers of the $|v_{g,u}(q_{g,u})\rangle$ and $|n_{i,g,u}(Q_{i,g,u})\rangle$ wavefunctions, respectively]. Evaluating the dipole selection rules, one finds that transitions between $|v(q_u)\rangle$ states are infrared active, whereas transitions between $|v_g(q_g)\rangle$ states are observed in the Raman band of the O–H stretching mode [65,67,68]. The $v_u = 1$ potential energy surface along the gerade $Q_{i,g}$ coordinate remains unaltered, whereas the excitonic coupling V_0 leads to a splitting of the $v_u = 1$ potential energy surface along the ungerade $Q_{i,u}$ coordinate by $2V_0$. The resulting line shape consists of two different progressions between the $n_{i,u} = 0$ level of the $Q_{i,u}$ mode in the $v_u = 0$ state and the $n_{i,u}^-$ levels in the $v_u = 1$ state as well as between the $n_{i,u} = 1$ level in the $v_u = 0$ state and the $n_{i,u}^+$ levels in the $v_u = 1$ state (Figure 2.4d). Simultaneously, the number of quanta in the $Q_{i,g}$ mode can be changed when exciting the system to the $v_{OH} = 1$ state, introducing an additional degeneracy of the lines in the respective progression. An individual molecule displays only one of those progressions, depending on whether the $n_{i,Q} = 0$ or $n_{i,Q} = 1$ level in the $v_u = 0$ state is populated. In an ensemble of molecules at a sufficiently high vibrational temperature, both levels are populated, and "consequently" both series of lines contribute to the overall vibrational band.

The excitonic coupling strength of the O–H stretching oscillators in carboxylic acid dimers has remained uncertain. Early work [65] suggested values of

$V_0 = -85$ cm^{-1}, whereas a later analysis assumed much smaller values. For the C=O stretching oscillators of acetic acid dimers, a coupling strength of approximately 50 cm^{-1} has been reported [70].

Fermi resonances between the $v_{OH} = 1$ state of the stretching mode and overtones or combination bands of modes at lower frequency lead to an additional splitting of the O–H stretching transition into different components, with a relative separation determined by twice the respective coupling. For large couplings, Fermi resonances have a strong influence on the line shape of the O–H stretching band. In the O–H . . . O hydrogen bonds of carboxylic acid dimers, for instance, Fermi resonances occur between the $v_{OH} = 1$ level and the $\delta_{OH} = 2$ bending level, as well as between $v_{OH} = 1$ and combination tones of δ_{OH} with v_{C-O} and $v_{C=O}$ stretching modes [66]. Recent theoretical work suggests an absolute value of the third-order coupling for such modes on the order of 100 cm^{-1} and attributes the coarse shape of the O–H stretching band to Fermi resonances, without, however, allowing for a full quantitative understanding [71, 72]. In addition to influencing vibrational line shapes, Fermi resonances open pathways for the population relaxation of the $v_{OH} = 1$ state of the stretching mode. Here, also couplings to hydrogen bond modes may play a role, as has been discussed in detail in a review by Hynes et al. [73].

In a liquid, a hydrogen bond is subject to the fluctuating forces exerted by the solvent bath, resulting in modulation of the vibrational transition frequency and effects such as vibrational dephasing and spectral diffusion. Depending on the strength and time scale of such modulation, the spectra may vary between a distribution of transition frequencies corresponding to different hydrogen bond configurations (inhomogeneous broadening) or a single motionally narrowed transition (homogeneous broadening). There are different theoretical approaches to describe this problem. The first approach considers a direct interaction of the dipole moment of the hydrogen stretching mode with the local electric field exerted by the fluctuating solvent [74]. The second type of model introduces a stochastic motion of the low-frequency hydrogen bond modes that couple anharmonically to the fast-stretching motion [75, 76]. In recent work, the solvent dynamics has been modeled on the basis on molecular dynamics simulations [77–80], and the resulting dephasing and spectral diffusion processes have been analyzed for the system HDO in D_2O.

The couplings (i) to (iii) transform the hydrogen stretching oscillator into a vibrational multilevel system with a multitude of transition lines. The interaction with the fluctuating surrounding [coupling (iv)] leads to an additional broadening of the individual lines, in many cases resulting in spectra without much fine structure. For most systems, both frequency domain measurements of linear vibrational spectra and theoretical calculations of the line shape have not allowed an unambiguous quantitative analysis of the line shapes and the microscopic couplings. Nonlinear vibrational spectroscopy allows separating the different couplings in the nonlinear time-resolved response following femtosecond vibrational excitation. In particular, the coherent vibrational dynamics can be isolated from processes of population relaxation and energy redistribution.

2.4.2 Low-Frequency Wavepacket Dynamics in Hydrogen Bonds

The multilevel character of X–H stretching excitations in hydrogen bonds directly influences the ultrafast vibrational dynamics. Recent pump-probe experiments with femtosecond time resolution have demonstrated coherent nuclear motions along low-frequency modes that are initiated by femtosecond broadband excitation of O–H/O–D stretching modes. Such excitation creates a (quantum) coherent superposition of several vibrational levels of anharmonically coupled low-frequency modes that gives rise to oscillatory wavepacket motions. In the following, recent results for medium-strong intra- and intermolecular hydrogen bonds are presented.

2.4.2.1 Intramolecular Hydrogen Bonds

Phthalic acid monomethyl ester with an intramolecular O–H...O (PMME–H) or O–D...O (PMME–D) hydrogen bond (Figure 2.5) and 2-(2'-hydroxyphenyl)benzothiazole with an O–D...N bond (HBT–D; Figure 2.6) represent model systems with a well-defined bonding geometry when dissolved in nonpolar non-hydrogen-bonding solvents. In the pump-probe experiments, mid-infrared pulses of 100 fs duration generated a vibrational excitation on the O–H or O–D stretching band (Figures 2.5 and 2.6) and the resulting change of the O–H/O–D stretching absorption were measured by weak probe pulses [81,82]. In Figure 2.7, the change of vibrational absorbance $\Delta A = -\log_{10}(T/T_0)$ of PMME–H is plotted as a function of pump-probe delay for different center frequencies of the pulses

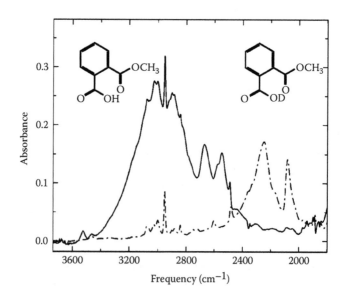

FIGURE 2.5 Molecular structures of PMME-H and PMME-D together with the linear O–H (solid line) and O–D stretching (dash-dotted line) bands (solvent C_2Cl_4).

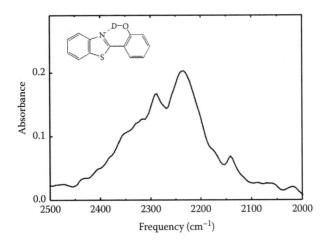

FIGURE 2.6 Molecular structure of HBT-D together with the linear O–D stretching band.

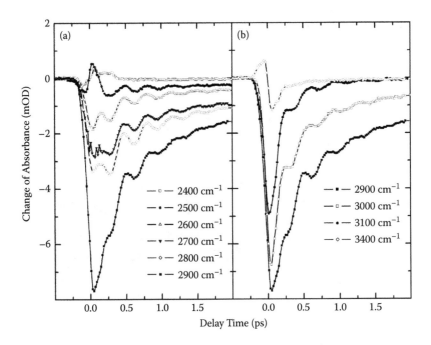

FIGURE 2.7 Pump-probe transients measured in the range of the O–H stretching band of PMME-H dissolved in C_2Cl_4. The change of vibrational absorbance $\Delta A = -\log(T/T_0)$ is plotted as a function of the delay time between the pump pulse pump and probe pulses (T, T_0, sample transmission with and without excitation). Pump and probe are centered at the same frequency position and the spectrally integrated probe is detected.

that cover a major part of the O–H stretching band (T, T_0, transmission with and without excitation, respectively). In such measurements, the probe pulses were detected in a spectrally integrated way. All transients display incoherent rate-like kinetics that are superimposed by strong oscillatory contributions to the absorbance change. The rate-like components that reflect a decrease of vibrational absorption due to the depletion of the $v_{OH} = 0$ state and stimulated emission from the $v_{OH} = 1$ state are determined by the relaxation of vibrational nonequilibrium populations of the O–H stretching oscillator and energy transfer processes to other intramolecular modes and the surrounding solvent C_2Cl_4. The lifetime of the $v = 1$ state of the O–H stretching mode is 220 fs, whereas energy transfer to the solvent occurs on a slower picosecond time scale. Such processes and their impact on the nonlinear vibrational absorption have been discussed elsewhere [82, 83].

The oscillatory pump-probe signals in Figure 2.7 display a frequency that is independent of the spectral position of the probe. The frequency spectrum of the oscillations peaks at $100 \, cm^{-1}$ with a FWHM of approximately $10 \, cm^{-1}$. Measurements on the O–D stretching absorption of the deuterated compound PMME-D reveal the same oscillation frequency as is evident from the data in Figure 2.8 taken with a spectrally resolved detection of the probe [81]. Such behavior demonstrates that the oscillation frequency is not related to the separation of the coarse spectral features in the line shape of the O–H/O–D stretching bands, most probably

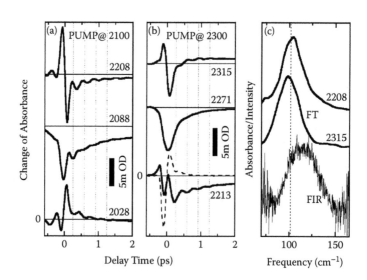

FIGURE 2.8 (a) and (b) Time-resolved change of O–D stretching absorbance of PMME-D after excitation with pump pulses centered at 2100 and $2300 \, cm^{-1}$. The change of absorbance measured at different probe frequencies is plotted versus pump-probe delay (solid lines). The dashed line in (b) represents the response of the solvent C_2Cl_4. (c) Fourier spectra (FT) of the oscillatory absorbance changes for two different probe frequencies and far-infrared (FIR) absorption spectrum of PMME-D.

originating from Fermi resonances. Instead, the oscillations are due to the generation of vibrational wavepackets along a low-frequency mode that is anharmonically
coupled to the high-frequency hydrogen or deuterium stretching mode.

The electric field envelope of the femtosecond pump pulse covers a frequency
range much broader than the energy spacing $\hbar\Omega$ of individual levels of the low-
frequency mode. In other words, the pump spectrum overlaps with several lines of
the vibrational progression depicted in Figure 2.4b. As a result, impulsive dipole
excitation from the $v_{OH} = 0$ to 1 state creates a nonstationary superposition of the
wavefunctions of low-frequency levels in the $v_{OH} = 1$ state with a well-defined
mutual phase. This (quantum-coherent) wavepacket oscillates in the $v_{OH} = 1$ state
with a frequency Ω and leads to a modulation of O–H stretching absorption that
is measured by the probe pulses. In addition to the wavepacket in the $v_{OH} = 1$
state, impulsive Raman excitation within the spectral envelope of the pump pulse
creates a wavepacket in the $v_{OH} = 0$ state, also oscillating with Ω and modulating
the O–H stretching absorption.

The oscillatory absorbance change is observed over a period of 1 to 2 ps,
pointing to a comparably slow vibrational dephasing (i.e., loss of mutual phase of
the wavefunctions contributing to the oscillations). The wavepacket in the $v_{OH} =$
1 state is damped efficiently by the $v_{OH} = 1$ population decay with a 220 fs
lifetime [83]. Thus, the oscillations observed at late delay times originate mainly
from the wavepacket in the $v_{OH} = 0$ state generated through the Raman process.
Normal-mode calculations for PMME-H/D discussed in Ref. [82] give six modes
below a frequency of 200 cm^{-1}. In this group of modes, there is a single out-
of-plane torsion of the groups forming the hydrogen bond that modulates the
length of the hydrogen bond appreciably. This mode of a calculated frequency of
approximately 70 cm^{-1} displays a negligible frequency shift on deuteration. The
oscillations observed in the experiment that occur at a somewhat higher frequency
of 100 cm^{-1} are attributed to this mode and thus reflect a periodic modulation of
the length and the strength of the intramolecular hydrogen bond. It should be noted
that theoretical studies of vibrational couplings and quantum dynamics give other
modes of higher frequency that also couple to the O–H/O–D stretching mode and
affect the hydrogen bond geometry strongly [84, 85]. The spectral bandwidth of
the 100-fs pulses applied in the experiments, however, is insufficient to generate
wavepackets along such modes, and consequently the oscillations are dominated
by the low-frequency mode at 100 cm^{-1}.

Coherent low-frequency oscillations of the intramolecular hydrogen bond have
also been observed with HBT-D [86]. In Figure 2.9, results of a pump-probe
study of HBT-D dissolved in toluene are summarized. The spectrally resolved
change of vibrational absorption (Figure 2.9a) exhibits bleaching due to $v_{OD} = 0$
bleaching and stimulated emission from the $v_{OD} = 1$ state as well as a blue-shifted
increase of absorption persisting for longer delay times. The latter contribution
is caused by the vibrationally "hot" ground state in which the O–D stretching
mode is in the $v_{OD} = 0$ state, that is, the $v_{OD} = 1$ state has been depopulated,
and other anharmonically coupled vibrations have accepted the excess energy
supplied by pump pulse. The blue-shifted absorption points to a weakening of the

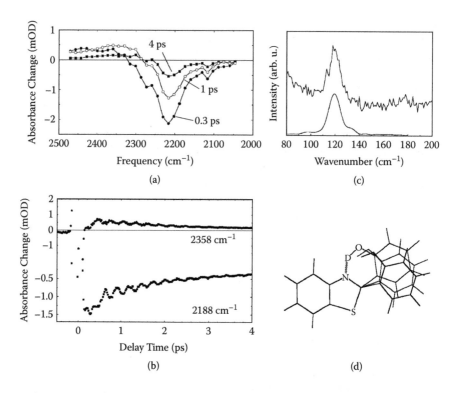

FIGURE 2.9 (a) Transient O–D absorption spectra of HBT-D for three different delay times. The change of absorbance is plotted as a function of probe frequency. (b) Time-resolved absorbance changes measured at probe frequencies of 2188 and 2358 cm^{-1}. Pronounced oscillatory signals are observed. (c) Fourier transform of the oscillatory absorbance changes of (b) (lower trace) and resonance Raman spectrum of HBT displaying a strong low-frequency band at 120 cm^{-1}. (d) Microsopic elongations connected with a wavepacket motion along the 120 cm^{-1} in-plane mode.

hydrogen bond in the hot molecules and decays by vibrational energy transfer to the surrounding solvent on a 30- to 50-ps time scale [87].

Oscillatory absorption changes are observed throughout the O–D stretching band. Two representative transients are shown in Figure 2.9b. The signals at negative delay times and around delay zero are dominated by the perturbed free induction decay of the vibrational polarization and the coherent pump-probe coupling. At positive delay times (i.e., for a sequential interaction of the molecules with pump and probe pulses), the transients exhibit strong oscillations with a period of 280 fs. The corresponding Fourier transform (Figure 2.9c) peaks at 120 cm^{-1}, a frequency in perfect agreement with that of a Raman active in-plane mode (Raman band: upper trace in Figure 2.9c) modulating the length of the hydrogen bond strongly (Figure 2.9d). As the lifetime of the $v_{OD} = 1$ state of approximately 200 fs is comparable to the period of the wavepacket oscillation, the wavepacket in the $v_{OD} = 1$ state is strongly damped, and the oscillations observed for picoseconds are due to the wavepacket generated via Raman excitation in the $v_{OD} = 0$ state. The phase of the oscillatory pump-probe signal displays a change by approximately π at the maximum of the O–D stretching band, even for pump pulses centered in the wing of the linear absorption band [86]. This finding demonstrates a resonant enhancement of the Raman process by the O–D stretching transition moment, pointing again to a strong anharmonic coupling of the low- and high-frequency modes.

The impulsive excitation scheme of low-frequency wavepackets applied here is not mode specific. In principle, all modes displaying a finite anharmonic coupling to the O–H/O–D stretching mode and a vibrational frequency that is smaller than or comparable to the pump bandwidth are excited. This subset of modes can include vibrations not affecting the hydrogen bond geometry directly. Underdamped modes in this subset give rise to oscillatory motions with dephasing times in the subpico- to picosecond time domain. In contrast to such slow dephasing, coherence on the O–H stretching transition can decay much faster, as discussed in Section 2.5. In the systems presented so far, low-frequency wavepacket motion is dominated by a single mode modulating the length and strength of the hydrogen bond. In the next section, a system displaying coherent motions along several low-frequency modes is discussed.

2.4.2.2 Hydrogen Bonded Dimers

Cyclic dimers of carboxylic acids represent important model systems forming two coupled intermolecular hydrogen bonds. The linear vibrational spectra of carboxylic acid dimers have been studied in detail, in both the gas and the liquid phase, and a substantial theoretical effort has been undertaken to understand the line shape of their O–H or O–D stretching bands (see Section 2.4.1). In contrast, there have been only a few experiments on the nonlinear vibrational response. The coupling of the two carbonyl oscillators in acetic acid dimers has been investigated by femtosecond pump-probe and photon echo measurements [70], and vibrational relaxation following O–H stretching excitation has been addressed in picosecond pump-probe studies [88]. In the following, recent extensive pump-probe studies of cyclic acetic acid dimers in the femtosecond time domain are presented [69,89,90].

Dimer structures containing two O–H . . . O (OH/OH dimer) or two O–D . . . O (OD/OD dimer) hydrogen bonds as well as dimers with one O–H . . . O and one O–D . . . O hydrogen bond (mixed dimers; Figure 2.10) were dissolved in CCl_4 with concentrations between 0.2 and 0.8 M. Two-color pump-probe experiments with independently tunable pump and probe pulses were performed with a 100-fs time resolution. Approximately 1% of the dimers present in the sample volume were excited by the 1-μJ pump pulse. After interaction with the sample, the probe pulses were spectrally dispersed to measure transient vibrational spectra with a spectral resolution of 6 cm^{-1}.

The steady-state and the transient O–H stretching absorption spectra of OH/OH dimers are displayed in Figures 2.11a and 2.11b, respectively. The transient spectra show strong bleaching in the central part of the steady-state band and enhanced absorption on the red and blue wings. The bleaching, which consists of a series of

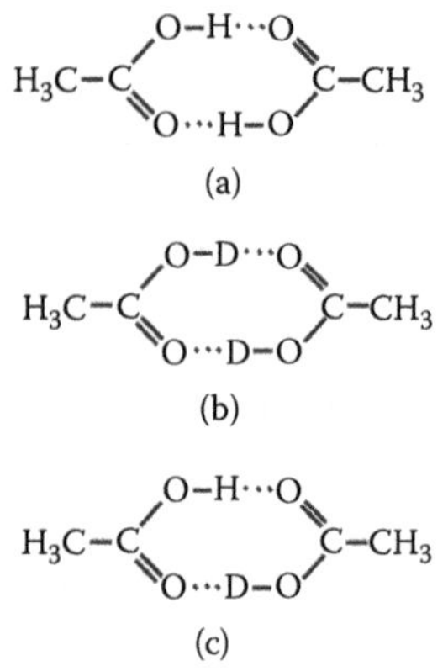

$$H_3C-C \overset{O-H\cdots O}{\underset{O\cdots H-O}{\diagup\diagdown}} C-CH_3$$

(a)

$$H_3C-C \overset{O-D\cdots O}{\underset{O\cdots D-O}{\diagup\diagdown}} C-CH_3$$

(b)

$$H_3C-C \overset{O-H\cdots O}{\underset{O\cdots D-O}{\diagup\diagdown}} C-CH_3$$

(c)

FIGURE 2.10 Acetic acid dimer structures: (a) cyclic dimer with two O–H . . . O hydrogen bonds (I), (b) cyclic dimer with two O–D . . . O hydrogen bonds (II), and (c) mixed dimer with one O–H . . . O and one O–D . . . O hydrogen bond.

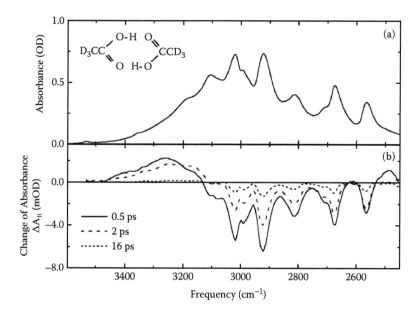

FIGURE 2.11 (a) Linear O–H stretching band of cyclic acetic acid dimers. (b) Transient vibrational absorption spectra measured for different pump-probe delays. The change of vibrational absorbance $\Delta A_{\parallel} = -\log(T/T_0)$ for pump and probe pulses of parallel linear polarization is plotted as a function of the probe frequency (T, T_0, sample transmission with and without excitation, respectively).

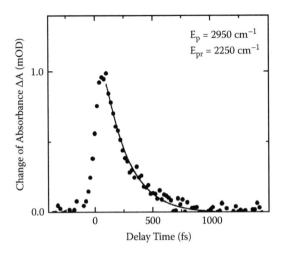

FIGURE 2.12 Nonlinear O–H stretching absorption of acetic acid dimers in the range of the $v_{OH} = 1 \rightarrow 2$ transition. The change of absorbance at $E_{pr} = 2250$ cm^{-1} is plotted as a function of pump-probe delay (points, pump pulses centered at $E_p = 2950$ cm^{-1}). Solid line: monoexponential decay with a 200-fs time constant.

comparably narrow spectral dips, originates from the depopulation of the $v_{OH} = 0$ state and stimulated emission from the $v_{OH} = 1$ state. The enhanced absorption at small frequencies is due to the $v_{OH} = 1 \rightarrow 2$ transition and decays by depopulation of the $v_{OH} = 1$ state with a lifetime of approximately 200 fs. The latter is evident from time-resolved measurements at a probe frequency of 2250 cm^{-1} (Figure 2.12). The enhanced absorption on the blue side is caused by the vibrationally hot ground state formed by relaxation of the $v_{OH} = 1$ state, similar to the behavior discussed for HBT–D (cf. Section 2.4.2.1). This transient absorption decays by vibrational cooling on a 10- to 50-ps time scale. Transient spectra have also been measured for the OD/OD and the mixed dimers—both on the O–H and O–D stretching bands—and display similar behavior.

The time evolution of the nonlinear O–H stretching absorption shows pronounced oscillatory signals for all types of dimers studied. In Figure 2.13, data for OD/OD dimers are presented that were recorded at three different spectral positions in the O–D stretching band. For positive delay times, one finds rate-like kinetics, which is due to population and thermal relaxation of the excited dimers. Such rate-like kinetics are superimposed by very strong oscillatory absorption changes. In contrast to the intramolecular hydrogen bonds discussed, the time-dependent amplitude of the oscillations displays features of a beatnote. This result demonstrates the presence of more than one oscillation frequency. In Figure 2.14a, the Fourier transforms of the oscillatory signals are plotted for the three spectral positions. There are three prominent frequency components, a strong doublet with maxima at 145 and 170 cm^{-1} and a much weaker component around 50 cm^{-1}. Comparative pump-probe studies of OH/OH dimers reveal a similar doublet at 145

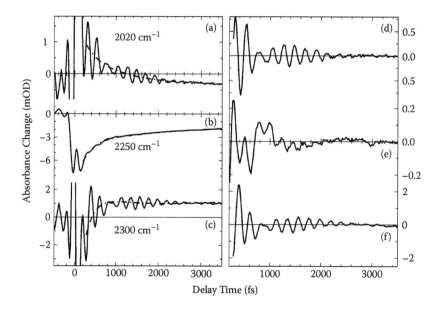

FIGURE 2.13 (a–c) Time-resolved change of O–D stretching absorbance as a function of the pump-probe delay for three different probe frequencies (solid lines). Around delay zero, coherent pump-probe coupling leads to a strong signal. The absorbance changes for positive delay times (i.e., sequential pump-probe interaction) consist of rate-like components reflecting population dynamics of the O–D stretching oscillator and oscillatory signal contributions. Dash-dotted lines: Numerical fits of the rate-like signals. (d)–(f) Oscillatory signals after subtraction of the rate-like components. The oscillations are due to coherent wavepacket motions along several low-frequency modes.

and 170 cm^{-1} with slightly changed relative intensities of the two components. The 50-cm^{-1} component is even weaker for the OH/OH than for the OD/OD case.

The two stretching oscillators in the OH/OH and OD/OD dimers should display an excitonic coupling, resulting in a splitting of their v = 1 states, on top of the anharmonic coupling to low-frequency modes. In the linear absorption spectrum of the ensemble of dimers, this results in two separate low-frequency progressions originating from the $v_Q = 0$ and $v_Q = 1$ levels in the v = 0 state of the stretching vibrations. In thermal equilibrium, a particular dimer populates only one of the v_Q levels at a certain instant in time, and thus only one of the progressions can be excited. Consequently, a quantum coherent nonstationary superposition of the splitted $v_{OH} = 1$ states of the stretching mode cannot be excited in an individual dimer, and quantum beats due to excitonic coupling are absent in the pump-probe signal. This behavior is evident from a comparison of transients recorded with OD/OD and OH/OD dimers, the latter displaying negligible excitonic coupling because of the large frequency mismatch between the O–H and the O–D stretching oscillator. The time-resolved change of O–D stretching absorption of the two types

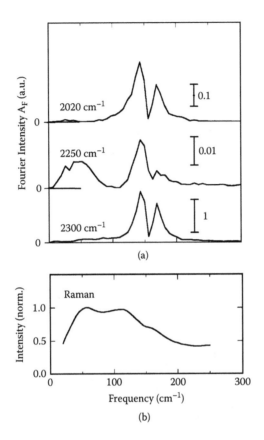

(a)

(b)

FIGURE 2.14 (a) Fourier spectra of the oscillatory absorbance changes of Figures 13 d–f. The spectra are scaled relative to each other and display three low-frequency modes. (b) Low-frequency spontaneous Raman spectrum of acetic acid (taken from Ref. [91]).

of dimers shows a very similar time evolution (Figure 2.15), and the Fourier spectra agree within the experimental accuracy.

A contribution of quantum beats between states split by Fermi resonances can be ruled out as well. There are different Fermi resonances within the O–H and O–D stretching bands. Depending on the spectral positions of pump and probe, this should lead to a variation of the oscillation frequencies, particularly when comparing O–H and O–D stretching excitations. Such behavior is absent in the experiment demonstrating identical oscillation frequencies for O–H and O–D stretching excitation, which remain unchanged for pumping throughout the respective stretching band.

The oscillatory absorption changes are due to coherent wavepacket motions along several low-frequency modes that anharmonically couple to the stretching modes. Wavepackets in the $v = 0$ state of the O–H or O–D stretching oscillators, which are generated through an impulsive resonantly enhanced Raman process,

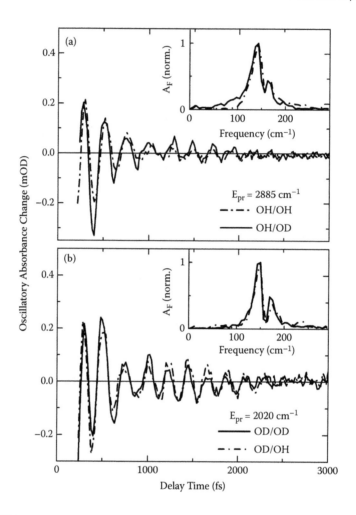

FIGURE 2.15 Oscillatory absorption changes of acetic acid dimers. (a) Comparison of the transient O–H stretching absorbance measured with OH/OH and mixed OH/OD dimers at a probe frequency of $E_{pr} = 2885$ cm^{-1} (pump pulses centered at 2950 cm^{-1}). (b) Oscillatory changes of O–D stretching absorbance in pure and mixed dimers (pump pulses centered at 2100 cm^{-1}). Insets: Fourier transforms of the transients.

govern the oscillatory response, whereas wavepackets in the $v = 1$ states are strongly damped by the fast depopulation processes.

Low-frequency modes of acetic acid have been studied in a number of Raman experiments. The spectrum in Figure 2.14b was taken from Ref. [91] and displays three maxima around 50, 120, and 160 cm^{-1}. The number of subbands in such strongly broadened spectra and their assignment has remained controversial [92]. The character of the different low-frequency modes and their anharmonic coupling to the O–H stretching mode have been studied in normal mode calculations based on density functional theory [69]. In Figure 2.16a, the calculated Raman transitions

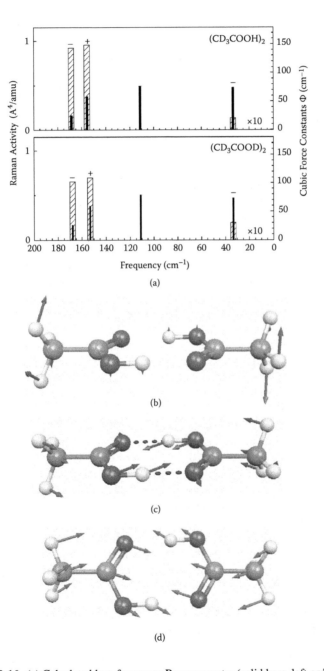

FIGURE 2.16 (a) Calculated low-frequency Raman spectra (solid bars, left ordinate) and cubic force constants Φ describing the coupling to the O–H or O–D stretching modes (hatched bars, right ordinate scale). Plus and minus signs indicate the sign of the force constants. (b–d) Microscopic elongations of the methyl torsion at 50 cm^{-1}, the dimer in-plane bending at 155 cm^{-1}, and the dimer stretching at 170 cm^{-1}.

(solid bars) and the respective cubic force constants for coupling to the hydrogen stretching mode (hatched bars) are shown for the OH/OH and OD/OD dimers. There are four vibrations; the methyl torsion at $44 \, \text{cm}^{-1}$ (Figure 2.16b), the out-of-plane wagging mode at $118 \, \text{cm}^{-1}$, the in-plane bending mode around $155 \, \text{cm}^{-1}$ (Figure 2.16c), and the dimer stretching mode at $174 \, \text{cm}^{-1}$ (Figure 2.16d). In this group, the in-plane bending and the dimer stretching modes couple strongly to the hydrogen/deuterium stretching mode via a third-order term in the vibrational potential that dominates compared with higher-order terms. The coupling of the methyl torsion is much weaker and that of the out-of-plane wagging mode is even negligible. Such theoretical results are in good agreement with the experimental findings: the strong doublet in the Fourier spectra (Figure 2.14a) is assigned to the in-plane bending and the dimer stretching and the weak band around $50 \, \text{cm}^{-1}$ to the methyl torsion. The out-of-plane wagging is not observed. It should be noted that the spectra derived from the oscillatory pump-probe signals (i.e., time domain data) allow for a much better separation of the low-frequency mode coupling than the steady-state spontaneous Raman spectra. The calculated anharmonic couplings Φ of the O–H/O–D stretching vibrations and the three low-frequency mode observed are in the same order of magnitude as the couplings calculated for Fermi resonances between the $v_{OH} = 1$ state and combination and overtones of the O–H bending and other fingerprint modes [71].

In conclusion, the results presented here demonstrate how nonlinear pump-probe spectroscopy allows for a separation of different microscopic couplings present in hydrogen-bonded systems. The anharmonic coupling of high-frequency hydrogen/deuterium stretching modes with low-frequency vibrations underlies oscillatory wavepacket motions contributing to the pump-probe signals, whereas excitonic couplings and Fermi resonances play a minor role. The results for cyclic acetic acid dimers demonstrate coherent intermolecular motions for several picoseconds. This should allow for the generation of tailored vibrational wavepackets by excitation with phase-shaped infrared pulses. Although cyclic carboxylic acid dimers are nonreactive in a sense that the O–H . . . O hydrogen bonding geometry does not change on O–H stretching excitation, there are other dimer systems that can undergo hydrogen transfer along intermolecular hydrogen bonds. For instance, vibrationally induced concerted hydrogen transfer along the N–H . . . N and N–H . . . O bonds in the adenine-thymine base pair of DNA seems feasible [93] when considering the calculated ground-state energies of the different species.

2.5 DEPHASING AND RELAXATION OF HYDROGEN STRETCHING EXCITATIONS

2.5.1 PHASE RELAXATION OF COHERENT O–H STRETCHING EXCITATIONS

The results presented in Section 2.4 demonstrate a subpico- to picosecond dephasing of low-frequency excitations that are induced via the anharmonically coupled hydrogen or deuterium stretching modes. Such dephasing manifests itself in damping of the coherent wavepacket motions (i.e., a loss of mutual phase coherence of the wavefunctions contributing to the wavepackets). In this section, dynamics and mechanisms of dephasing of O–H stretching excitations (i.e., of the high-frequency mode) are discussed. Such processes affect the quantum mechanical phase between the $v_{OH} = 0$ and $v_{OH} = 1$ wavefunctions and influence the line shape of the O–H stretching band directly, in the simplest case resulting in a homogeneously broadened band.

Dephasing of O–H stretching excitations in hydrogen bonds has been studied by different techniques of photon echo spectroscopy using resonant vibrational excitation. The experiments have concentrated on two different types of systems: (i) single or double hydrogen bonds with a well-defined intramolecular or dimer geometry [51, 52, 94] and (ii) extended disordered networks of intermolecular hydrogen bonds as in water or alcohols [50, 53–56]. In the first case, structural inhomogeneity, and thus inhomogeneous broadening and spectral diffusion, should play a minor role, whereas the different fluctuating binding geometries or strengths of hydrogen bonds in the second case should give rise to spectral diffusion on ultrafast time scales. In the following, the focus is on the first type of systems.

Coherent O–H stretching excitations were studied in 3PE experiments with PMME-H [51], a model system forming an intramolecular O–H...O hydrogen bond of well-defined geometry and displaying coherent low-frequency oscillations on excitation of the O–H stretching mode (cf. Section 2.4.2.1). The homodyne detected photon echo signal measured as a function of the coherence time τ (population time $T = 0$ corresponding to a 2PE) displays a very fast decay with a time constant on the order of 30 fs, similar to the signal from the pure solvent CCl_4 (Figure 2.17). This points to an extremely fast decay of the macroscopic coherent O–H stretching polarization, which is much faster than for stretching vibrations of free O–H groups. For the latter, a lower limit of the dephasing time of approximately 300 fs is estimated from the linewidth of the O–H stretching band. The 3PE peak shift of PMME-H exhibits an oscillatory behavior with very small absolute values of less than 10 fs [51]. The oscillation period of 330 fs corresponds to a frequency of 100 cm^{-1}, identical to that of the anharmonically coupled low-frequency mode that has been observed in the pump-probe experiments discussed above. The very small peak shift points to a negligible inhomogeneous broadening of the O–H stretching band.

Photon echo studies of hydrogen-bonded cyclic dimers of acetic acid reveal the multilevel character of vibrational coherences generated by excitation in the O–H stretching band [52]. Two-pulse photon echo data recorded with homodyne detection in the center of the O–H stretching band are presented in Figure 2.18 for mixed dimers with one O–H and one O–D oscillator (solid line) and for dimers with two O–H oscillators (dash-dotted line). After an initial very fast decay over several orders of magnitude, which is similar to the behavior of PMME-H, the signal shows pronounced recurrences with a separation of approximately 250 fs, pointing to longer-lived coherences created via the O–H stretching excitation. The results for the dimers with two O–H groups and the mixed O–H/O–D dimers are similar, suggesting a minor role of excitonic coupling for the observed nonlinear response. The 3PE peak shift for the mixed dimers (Figure 2.18b) decreases to a

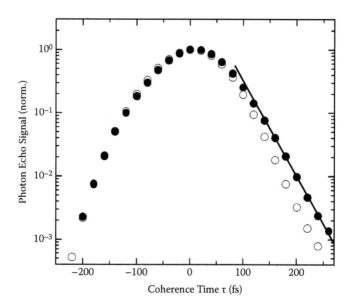

FIGURE 2.17 Homodyne detected two-pulse photon echo signal from PMME-H as a function of the coherence time τ (solid circles).The macroscopic polarization shows a decay with a time constant of 30 fs (solid line, monoexponential decay), close to the instrumental response function (open circles).

value below 10 fs during the pulse overlap and displays oscillatory features on the time scale of the recurrences of the photon echo signal.

It is interesting to relate this behavior to the pump-probe spectra of Figure 2.11b. The position and width of the spectral dips in the nonlinear absorption spectrum remain essentially unchanged for delay times up to 50 ps. This result provides direct evidence for a negligible spectral diffusion in this time domain and a predominant homogeneous broadening of the band, in agreement with the very small 3PE peak shift shown in Figure 2.18b. From the spectral width of the individual dips in the nonlinear absorption spectrum of approximately $50\,cm^{-1}$, one estimates a lower limit of the dephasing time of the O–H stretching polarization of 200 fs, attributing the dip to a single homogeneously broadened vibrational transition [52].

The recurrences of the photon echo signal reflect the long-lived ($\approx$ 1-ps) coherence in the anharmonically coupled low-frequency modes that give rise to the vibrational progressions discussed in Section 2.4.1. The broadband infrared pulses generate coherences on a subset of vibrational progression lines overlapping with the spectral envelope of the driving electric field. In this way, low-frequency coherences are induced and read out in the interaction with the third pulse in the photon echo scheme. As coherences between different low-frequency levels live substantially longer than the quantum mechanical phase between the $\nu_{OH} = 0$ and 1 states is preserved, recurrences of the photon echo occur at late coherence

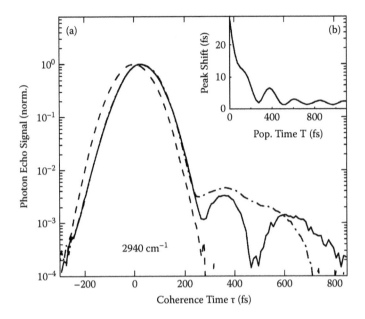

FIGURE 2.18 O–H stretching photon echo results for acetic acid dimers. (a) Homodyne detected two-pulse photon echo signal as a function of the coherence time τ for mixed (solid line) and pure (dash-dotted line) dimers. Both transients display pronounced recurrences of the signal. Dashed line: instrumental response function. (b) Three-pulse photon echo peak shift for pure acetic acid dimers.

times. This multilevel quantum beat in the vibrational manifold corresponds to an interference of different nonlinear interaction pathways that involve the different low-frequency modes.

A theoretical analysis of such behavior requires modeling of the anharmonically coupled multilevel system, a treatment that is beyond most existing models for vibrational dephasing [56, 95, 96]. Theoretical calculations based on a model Hamiltonian including—for each low-frequency mode—a third-order coupling linear in the low-frequency coordinate and quadratic in the stretching coordinate are presented in Ref. [52]. Third-order nonlinear response functions were calculated in the time domain using a sum-over-states approach, including the $v_{OH} = 0$ and 1 states and up to 20 levels of the coupled low-frequency modes. The total nonlinear response function represents a sum over Liouville space pathways, individually weighted by the relative thermal population of the initial state (temperature 300 K). Figure 2.19 displays the calculated photon echo signal (solid line). The calculation includes a 50- and a 150-cm^{-1} low-frequency mode of a 1-ps dephasing time and an instantaneous signal component around $\tau = 0$ to account for the solvent contribution. The photon echo data are reproduced quite well by this simulation. The intensity ratio of the photon echo signal at $\tau = 0$ and the recurrences set an upper limit to the dephasing time of the O–H stretching mode

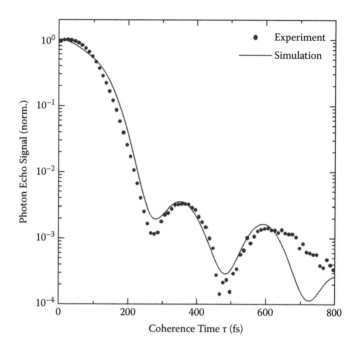

FIGURE 2.19 Calculated homodyne two-pulse photon echo signal from mixed acetic acid dimers (solid line) and experimental data (points). The calculation includes two low-frequency modes at 50 and 150 cm^{-1} that are anharmonically coupled to the O–H stretching mode. Dephasing times of 1 ps and 200 fs were assumed for the low-frequency modes and the O–H stretching vibration, respectively.

of 200 fs, a value very similar to the lower limit of approximately 200 fs that was derived from the transient absorption spectra in Figure 2.11.

Substantially more detailed information on the relevant vibrational couplings can be derived from heterodyne-detected 3PEs in combination with theoretical calculations, including ab initio vibrational potentials and the nonlinear response functions. Recently, first results have been obtained along such lines [97]. In addition to the anharmonic coupling to low-frequency modes, couplings of the $v_{OH} = 1$ state with combination and overtones of the O–H bending, the C=O and the C–O stretching are evident from such heterodyned photon echoes. Taking into account Fermi resonances and the coupling to low-frequency modes, both the photon echo results and the spectral substructure of the linear O–H stretching band were modeled in a quantitative way.

Population relaxation of the $v_{OH} = 1$ state plays a limited role for the fast dephasing of the coherent O–H stretching excitations in hydrogen-bonded PMME-H and cyclic acetic acid dimers. Even for a short lifetime of 200 fs in the acetic acid dimers, the corredponding dephasing time of $T_2 = 2T_1 = 400$ fs is substantially longer than the measured dephasing time of 200 fs. Thus, dephasing is dominated by the interaction of the hydrogen-bonded O–H stretching oscillator with the fluctuating environment [74–76,98]. It is important to note that dephasing in a hydrogen bond is much faster than for free O–H groups. Although there are no photon echo measurements for free O–H groups, one may estimate a lower limit for the dephasing time from the spectral width of the respective O–H stretching band, assuming predominant homogeneous broadening. Typical values are around 20 cm^{-1} (e.g., for phenol monomers in a nonpolar solvent such as C_2Cl_4), translating into a dephasing time $T_2 \geq 500$ fs. The fast dephasing of O–H stretching excitations in hydrogen bonds can be related to the marked anharmonicity of the O–H oscillator as discussed in Ref. [50]. The oscillator is subject to the fluctuating forces of the surrounding liquid, resulting in a modulation of its transition frequency. A basic approach of vibrational dephasing theory considers an expansion of the frequency fluctuation $\delta\omega(t)$ with forces according to

$$\hbar\delta\omega(t) = (q_{11} - q_{00})F_1(t) + ((q^2)_{11} - (q^2)_{00})F_2(t) + \ldots$$

where q_{ii} and $(q^2)_{ii}$ ($i = 0, 1$) are matrix elements of the stretching coordinate to be taken in the $v_{OH} = 0$ and 1 states, and $F_{1,2}(t)$ are the fluctuating forces of the solvent [98]. The first term describing the influence of the linear force $F_1(t)$ vanishes for a harmonic oscillator as $(q_{11} - q_{00}) = 0$. For finite anharmonicity, however, this term is nonzero and may dominate vibrational dephasing [95,99]. In this limit, the dephasing rate is proportional to the anharmonicity times the time integral of the correlation function of the linear force $F_1(t)$. Thus, this picture accounts in a qualitative way for the enhanced dephasing rates of O–H stretching excitations in hydrogen bonds.

2.5.2 POPULATION RELAXATION OF HYDROGEN-BONDED O–H STRETCHING OSCILLATORS

Processes of vibrational population relaxation and energy redistribution in hydrogen bonds have been studied extensively (for a review, see Refs. [5, 73, 100]). Although such topics are beyond the main scope of this chapter, a few comments should be made on the relaxation of O–H/O–D stretching excitations in hydrogen bonds. Different one- and two-color pump-probe schemes as well as anti-Stokes Raman scattering after infrared excitation have been applied to measure the kinetics of the $v_{OH} = 1$ populations and derive vibrational lifetimes. In addition to the kinetics of vibrational populations, spectral diffusion processes can contribute to the measured transients. This issue is particularly relevant in systems forming an extended fluctuating network of hydrogen bonds (e.g., HDO in D_2O or H_2O) and displaying spectral diffusion of the $v_{OH} = 0 \rightarrow 1$ transition on femto- to picosecond time scales. For HDO in D_2O, both spectral diffusion and population relaxation have been analyzed in detail, and a $v_{OH} = 1$ lifetime of approximately 700 fs has been derived from different experiments, a value somewhat depending on the spectral position within the O–H stretching band [101–105]. The Raman experiments of Ref. [106] have shown a substantial population of the O–H bending modes of HDO and of the surrounding D_2O through relaxation of the O–H stretching excitation. It should be noted that the $v_{OH} = 1$ population decay is much slower than the dephasing of coherent O–H stretching polarizations on a 100-fs time scale; that is, population relaxation makes a minor contribution to dephasing.

For pure H_2O, the lifetime of the $v_{OH} = 1$ state has remained controversial. The pump-probe data in Ref. [107] have been interpreted in terms of a spectral diffusion below 100 fs, a 260-fs lifetime of the $v_{OH} = 1$ state, and a 550-fs lifetime of an intermediate state populated through the relaxation of the O–H stretching oscillator. In contrast, the 300- and 600-fs kinetic components found in the time-resolved anti-Stokes Raman signals have been attributed to spectral diffusion and the $v_{OH} = 1$ lifetime, respectively [108–110].

Spectral diffusion in the subpicosecond time domain is essentially absent in intramolecular hydrogen bonds and hydrogen-bonded dimers with a well-defined geometry that remains unchanged on O–H stretching excitation. For cyclic acetic acid dimers in nonpolar solution, an O–H stretching lifetime of 200 fs has been measured (Figure 2.12). Two-color pump-probe studies gave evidence of a strong anharmonic coupling of the O–H stretching vibration to the O–H bending (δ_{OH}) and the C–O stretching (v_{CO}) mode, on top of the coupling to the low-frequency modes discussed in Section 2.4.2.2. There is, however, no evidence for a transfer of vibrational population from the $v_{OH} = 1$ state into the $v = 1$ states of the infrared active δ_{OH} and v_{CO} modes [46].

Different pathways of population relaxation have been considered for the $v_{OH} = 1$ state of hydrogen-bonded O–H groups. Energy transfer from the $v_{OH} = 1$ state into high-lying levels of the low-frequency hydrogen bond modes is considered in the vibrational predissociation model, making use of an adiabatic separation of time scales of the high- and low-frequency modes [111]; (cf. Section 2.4.1). In Ref. [111], a single O–H . . . O hydrogen bond was considered. Using empirical Lippincott–Schroeder potentials for the low-frequency O . . . O mode in the $v_{OH} = 0$ and 1 states, it was shown that $v_{OH} = 1$ lifetimes on the order of 10 ps result from the pronounced nonadiabatic coupling of the two modes and the substantial overlap of bound and continuum wavefunctions of the low-frequency mode at the inner turning point of the $v_{OH} = 1$ potential. This picture has also been invoked to explain the much shorter (0.7-ps) lifetime of the O–H stretching vibration of HDO in D_2O [100]. On the other hand, population transfer from the $v_{OH} = 1$ state into the $v = 2$ state of the O–H bending mode δ_{OH} is considered a major pathway of relaxation. Depending on the particular system, there may be a Fermi resonance between the $v_{OH} = 1$ and the $v = 2$ state of δ_{OH}, resulting in new coupled states or a substantial energy mismatch that has to be taken up by other modes (e.g., those of the solvent). Recent theoretical calculations of the $v_{OH} = 1$ lifetime of HDO in D_2O [79, 112, 113] give values between 2.3 and 8 ps, depending on the particular theoretical approach, as discussed in a recent review [73]. Such calculated lifetimes are substantially longer than the experimental value of 0.7 ps. This discrepancy suggests several relaxation pathways contributing to the $v_{OH} = 1$ decay. Further experimental and theoretical work is required to unravel the relaxation schemes of HDO in D_2O, H_2O, acetic acid dimers, and other hydrogen-bonded systems.

The O–H stretching relaxation in the intramolecular hydrogen bond of PMME-H has been investigated, combining two-color pump-probe experiments with quantum dynamical calculations [83]. The lifetimes of the $v = 1$ states of the O–H stretching and bending modes are 220 and 800 fs, respectively. There is clear experimental evidence for a population of the $v = 1$ state of δ_{OH} by relaxation of the O–H stretching excitation. In addition to the O–H stretching and bending vibrations, the theoretical model applied includes the O . . . O hydrogen bond mode, as well as two O–H out-of-plane deformation modes at 700 and 800 cm^{-1}, displaying a strong anharmonic coupling to the O–H stretching mode. The calculations suggest a population of the $v = 1$ state of δ_{OH} through combination tones made up of a δ_{OH} excitation plus two quanta of the out-of-plane modes. In contrast, depopulation of the O–H stretching mode through high-lying levels of the O . . . O hydrogen bond mode plays a minor role. The out-of-plane modes are also essential for the 800-fs decay of the bending population.

After depopulation of the O–H/O–D stretching vibration, the vibrational excess energy is redistributed over a larger number of modes (i.e., a hot ground state is created). This state gives rise to a blue-shifted absorption on the $v_{OH} = 0 \rightarrow 1$ transition, as discussed in Section 2.4.2. The comparably slow disappearance of the blue shift on a time scale of several tens of picoseconds reflects the transfer of vibrational energy to the surrounding solvent [87].

In conclusion, population relaxation of O–H stretching excitations in hydrogen bonds represents an important field of ultrafast vibrational spectroscopy. Although an increasing amount of experimental information has become available, the time scales and particularly the pathways of vibrational relaxation are not fully understood. Multicolor pump-probe and Raman experiments have the potential to follow population transfers into different anharmonically coupled modes and thus to unravel the different channels of population and energy redistribution.

2.6 CONCLUSIONS

Nonlinear vibrational spectroscopy in the femtosecond time domain provides direct insight into both the dynamics and couplings of vibrational excitations of hydrogen bonds, information that is difficult to derive from linear vibrational spectra. Femtosecond excitation of the O–H or O–D stretching vibrations in hydrogen bonds creates different types of quantum coherence that reflect the anharmonic coupling to modes at lower frequencies (e.g., fingerprint vibrations or low-frequency modes of the hydrogen bonds) as well as the role of the fluctuating liquid environment for vibrational dephasing. Coherent wavepacket motions along low-frequency modes (i.e., a nonstationary quantum coherent superposition of wavefunctions of different low-frequency levels) were observed in intramolecular hydrogen bonds and hydrogen-bonded dimers, in the latter case displaying motions along several underdamped modes. The coherent low-frequency response is dominated by wavepackets in the $v_{OH} = 0$ state, which are created through a Raman process resonantly enhanced by the O–H stretching transition dipole. The comparably slow dephasing of such motions on a time scale of 1 to 2 ps demonstrates the underdamped character of the coupled modes, even in a liquid environment, and their quantized nature in contrast to the frequently assumed classical description of slow vibrational coordinates. The occurrence of wavepacket motions also confirms the much-debated picture of vibrational low-frequency progressions within the strongly broadened O–H stretching bands, as introduced in the early theoretical literature on linear vibrational spectra.

So far, such oscillations have not been observed in extended fluctuating networks of hydrogen bonds as in water. There are, however, indications of a slightly underdamped intermolecular O....O motion of HDO in D_2O, from both photon echo studies and molecular dynamics simulations. The picosecond decay of low-frequency coherences allows for generating and manipulating vibrational motion with phase-shaped infrared pulses. This may be of particular interest for reactive systems in which processes of hydrogen transfer along a preexisting hydrogen bond occur and may become accessible for optical control.

The decay of quantum coherence on the O–H stretching transition of high frequency is much faster, reflecting the loss of quantum coherence between the wavefunctions of the $v_{OH} = 0$ and 1 states. Here, coupling of the stretching oscillator to the fluctuating environment plays the dominant role. Dephasing rates are enhanced by the pronounced anharmonicity of the O–H stretching vibration in hydrogen bonds, making the oscillator much more sensitive to the fluctuating linear force of the surrounding. In a number of systems with well-defined hydrogen-bonding geometries, one finds an essentially homogeneous broadening of the O–H stretching band, whereas O–H stretching excitations in fluctuating hydrogen-bonded networks display spectral diffusion on a multitude of time scales, directly influencing the vibrational line shapes.

ACKNOWLEDGMENTS

I would like to acknowledge the important contributions of my present and former coworkers Jens Stenger, Dorte Madsen, Nils Huse, Karsten Heyne, Jens Dreyer, Peter Hamm, and Erik Nibbering to the work reviewed in this chapter. It is my pleasure to thank Casey Hynes for many interesting discussions on hydrogen bonding. I also thank the Deutsche Forschungsgemeinschaft and the Fonds der Chemischen Industrie for financial support.

REFERENCES

1. P. Schuster, G. Zundel, and C. Sandorfy (Eds.). *The Hydrogen Bond: Recent Developments in Theory and Experiment*, Vols. 1–3. North Holland, Amsterdam, 1976.
2. T. Elsaesser and H. J. Bakker (Eds.). *Ultrafast Hydrogen Bonding Dynamics and Proton Transfer Processes in the Condensed Phase*, Kluwer, Dordrecht, 2002.

3. S. Mukamel, *Principles of Nonlinear Optical Spectroscopy*, University Press, Oxford, 1995.

4. M. D. Fayer (Ed.). *Ultrafast Infrared and Raman Spectroscopy*, Dekker, New York, 2001.

5. E. T. J. Nibbering and T. Elsaesser. Ultrafast vibrational dynamics of hydrogen bonds in the condensed phase. *Chem. Rev.*, 104:1887–1914, 2004.

6. O. Svelto. *Principles of Lasers*, 4th ed. Plenum Press, New York, 1998.

7. Y. R. Shen. *The Principles of Nonlinear Optics*, Chap. 5, Wiley, New York, 1984.

8. V. G. Dmitriev, G. G. Gurzadyan, and D. N. Nikogosyan. *Handbook of Nonlinear Optical Crystals*, Springer-Verlag, Berlin, 1991.

9. A. M. Weiner. Effect of group-velocity mismatch on the measurement of ultrashort pulse via second harmonic generation. *IEEE J. Quantum Electron.*, 19:1276–1283, 1983.

10. M. Joffre, A. Bonvalet, A. Migus, and J. L. Martin. Femtosecond diffracting Fourier-transform infrared interferometer. *Opt. Lett.*, 21:964–967, 1996.

11. A. Bonvalet, M. Joffre, J. L. Martin, and A. Migus. Generation of ultrabroadband femtosecond pulses in the mid-infrared by optical rectification of 15 fs light pulses at 100 MHz repetition rate. *Appl. Phys. Lett.*, 67:2907–2909, 1995.

12. R. A. Kaindl, D. C. Smith, M. Joschko, M. P. Hasselbeck, M. Woerner, and T. Elsaesser. Femtosecond infrared pulses tunable from 9 to 18 μm at 88 MHz repetition rate. *Opt. Lett.*, 23:861–863, 1998.

13. R. A. Kaindl, F. Eickemeyer, M. Woerner, and T. Elsaesser. Broadband phase-matched difference frequeny mixing of femtosecond pulses in GaSe: experiment and theory. *Appl. Phys. Lett.*, 75:1060–1062, 1999.

14. R. Huber, A. Brodschelm, F. Tauser, and A. Leitenstorfer. Generation and field-resolved detection of femtosecond electromagnetic pulses tunable up to 41 THz. *Appl. Phys. Lett.*, 76:3191–3193, 2000.

15. E. S. Wachman, D. C. Edelstein, and C. L. Tang. Continuous-wave mode-locked and dispersion-compensated femtosecond optical parametric oscillator. *Opt. Lett.*, 15:136–138, 1990.

16. Q. Fu, G. Mak, and H. M. van Driel. High-power, 62-fs infrared optical parametric oscillator synchronously pumped by a 76-MHz Ti:sapphire laser. *Opt. Lett.*, 17:1006–1008, 1992.

17. K. C. Burr, C. L. Tang, M. A. Arbore, and M. M. Fejer. Broadly tunable mid-infrared femtosecond optical parametric oscillator using all-solid-state-pumped periodically poled lithium niobate. *Opt. Lett.*, 22:1458–1460, 1997.

18. A. Lohner, P. Kruck, and W. W. Rühle. Generation of 200 fs pulses tunable between 2.5 and 5.5 μm. *Appl. Phys. B*, 59:211–213, 1994.

19. J. M. Fraser, D. Wang, A. Hache, G. R. Allan, and H. M. van Driel. Generation of high-repetition-rate femtosecond pulses from 8 to 18 μm. *Appl. Opt.*, 36:5044–5047, 1997.

20. S. Ehret and H. Schneider. Generation of subpicosecond infrared pulses tunable between 5.2 μm and 18 μm at a repetition rate of 76 MHz. *Appl. Phys. B*, 66:27–30, 1998.

21. T. Elsaesser and M. C. Nuss. Femtosecond pulses in the mid-infrared generated by down-conversion of a traveling wave dye laser. *Opt. Lett.*, 16:411–413, 1991.

22. P. Hamm, C. Lauterwasser, and W. Zinth. Generation of tunable subpicosecond light pulses in the midinfrared between 4.5 and 11.5 μm. *Opt. Lett.*, 18:1943–1945, 1993.

23. W. R. Bosenberg and R. C. Eckardt (Eds.). Special issue on parametric devices. *J. Opt. Soc. Am. B*, vol. 12, 1995.
24. V. Petrov, F. Rotermund, and F. Noack. Generation of high-power femtosecond light pulses at 1 kHz in the mid-infrared spectral range between 3 and 12 μm by second-order nonlinear processes in crystals. *J. Opt. A: Pure Appl. Opt.*, 3:R1–R19, 2001.
25. F. Seifert, V. Petrov, and M. Woerner. Solid-state laser system for the generation of midinfrared femtosecond pulses tunable from 3.3 to 10 μm. *Opt. Lett.*, 19:2009–2011, 1994.
26. R. A. Kaindl, M. Wurm, K. Reimann, P. Hamm, A. M. Weiner, and M. Woerner. Generation, shaping, and characterization of intense femtosecond pulses tunable from 3 to 20 μm. *J. Opt. Soc. Am. B*, 17:2086–2094, 2000.
27. P. Hamm, R. A. Kaindl, and J. Stenger. Noise suppression in femtosecond mid-infrared light sources. *Opt. Lett.*, 25:1798–1800, 2000.
28. K. Reimann, R. P. Smith, A. M. Weiner, T. Elsaesser, and M. Woerner. Direct field-resolved detection of terahertz transients with megavolt/cm amplitudes. *Opt. Lett.*, 28:471–473, 2003.
29. Q. Wu and X. C. Zhang. Free-space electro-optics sampling of mid-infrared pulses. *Appl. Phys. Lett.*, 71:1285–1287, 1997.
30. J. A. Gruetzmacher and N. P. Scherer. Few-cycle mid-infrared pulse generation, characterization, and coherent propagation in optically dense media. *Rev. Sci. Instrum.*, 73:2227–2236, 2002.
31. B. A. Richman, K. W. DeLong, and R. Trebino. Temporal characterization of the Stanford mid-IR FEL micropulses by "FROG." *Nucl. Instrum. Meth. A*, 358:268–271, 1995.
32. D. T. Reid, P. Loza-Alvarez, C. T. A. Brown, T. Beddard, and W. Sibbett. Amplitude and phase measurement of mid-infrared femtosecond pulses by using cross-correlation frequency-resolved optical gating. *Opt. Lett.*, 25:1478–1480, 2000.
33. F. Eickemeyer, R. A. Kaindl, M. Woerner, T. Elsaesser, and A. M. Weiner. Controlled shaping of ultrafast electric field-transients in the mid-infrared spectral range. *Opt. Lett.*, 25:1472–1474, 2000.
34. T. Witte, D. Zeidler, D. Proch, K. L. Kompa, and M. Motzkus. Programmable amplitude- and phase-modulated femtosecond laser pulses in the mid-infrared. *Opt. Lett.*, 27:131–133, 2002.
35. N. Belabas, J. P. Likforman, L. Canioni, B. Bousquet, and M. Joffre. Coherent broadband pulse shaping in the mid infrared. *Opt. Lett.*, 26:743–745, 2001.
36. H. S. Tan, E. Schreiber, and W. S. Warren. High-resolution indirect pulse shaping by parametric transfer. *Opt. Lett.*, 27:439–441, 2002.
37. T. Witte, T. Hornung, L. Windhorn, D. Proch, R. de Vivie-Riedle, M. Motzkus, and K. L. Kompa. Controlling molecular ground-state dissociation by optimizing vibrational ladder climbing. *J. Chem. Phys.*, 118:2021–2024, 2003.
38. G. M. H. Knippels, R. F. X. A. M. Mols, A. F. G. van der Meer, D. Oepts, and P. W. van Amersfoort. Intense far-infrared free-electron laser pulses with a length of six optical cycles. *Phys. Rev. Lett.*, 75:1755–1758, 1995.
39. G. M. H. Knippels, M. J. van de Pol, H. P. M. Pellemans, P. C. M. Planken, and A. F. G. van der Meer. Two-color facility based on a broadly tunable infrared free-electron laser and a subpicosecond-synchronized 10-fs-Tisapphire laser. *Opt. Lett.*, 23:1754–1756, 1998.

40. D. D. Dlott. Vibrational energy redistribution in polyatomic liquids: 3D infrared Raman spectroscopy. *Chem. Phys.*, 266:149–166, 2001.

41. M. Woerner, A. Seilmeier, and W. Kaiser. Reshaping of infrared picosecond pulses after passage through atmospheric CO_2. *Opt. Lett.*, 14:636–638, 1989.

42. W. T. Pollard and R. A. Mathies. Analysis of femtosecond dynamic absorption spectra of nonstationary states. *Annu. Rev. Phys. Chem.*, 43:497–523, 1992.

43. P. Hamm and R. M. Hochstrasser. Structure and dynamics of proteins and peptides: femtosecond two-dimensional infrared spectroscopy. In *Ultrafast Infrared and Raman Spectroscopy* (MD, Fayer, Ed.). Dekker, New York, 2001.

44. S. Mukamel. Multidimensional femtosecond correlation spectroscopies of electronic and vibrational excitations. *Annu. Rev. Phys. Chem.*, 51:691–729, 2000.

45. A. M. Moran, S. M. Park, J. Dreyer, and S. Mukamel. Linear and nonlinear infrared signatures of local α- and 3_{10}-helical structures in alanine polypeptides. *J. Chem. Phys.*, 118:3651–3659, 2003.

46. K. Heyne, N. Huse, E. T. J. Nibbering, and T. Elsaesser. Ultrafast relaxation and anharmonic coupling of O–H stretching and bending excitations in cyclic acetic acid dimers. *Chem. Phys. Lett.*, 382:19–25, 2003.

47. G. R. Fleming and M. Cho. Chromophore-solvent dynamics. *Annu. Rev. Phys. Chem.*, 47:109–134, 1996.

48. W. P. de Boeij, M. S. Pshenichnikov, and D. A. Wiersma. Ultrafast solvation dynamics explored by femtosecond photon echo spectroscopies. *Annu. Rev. Phys. Chem.*, 49:99–123, 1998.

49. D. M. Jonas. Two-dimensional femtosecond spectroscopy. *Annu. Rev. Phys. Chem.*, 54:425–463, 2003.

50. J. Stenger, D. Madsen, P. Hamm, E. T. J Nibbering, and T. Elsaesser. Ultrafast vibrational dephasing of liquid water. *Phys. Rev. Lett.*, 87:027401/1–4, 2001.

51. J. Stenger, D. Madsen, J. Dreyer, P. Hamm, E. T. J. Nibbering, and T. Elsaesser. Femtosecond mid-infrared photon echo study of an intramolecular hydrogen bond. *Chem. Phys. Lett.*, 354:256–263, 2002.

52. N. Huse, K. Heyne, J. Dreyer, E. T. J. Nibbering, and T. Elsaesser. Vibrational multi-level quantum coherence due to anharmonic couplings in intermolecular hydrogen bonds. *Phys. Rev. Lett.*, 91:197401/1–4, 2003.

53. J. Stenger, D. Madsen, P. Hamm, E. T. J. Nibbering, and T. Elsaesser. A photon echo peak shift study of liquid water. *J. Phys. Chem. A*, 106:2341–2350, 2002.

54. S. Yeremenko, M. S. Pshenichnikov, and D. A. Wiersma. Hydrogen-bond dynamics in water explored by heterodyne detected photon echo. *Chem. Phys. Lett.*, 369:107–113, 2003.

55. C. J. Fecko, J. D. Eaves, J. J. Loparo, A. Tokmakoff, and P. L. Geissler. Ultrafast hydrogen-bond dynamics in the infrared spectroscopy of water. *Science*, 301:1698–1702, 2003.

56. J. B. Asbury, T. Steinel, C. Stromberg, S. A. Corcelli, C. P. Lawrence, J. L. Skinner, and M. D. Fayer. Water dynamics: vibrational echo correlation spectroscopy and comparison to molecular dynamics simulations. *J. Phys. Chem. A*, 108:1107–1119, 2004.

57. R. Agarwal, B. S. Prall, A. H. Rizvi, M. Yang, and G.R. Fleming. Two-color three pulse photon echo peak shift spectroscopy. *J Chem. Phys.*, 116:6243–6252, 2002.

58. C. Scheurer, A. Piryatinski, and S. Mukamel. Signatures of beta-peptide un-
    folding in two-dimensional vibrational echo spectroscopy. *J. Am. Chem. Soc.*,
    123:3114–3124, 2001.

59. I. V. Rubtsov, J. P. Wang, and R. M. Hochstrasser. Dual-frequency 2D-IR
    spectroscopy: heterodyned photon echo of the peptide bond. *Proc. Natl. Acad. Sci.
    U.S.A.*, 100:5601–5606, 2003.

60. T. Joo, Y. Jia, J. Y. Yu, M. J. Lang, and G. R. Fleming. Third-order nonlinear time
    domain probes of solvation dynamics. *J. Chem. Phys.*, 104:6089–6108, 1996.

61. W. P. de Boeij, M. S. Pshenichnikov, and D. A. Wiersma. On the relation between
    the echo-peak shift and Brownian-oscillator correlation function. *Chem. Phys.
    Lett.*, 253:53–60, 1996.

62. C. Sandorfy. The near-infrared—a reminder. *Bull. Pol. Acad. Sci., Chem.*, 43:7–24,
    1995.

63. W. Mikenda and S. Steinbock. Stretching frequency versus bond distance corre-
    lation of hydrogen bonds in solid hydrates: a generalized correlation function. *J.
    Mol. Struct.*, 384:159–163, 1996.

64. B. I. Stepanov. Interpretation of the regularities in the spectra of molecules
    forming the intermolecular hydrogen bond by the predissociation effect. *Nature*,
    157:808–810, 1946.

65. Y. Maréchal and A. Witkowski. Infrared spectra of H-bonded systems. *J. Chem.
    Phys.*, 48:3697–3705, 1968.

66. Y. Maréchal. IR spectra of carboxylic acids in the gas phase: a quantitative
    reinvestigation. *J. Chem. Phys.*, 87:6344–6353, 1987.

67. O. Henri-Rousseau, P. Blaise, and D. Chamma. Infrared lineshapes of weak hydro-
    gen bonds: recent quantum developments. *Adv. Chem. Phys.*, 121:241–309, 2002.

68. D. Chamma and O. Henri-Rousseau. IR theory of weak hydrogen bonds: Davydov
    coupling, Fermi resonances and direct relaxations. I. Basic equations within the
    linear response theory. *Chem. Phys.*, 248:53–70, 1999.

69. K. Heyne, N. Huse, J. Dreyer, E. T. J. Nibbering, and T. Elsaesser. Coherent
    low-frequency motions of hydrogen bonded acetic acid dimers in the liquid phase.
    J. Chem. Phys., 121:902–913, 2004.

70. M. Lim and R. M. Hochstrasser. Unusual vibrational dynamics of the acetic acid
    dimer. *J. Chem. Phys.*, 115:7629–7643, 2001.

71. G. M. Florio, T. S. Zwier, E. M. Myshakin, K. D. Jordan, and E. L. Sibert, III.
    Theoretical modeling of the OH stretch infrared spectrum of carboxylic acid dimers
    based on first-principles anharmonic couplings. *J. Chem. Phys.*, 118:1735–1746,
    2003.

72. C. Emmeluth, M. A. Suhm, and D. Luckhaus. A monomers-in-dimers model for
    carboxylic acid dimers. *J. Chem. Phys.*, 118:2242–2255, 2003.

73. R. Rey, K. B. Moller, and J. T. Hynes. Ultrafast vibrational population dynamics of
    water and related systems: a theoretical perspective. *Chem. Rev.*, 104:1915–1928,
    2004.

74. N. Rösch and M. Ratner. Model for the effects of a condensed phase on the infrared
    spectra of hydrogen-bonded systems. *J. Chem. Phys.*, 61:3344–3351, 1974.

75. S. Bratos. Profiles of hydrogen stretching ir bands of molecules with hydrogen
    bonds: a stochastic theory. I. Weak and medium-strong hydrogen bonds. *J. Chem.
    Phys.*, 63:3499–3509, 1975.

76. G. N. Robertson and J. Yarwood. Vibrational relaxation of hydrogen-bonded
    species in solution. I. Theory. *Chem. Phys.*, 32:267–282, 1978.

77. M. Diraison, Y. Guissani, J. C. Leicknam, and S. Bratos. Femtosecond solvation dynamics of water: solvent response to vibrational excitation of the solute. *Chem. Phys. Lett.*, 258:348–351, 1996.

78. R. Rey, K. B. Moller, and J. T. Hynes. Hydrogen bond dynamics in water and ultrafast infrared spectroscopy. *J. Phys. Chem. A*, 106:11993–11996, 2002.

79. C. P. Lawrence and J. L. Skinner. Vibrational spectroscopy of HOD in liquid D_2O. I. Vibrational energy relaxation. *J. Chem. Phys.*, 117:5827–5838, 2002.

80. C. P. Lawrence and J. L. Skinner. Vibrational spectroscopy of HOD in liquid D_2O. II. Infrared line shapes and vibrational Stokes shift. *J. Chem. Phys.*, 117:8847–8854, 2002.

81. J. Stenger, D. Madsen, J. Dreyer, E. T. J. Nibbering, P. Hamm, and T. Elsaesser. Coherent response of hydrogen bonds in liquids probed by ultrafast vibrational spectroscopy. *J. Phys. Chem. A*, 105:2929–2932, 2001.

82. D. Madsen, J. Stenger, J. Dreyer, P. Hamm, E. T. J. Nibbering, and T. Elsaesser. Femtosecond mid-infrared pump-probe study of wave packet motion in a medium-strong intramolecular hydrogen bond. *Bull. Chem. Soc. Jpn.*, 75:909–917, 2002.

83. K. Heyne, E. T. J. Nibbering, T. Elsaesser, M. Petkovic, and O. Kühn. Cascaded energy redistribution upon O–H stretching excitation in an intramolecular hydrogen bond. *J. Phys. Chem. A*, 108:6083–6086, 2004.

84. H. Naundorf, G. A. Worth, H. D. Meyer, and O. Kühn. Multiconfiguration time-dependent Hartree dynamics on an ab initio reaction surface: ultrafast laser-driven proton motion in phthalic acid monomethylester. *J. Phys. Chem. A*, 106:719–724, 2002.

85. O. Kühn and H. Naundorf. Dissipative wave packet dynamics of the intramolecular hydrogen bond in *o*-phthalic acid monomethylester. *Phys. Chem. Chem. Phys.*, 5:79–86, 2003.

86. D. Madsen, J. Stenger, J. Dreyer, E. T. J. Nibbering, P. Hamm, and T. Elsaesser. Coherent vibrational ground state dynamics of an intramolecular hydrogen bond. *Chem. Phys. Lett.*, 341:56–61, 2001.

87. T. Elsaesser and W. Kaiser. Vibrational and vibronic relaxation of large polyatomic molecules in liquids. *Annu. Rev. Phys. Chem.*, 42:83–108, 1991.

88. G. Seifert, T. Patzlaff, and H. Graener. Ultrafast vibrational dynamics of doubly hydrogen bonded acetic acid dimers in liquid solution. *Chem. Phys. Lett.*, 333:248–254, 2001.

89. K. Heyne, N. Huse, E. T. J. Nibbering, and T. Elsaesser. Ultrafast coherent nuclear motions of hydrogen bonded carboxylic acid dimmers. *Chem. Phys. Lett.*, 369:591–596, 2003.

90. K. Heyne, N. Huse, E. T. J. Nibbering, and T. Elsaesser. Coherent vibrational dynamics of intermolecular hydrogen bonds in acetic acid dimers studied by ultrafast mid-infrared spectroscopy. *J. Phys.: Condens. Matter*, 15:S129–S135, 2003.

91. O. F. Nielsen and P. A. Lund. Intermolecular Raman active vibrations of hydrogen bonded acetic acid dimers in the liquid state. *J. Chem. Phys.*, 78:652–655, 1983.

92. T. Nakabayashi, K. Kosugi, and N. Nishi. Liquid structure of acetic acid studied by Raman spectroscopy and ab initio molecular orbital calculations. *J. Phys. Chem. A*, 103:8595–8603, 1999.

93. J. Florian, V. Hrouda, and P. Hobza. Proton transfer in the adenine-thymine base pair. *J. Am. Chem. Soc.*, 116:1457–1460, 1994.

94. A. T. Krummel, N. Mukherjee, and M. T. Zanni. Inter- and intrastrand vibrational couplings in DNA studied with heterodyned 2D-IR spectroscopy. *J. Phys. Chem. B*, 107:9165–9169, 2003.

95. A. Piryatinski, C. P. Lawrence, and J. L. Skinner. Vibrational spectroscopy of HDO in liquid D_2O. IV. Infrared two-pulse photon echoes. *J. Chem. Phys.*, 118:9664–9671, 2003.

96. A. Piryatinski, C. P. Lawrence, and J. L. Skinner. Vibrational spectroscopy of HDO in liquid D_2O. V. Infrared three-pulse photon echoes. *J. Chem. Phys.*, 118:9672–9679, 2003.

97. N. Huse, B. D. Bruner, M. L. Cowan, J. Dreyer, E. T. J. Nibbering, R. J. D. Miller, and T. Elsaesser. Anharmonic couplings underlying the ultrafast vibrational dynamics of hydrogen bonds in liquids. *Phy. Rev. Lett.*, 95:147402/1–4, 2005.

98. D. W. Oxtoby. Dephasing of molecular vibrations in liquids. *Adv. Chem. Phys.*, 40:1–48, 1979.

99. D. W. Oxtoby, D. Levesque, and J. J. Weis. Molecular dynamics simulation of dephasing in liquid nitrogen. *J. Chem. Phys.*, 68:5528–5533, 1978.

100. H. J. Bakker. Femtosecond mid-infrared spectroscopy of water. In: *Ultrafast hydrogen bonding dynamics and proton transfer processes in the condensed phase* (T. Elsaesser and H. J. Bakker, Eds.), Kluwer, Dordrecht, 2002:31–72.

101. R. Laenen, C. Rauscher, and A. Laubereau. Dynamics of local substructures in water observed by ultrafast infrared hole burning. *Phys. Rev. Lett.*, 80:2622–2625, 1998.

102. G. M. Gale, G. Gallot, F. Hache, N. Lascoux, S. Bratos, and J. C. Leicknam. Femtosecond dynamics of hydrogen bonds in liquid water: a real time study. *Phys. Rev. Lett.*, 82:1068–1071, 1999.

103. G. M. Gale, G. Gallot, and N. Lascoux. Frequency-dependent vibrational population relaxation time of the OH stretching mode in liquid water. *Chem. Phys. Lett.*, 311:123–125, 1999.

104. S. Woutersen, U. Emmerichs, H. K. Nienhuys, and H. J. Bakker. Anomalous temperature dependence of vibrational lifetimes in water and ice. *Phys. Rev. Lett.*, 81:1106–1109, 1998.

105. H. K. Nienhuys, S. Woutersen, R. A. van Santen, and H. J. Bakker. Mechanism for vibrational relaxation in water investigated by femtosecond infrared spectroscopy. *J. Chem. Phys.*, 111:1494–1500, 1999.

106. J. C. Deak, S. T. Rhea, L. K. Iwaki, and D. D. Dlott. Vibrational energy relaxation and spectral diffusion in water and deuterated water. *J. Phys. Chem. A*, 104:4866–4875, 2000.

107. A. J. Lock and H. J. Bakker. Temperature dependence of vibrational relaxation in liquid H_2O. *J. Chem. Phys.*, 117:1708–1713, 2002.

108. A. Pakoulev, Z. Wang, and D. D. Dlott. Vibrational relaxation and spectral evolution following ultrafast OH stretch excitation of water. *Chem. Phys. Lett.*, 371:594–600, 2003.

109. H. J. Bakker, A. J. Lock, and D. Madsen. Comment on "Vibrational relaxation and spectral evolution following ultrafast OH stretch excitation of water." *Chem. Phys. Lett.*, 385:329–331, 2004.

110. A. Pakoulev, Z. Wang, Y. Pang, and D. D. Dlott. Reply to comment on "Vibrational relaxation and spectral evolution following ultrafast OH stretch excitation of water" by H. J. Bakker, A. J. Lock, and D. Madsen. *Chem. Phys. Lett.*, 385:332–335, 2004.

111. A. Staib and J. T. Hynes. Vibrational predissociation in hydrogen-bonded OH...O complexes via OH stretch-OO stretch energy transfer. *Chem. Phys. Lett.*, 204:197–205, 1993.

112. R. Rey and J. T. Hynes. Vibrational energy relaxation of HOD in liquid D_2O. *J. Chem. Phys.*, 104:2356–2368, 1996.

113. C. P. Lawrence and J. L. Skinner. Vibrational spectroscopy of HOD in liquid D_2O. VI. Intramolecular and intermolecular vibrational energy flow. *J. Chem. Phys.*, 119:1623–1633, 2003.

3 Coherent Phonon Dynamics in π-Conjugated Chains

Guglielmo Lanzani

CONTENTS

3.1 INTRODUCTION

3.1.1 Vibrational Dynamics in Conjugated Polymers

The object of this chapter is the time domain investigation of phonon dynamics in carbon-based π-conjugated materials, including polyenes, carotenoids, and semiconducting polymers, which are photo-active in a number of systems or devices. All have in common π-electron delocalization along the carbon skeleton, which is responsible for the electronic properties and the coupling to the chain vibrations.

Many polyenes and carotenoids are found in living systems as photoreceptors; evolution has resulted in the matching of their typical transition energies with the environmental lighting condition (e.g., green-yellow for organisms exposed to daylight, blue for bacteria living in deep water). Their role in complex biological processes, the structure–property relationship, and the detailed electronic structures are subject to intensive and exciting research.

Conjugated polymers (CPs) are molecular semiconductors that have a large amount of technological impact on industrial areas such as lighting, display, automotive products, telecommunications, energy conversion, energy storage, and more. Their optoelectronic and photonics properties, again due to π-electrons, have been under investigation since 1980. Amid spectacular development in technology, with commercial products now on the shelf, basic understanding is somewhat lagging back and not yet satisfactory.

Here, we focus on the specific issue of electron–phonon coupling and vibrational dynamics in carbon-conjugated chains. Electron–phonon coupling governs excited-state (ES) relaxation, energy dissipation, thermal properties, and transport. It is large in a one-dimensional system due to ease of deformation on perturbation, and it has been largely investigated because of the intimate connection with the electronic structure, the ground-state (GS) and ES geometries, and the molecular functions. Research has involved applying conventional tools such as optical spectroscopy [1], infrared and Raman spectroscopy [2–4], and nonlinear, spectrally resolved techniques (e.g., Coherent anti-Stokes Roman spectroscopy [CARS]) [5–7]. Such studies highlight the role of a few "special" modes and of many that have strong coupling to the optical (π–π^*) transition. These are *breathing* modes, which do not lower the symmetry of the system and can be Raman active according to the Frank–Condon mechanism. Typically, conjugated chains have single and double carbon stretching modes, which get resonantly enhanced in the Raman shift region between 1000 and 2000 cm^{-1}. These vibrations play a crucial role in stabilizing the chain structure because they "drive" dimerization, leading to single and double bonds. The reason for that helps in understanding the origin of electron phonon coupling in such systems and deserves some attention.

Let us start with the ideal linear, infinite chain of equally spaced (i.e., nondimerized), frozen carbon atoms, each contributing one noninteracting π-electron. This would be a metal in the single-particle approximation (tight-binding or Hückel approximation) with half-filled band and zero gap. This is an oversimplified picture. Including electron correlation, for instance, causes metal–semiconductor transition. In addition, there is a geometrical argument.

In 1955, Peierls stated a theorem for the infinite chain [8]: a one-dimensional (1-D) solid with half-filled band is subject to a periodic distortion of the lattice that mixes energy levels close to the Fermi level. In a system with one π-electron per atom, as carbon chains typically are, the Fermi level is at $K_F = \pi/2a$, where a is the lattice constant, so that the maximum energy gain occurs for a distortion of period $2a$ (i.e., dimerization of the structure with doubling of the primitive cell). Pushing π-electron empty levels up and filled ones down, the mechanism is energetically favorable until balance is reached with sigma-bonds, which get squeezed and stretched, enhancing elastic energy. Those normal coordinates that are associated with such distortion modulate the electronic gap, hence the strong coupling with the optical transition. The experimental evidence for the existence of bond alternation, based on structural probes, supports the geometrical argument, even though electronic correlation and on-site repulsion also play a role, and may indeed contribute to enhance dimerization [9–11].

3.1.2 TIME-RESOLVED VIBRATIONAL STUDIES: A CRITICAL SURVEY ON AVAILABLE TECHNIQUES

Time resolution can be achieved in vibrational spectroscopy in several ways; for a review, see Ref. [12]. In standard pump-probe experiments, the probe beam can be used as quasi-monochromatic resonance Raman excitation for spontaneous scattering [13]. Raman spectra can thus be recorded at different pump-probe delays. The Stokes resonance Raman scattering (RRS) intensity is $\sim A(w)(n+1)$, where A is the electronic factor containing the energy denominators and the polarizability, and n is the phonon population. For both pump and probe in resonance with the optical transition, the differential signal is $\sim \Delta A(w)$ since the phonon population is negligible for large wavenumbers typically involved in polymers. This implies that the simplest outcome of the experiment is population dynamics monitored by GS RRS, used as a "chemical" selective probe [13]. ES Raman scattering can be obtained by tuning the probe in resonance with ES transitions. This has been done successfully in a number of compounds, such as β-carotene [14].

The technique suffers a number of constraints. Most important, spontaneous Raman scattering is rather weak, easily overwhelmed by photoluminescence. The signal is obtained by subtraction of overwhelming GS contribution. Sufficient spectral resolution requires quasi-monochromatic pulses with narrow bandwidth, typically of duration > 0.7 ps, so there is a limit in time resolution. The experiment does not give information on the vibrational dynamics (vibrational lifetime) and provides only partial information on vibrational dephasing, related to the Raman peak linewidth. To perform the experiment, a complete Raman spectrometer coupled to a picosecond optical table is needed, a rather complex setup.

An interesting technique that partially overcomes these limitations is based on stimulated Raman scattering (SRS). Here, a pump generates an excited state, then a pair of SRS pulses is used for inducing vibrational coherence in that state. A third beam, often identical to the first SRS pulse, generates the coherent Stokes or anti-Stokes emission (Coherent stokes Roman spectroscopy [CSRS] and CARS, respectively) [12]. ES modes can be detected in this way, and the associated dynamics is due to vibrational coherence loss (dephasing) for the two vibrational levels involved in the Raman process. The major advantages are that in this experiment the standard alignment and detection used in nonlinear spectroscopy are adopted; the signal propagates in a defined direction, is collimated, and is enhanced by Raman gain.

There is also an altered scheme of such a technique that adopts ultrashort broadband pulses as excitation and probe, in addition to a long, narrowband Raman pump [15, 16]. It has the advantage of preserving the time resolution as it is determined by the two broadband, short time pulses and, ultimately, by vibrational dephasing. It detects Raman spectra in the excited state, yet the dynamics is that of the underlying electronic resonance; no information on vibrational lifetime is obtained. In addition, it suffers the same limitation of the photoinduced spontaneous Raman scattering described above because the ES signal appears on top of the strong GS "baseline."

Direct probes of the phonon population are photoinduced absorption (PA) in the infrared spectral range and anti-Stokes (AS) RRS. AS-RRS intensity I_{AS} is $I_{AS} \sim A(\omega)n$, with the same meaning as above. A pump-probe difference technique allows measurement of $\Delta n/n$, where Δn is the vibrational population change induced by the pump [because $n \Delta A(\omega)$ is negligible]. The decay of I_{AS} is a measasure of $\Delta n(t)$, which provides the phonon lifetime T_1. The technique has been used extensively on inorganic semiconductors and small molecules [17] but only once in CPs [18]. Note in addition that in molecular semiconductors excess electronic energy is redistributed on a broad set of modes, most of them not optically active. This makes a difference with inorganic polar semiconductors, for which the Fröhlich interaction coupled excited carriers to specific longitudnal optical (LO) phonons.

3.1.3 VIBRATIONAL DYNAMICS IN THE TIME DOMAIN: GENERALITIES

The observation of vibrational wavepacket motion allows detection of molecular dynamics in real time. Such time domain investigation provides information on GS and ES vibrational frequencies, dephasing, anharmonicity, and nonadiabatic coupling [19–21]. The experiment consists of the standard pump-probe technique in which a short resonant pulse excites the samples, and a weaker, delayed pulse is used for probing pump-induced changes. Transmission changes are measured on changing pump-probe delay, which provides a time-dependent characterization. A coherent state is initiated whenever the pump pulse duration is shorter than some characteristic time constant in the response function. In our study, this is the vibrational period. Coherent phonons show up as an oscillation in the transient transmission signal. Often, the oscillation contains many frequencies, which can be singled out by Fourier transform or other numerical analysis, providing amplitudes, phase, and damping time constants for each mode. In the following, we introduce the sliding-window Fourier Transform (SWFT) algorithm, describing advantages and drawbacks of this technique for data analysis.

In general vibrational coherence is initiated in both excited and ground state, and the corresponding spectroscopical signatures overlap in a large spectral region. Separation of the two is not trivial. They are expected to have different phase, but this is not always easy to measure. Temperature dependence is different, yet the effect is very weak for phonon energy exceeding $k_B T$; in addition, a cryostat introduces dispersive optics, which requires further compensation and worse experimental conditions. Mode assignment based on theory is of little help given the difficulty in working out reliable ES normal modes. A few clear cases exist, however, in which assignment can be done quite reasonably:

(i) The pulse duration t_p is much shorter than the observed vibrational period τ_v ($\tau_v \approx 10\ t_p$). In this case, the delta-like (strictly impulsive) excitation condition is fulfilled, and coherent motion is initiated only in the excited state.

(ii) Relaxation processes quickly destroy ES coherence (e.g., internal conversion, exciton self-trapping). In this case, the coherent signal detected after relaxation will necessarily contain only GS contribution.

(iii) ES frequencies are quite different from GS ones and assigned (by other experiments or straightforward calculations). This situation, rarely encountered, implies that a simple comparison of the transient modes with those in the conventional Raman spectrum leads to the assignment.

As a remark on the last point, one should note that observation of well-known molecular dynamics in real time always provides additional information to conventional techniques, so that it may be worthwhile even in well-known situations. The clear case here is GS coherence. In principle, this is detected by standard, frequency domain Raman spectroscopy using monochromatic excitation. Yet, impulsive Raman scattering off the ground state has benefits. Dephasing as calculated from Raman linewidth is in regard to the 0–1 vibrational transition of the active mode. Impulsive Raman scattering contains coherence from many vibrational levels, up to higher quantum numbers. This provides an estimate of the lifetime of higher levels and about anharmonicity. Anharmonic coupling between modes can be observed as a modulation in the vibrational parameters. This effect, which is subtle and hard to detect, maybe be better resolved in the time domain [21]. Finally, low-frequency phonons, crucial in fields such as that of protein conformation, are much better and more easily detected in the time domain.

This chapter presents a series of cases encountered by applying time domain vibrational spectroscopy to CPs. First, we report on vibrational coherence initiated in slow modes, with the period much longer than the pulse duration, which can be safely assigned to ES motion. Then, we discuss high-frequency modes and introduce cases for which the electronic dynamics allow conclusive assignment. Finally, we discuss involved situations for which some degree of speculation is needed to rationalize the findings. All the experimental results shown were obtained recently in our laboratory using the experimental setup and detection techniques as discussed in detail in Chapter 1.

3.2 QUANTUM MECHANICAL MODELING OF TIME AND FREQUENCY DOMAIN RAMAN SCATTERING

3.2.1 QUALITATIVE COMPARISON BETWEEN RRS AND COHERENT PHONON SPECTRA

Resonance (spontaneous) Raman scattering (RRS) spectra and coherent phonon spectra (CPS), obtained by Fourier Transformation of the oscillating signal in time, bear close similarities, suggesting common origin. However, at closer look there are differences, for instance, in peak positions, relative amplitudes, and width; in addition, Raman spectra, which have a much larger signal-to-noise ratio, have far richer structure. It turns out that the precise relationship between the two is not straightforward, even if appropriate theoretical modeling can account for most of the observed features [22]. In Ref. [22], considering experiments in small molecules, it is pointed out that CPS represents a subset of the spontaneous RRS that has been projected out by the spectral and temporal characteristics of the pump and probe pulses. Different portions of an inhomogeneous distribution are added in the frequency domain in RRS and in the time domain in transient measurements, which may cause interference effects. Here, we aim to investigate this relationship further. To start, let us point out some simple statements on the comparison between RRS and CPS:

(i) CPS may contain both GS and ES contributions, while RRS concerns only the ground state.

(ii) CPS dynamics represents instants of propagation of the wavepacket onto the (multidimensional) potential energy surface. RRS is local.

(iii) CPS are affected by the pulse properties (bandwidth and duration), both in excitation and detection, while RRS depends only on the energy of the monochromatic excitation.

(iv) CPS may depend on electronic transition matrix elements connecting excited states other than the initial one; RRS depends only on the GS–ES transition involved by the excitation.

To enhance our understanding of the relationship between the two phenomena and better visualize them, we carried out quantum mechanical calculation of RRS and CPS spectra in a few study cases. In the following, the procedure and the results are outlined.

3.2.2 THE MODEL SYSTEM

We consider two electronic levels, coupled by a dipole-allowed transition μ, separated in frequency by Ω_{00} and with two vibrational modes, $\bar{\nu}_1 = 300$ cm^{-1} and $\bar{\nu}_2 = 1600$ cm^{-1}. A sketch of the model is given in Figure 3.1. Electron–phonon coupling is expressed by the corresponding dimensionless displacements of the equilibrium position in configurational space between the GS and ES energy

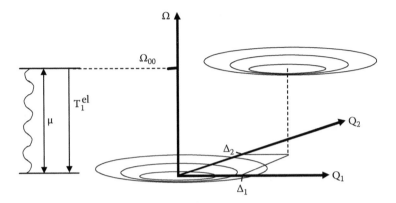

FIGURE 3.1 The model for the excited state structure used in the calculation.

surfaces, $\Delta_1 = 0.15$ and $\Delta_2 = 0.4$, respectively. GS and ES vibrational frequencies are assumed equal.

3.2.3 ABSORPTION LINESHAPE

The absorption cross-section is calculated by using the time-dependent method according to Ref. [23]. Assuming no changes between GS and ES vibrational frequencies and neglecting Duschinsky rotation, the required expression is

$$\sigma_A = \frac{4\pi \mu^2 \omega}{3\hbar c} i \int_0^\infty ds\, e^{i\omega s} K_g \qquad (3.1)$$

The analytic expression for the multimode absorption correlator function is $K_g = \exp(-i\Omega_{00}s - \Gamma s - g(s))$, with the phonon part, neglecting temperature dependence, given by

$$g(s) = \sum_{k=1}^{2} \frac{1}{2} \Delta_k^2 (1 - \exp(-i\omega_k s))$$

Line broadening is assumed homogeneous and is described by a single exponential decay with rate Γ.

3.2.4 RAMAN CROSS SECTION AND RAMAN SPECTRA

Similar to Eq. 3.1, the Raman scattering probability into vibrational mode m, with frequency ω_m, for excitation at ω_0, is given by

$$\sigma_R^m = \frac{8\pi \mu^4 \omega_S^3 \omega_0}{9\hbar^2 c^4} \left| \int_0^\infty ds\, e^{i\omega_0 s} K_R^m \right|^2 \qquad (3.2)$$

where $K_R^m = (e^{-i\omega_m s} - 1)K_g$.

The Raman spectrum for a given monochromatic excitation is built from $\sigma_R^m(\omega_0, \omega_m)$ assuming Lorentzian lineshape $g(\omega_s = \omega_0 - \omega, \omega_m, T_2^m)$ according to

$$I_{RS}(\omega_s) = \sum_{m=1}^{k} \sigma_R^m g\left(\omega_s, \omega_m, T_2^m\right) \tag{3.3}$$

where T_2^m is the vibrational dephasing time, corresponding to the Raman linewidth, which describes the decay of the vibrational coherent state.

3.2.5 VIBRATIONAL COHERENCE AMPLITUDE

The transmission difference signal that we detect in pump-probe experiments originates from the third-order polarization induced in the sample. Typically, plotted quantity is the normalized transmission difference $\Delta T / T$, coinciding for small signal with the transient absorption change $\Delta \alpha$. This is understood, in the frequency-resolved scheme, as

$$\Delta \alpha(\omega, \tau) \cdot L \approx -\frac{\Delta T}{T} \approx -\frac{n\omega L}{c\varepsilon_0} Im \frac{P^{(3)}(\omega, \tau)}{E_t(\omega)} \tag{3.4}$$

where n is the refractive index, L is the thickness of the sample (in the thin thickness approximation), and c and ε_0 have the usual meanings. If we assume that the delay τ is larger than the pulse duration (for $\tau > 20$ fs in our case), then the third-order susceptibility can be described within the effective linear approximation as the linear response of a nonstationary state according to

$$P^{(3)}(t, \tau) = \int dt' \Delta \chi(t, t') E_{pr}(t' - \tau) \tag{3.5}$$

where

$$\Delta \chi(t, t') = \int\limits_{-\infty}^{t'} dt'' \int\limits_{-\infty}^{t''} dt''' \chi^{(3)}(t, t', t'', t''') E_{pu}(t''') E_{pu}(t'') \tag{3.6}$$

is the nonstationary medium response, E_{pu} is the pump field, E_{pr} is the probe field, and $\chi^{(3)}$ is the third-order susceptibility. Equations (3.5) and (3.6) are the sequential terms of the complete third-order nonlinear response function, properly split into two integrals. $\Delta \chi$ is derived from the "jump" in the density matrix of the system, also called the doorway function $\delta \rho$ [24], induced by the second-order interaction with the pump field. This does not create electronic coherence, even in the field interaction, but contains vibrational coherence. To the lowest order, this is represented by the vibrational "displaced" state, $\delta \rho = \hat{D} \rho_0 \hat{D}^+$, where the displacement operator is $\hat{D} = \exp(\lambda_u \hat{a}^+ - \lambda_u^* \hat{a})$ and λ is the complex displacement, which contains the initial position and momentum of the wavepacket

[25] $\lambda_u = (Q_{0u} + i P_{0u})/\sqrt{2}$, with u for ground or excited states. Dynamics of the coherent state is embodied by the time dependence of $Q_u(t)$ and $P_u(t)$, which follow sinusoidal behavior, eventually multidimensional if there are several coupled vibrational modes. Finally, according to standard perturbation theory, we have:

$$\Delta\chi(t, \tau) = \frac{i}{\hbar} Tr(\delta\hat{\rho}(\tau)[\hat{\mu}(t), \hat{\mu}(\tau)]) \qquad (3.7)$$

Kumar et al. developed a convenient approach to the calculation of the pump-probe signal under this approximation [26], showing that the time-dependent susceptibility $\Delta\chi(t, \tau)$ stems from the perturbed correlator functions $K_u(t, \tau)$, that is, the oscillating lineshape that acquires time-dependent character from the underlying, classical wavepacket dynamics. This leads to an analytical expression for the modulation depth of the transmission signal in the zero dephasing approximation. From a practical point of view, inputs for the calculation are the pulse spectrum and the molecular absorption lineshape. The former is calculated as described above; the latter is analytically simulated by a super-Gaussian (sixth order) of appropriate width. With those inputs, the initial position and momentum of the wavepacket are worked out to second order in pump interaction for both GS and ES by considering a Raman-like interaction. Then, amplitude of the periodic modulation in ΔT is calculated by using Eq. (3.5). Note that this approach is fully equivalent to the extended nonlinear response function theory, usually implemented in the time domain by working out nested integrals, but this requires much less computing effort.

3.2.6 RRS VERSUS COHERENT AMPLITUDE EXCITATION PROFILES

The first result we report, in Figure 3.2, is the comparison between RRS excitation profiles and coherent phonon amplitude profiles. The latter represent the *modulation depth* of the transient transmission at different probe wavelengths (spectral component of the pulse broadband). The inset in Figure 3.2 shows the absorption and pulse spectra according to the simulation. The excitation condition, with the pulse spectrum only slightly overlapping the absorption one, is chosen after a number of real experimental situations. Coherent oscillations are observed at longer wavelength with respect to the absorption edge, reflecting modulation of stimulated emission (SE) and GS absorption for ES and GS coherence, respectively. Coherence in the excited state gives rise to much larger signals than in the ground state, particularly for the low-frequency mode. This observation stems from the mechanism that generates GS coherence. In the "Raman" process responsible for it, the wavepacket initially placed onto the excited state should propagate before coupling back to the ground state. If the travelled distance is negligibly small in the configurational space, then the transfer to the GS wavepacket also will

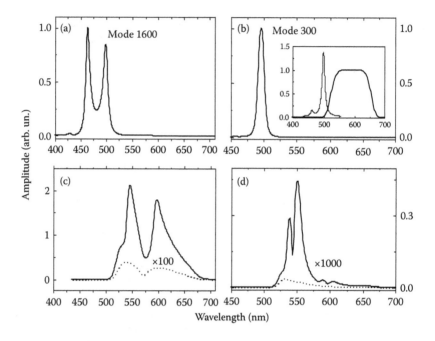

FIGURE 3.2 (a) Raman excitation profile for the vibrational mode at 1600 cm^{-1}; (b) same as (a) for the vibrational mode at 300 cm^{-1}; (c) Modulation depth in the transient transmission difference trace as a function of probe wavelength for the 1600 cm^{-1} mode; (d) same as (c) for the 300 cm^{-1} phonon. Solid line for excited state; Dotted line for ground state. Inset in (b) shows absorption and pulse.

be negligible. Of course, how much the wavepacket propagates depends on the vibrational period compared with the exciting pulse duration. For this reason, pulses with time duration much shorter than the vibrational period are largely ineffective for initiating GS coherence. A simple calculation of amplitude signal versus pulse duration (as full width at half maximum [FWHM] of the pulse electric field amplitude) is shown in Figure 3.3 for both ground and excited states. While ES contribution saturates to a maximum value for short pulses, after a maximum, GS coherence is reduced on shortening of pulse duration.

Let us now turn to the main issue of this discussion, the comparison in frequency domain of RRS and CPS spectra. To build such spectra from the oscillation amplitudes, we assume exponential damping. In the first simulation, dephasing times for GS coherence are the same used for Raman linewidth, $T_2(\nu_1) = 2000$ fs and $T_2(\nu_2) = 1000$ fs, while for excited state these values are halved. Obtained spectra are shown in Figure 3.4. Raman spectra are essentially out of resonance, so they display little difference on changing the excitation wavelength. CPS have much larger

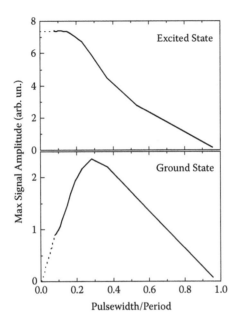

FIGURE 3.3 Behavior of the maximum modulation depth signal upon pulse duration, in ground (bottom) and excited (top) state. The pulse band is centered at the 0–0 electronic transition. The calculation was made for the 300 cm^{-1} (111 fs period) mode.

peak width and, notably for 550 nm, display a very large signal at low frequency. The probe wavelength dependence is clearly much more pronounced. For ultrafast dephasing of the vibrational coherence of mode 2 in the excited state, $T_2(\nu_2) = 25$ fs, the obtained CPS are shown in Figure 3.5, again compared with RRS. Here, the two sets are dramatically different. In particular, in CPS the low-frequency mode blows up with respect to the higher one, a situation often encountered in experiments.

Next, Figure 3.6 shows the comparison of RRS and GS coherence *only* for the latter set of parameters. In this case, the two sets of plots match quite well. Differences arise only from finite bandwidth of the probe pulse in the transient experiment. This elucidates the intuitive statement that only GS coherence should be compared with RRS.

When the exciting and probing pulses are in full resonance with the absorption, we get the situation depicted in Figure 3.7. Here, modulation depth is larger for both GS and ES, with the former having higher relative weight. The profile shape is different, with broader width, but covering essentially the same spectral region as in preresonance. Still clear is the persistent signal to the red of the absorption spectrum. Note that not only is this due to SE (ES contribution), but it also shows up for GS.

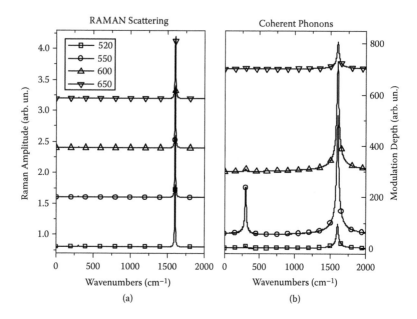

FIGURE 3.4 (a) Calculated Raman spectra and (b) coherent phonon Fourier transform spectra. Dephasing times for ground-state coherence are the same as used for Raman linewidth, $T_2(\nu_1) = 2000$ fs, $T_2(\nu_2) = 1000$ fs; for excited state, these values are halved.

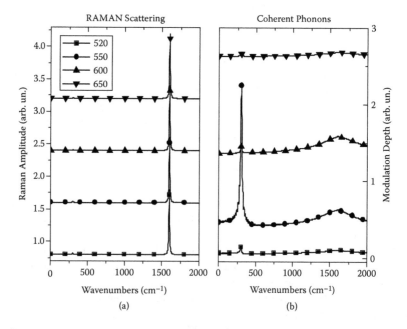

FIGURE 3.5 (a) Calculated Raman spectra and (b) coherent phonon Fourier transform spectra for ultrafast dephasing of the vibrational coherence of mode "2" in the excited state, $T_2(\nu_2) = 25$ fs.

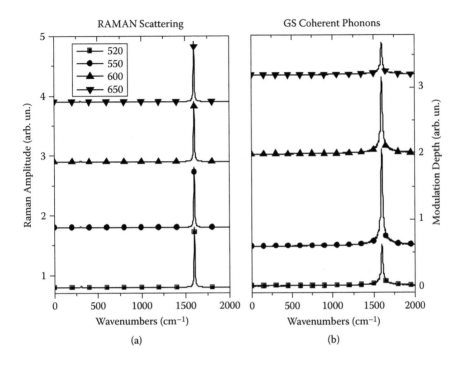

FIGURE 3.6 Comparison of RRS and GS coherence *only* for the set of parameters as in Figure 3.5.

Concluding, quantum mechanical simulation shows that RRS and CPS of the GS almost coincide, while CPS in ES are quite different from RRS and can be dominated by low-frequency modes, which typically have longer dephasing time. The "common sense" that GS coherence profiles follow the first derivative of absorption and those of ES follow the first derivative of emission is no longer true for multimode systems. Finally, when comparing experiments, one should keep in mind that most systems are inhomogeneously broadened, while here we considered the simple case of homogeneous Lorentzian lineshape. This affects dynamics and damping, expected to be faster, as well as the excitation profiles. Calculation, as examples will show, should be done by integrating a weighted distribution of homogeneous response functions.

3.3   VIBRATIONAL COHERENCE IN THE EXCITED STATE: THE IMPULSIVE EXCITATION LIMIT

We discuss the *impulsive regime* of excitation with an example for which the vibrational period of the excited mode largely exceeds the pulse duration. The sample under study is a thiophene oligomer, quinquetiophene-S, S-dioxide (T_5AO_2),

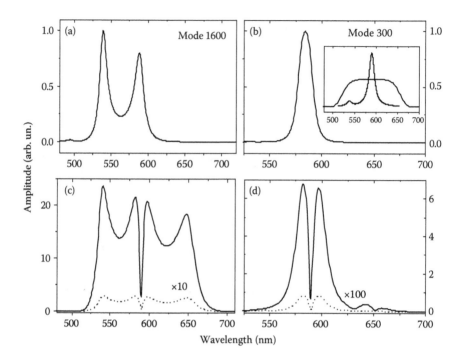

FIGURE 3.7 (a) Raman excitation profile for the vibrational mode at 1600 cm^{-1}; (b) same as (a) for the vibrational mode at 300 cm^{-1}; (c) modulation depth in the transient transmission difference trace as a function of probe wavelength for the 1600-cm^{-1} mode; (d) same as (c) for the 300-cm^{-1} phonon. Solid line for excited state; dotted line for ground state. Inset in (b) shows absorption and pulse spectra.

in cyclohexane solution at room temperature. The chemical structure of T_5AO_2 is shown in Figure 3.8. This is a member of a very interesting family of organic compounds with high technological impact and synthesized by the group of G. Barbarella; it has been used for realization of color light-emitting diodes (LEDs), white LEDs, and Lasers [27–32]. The molecule is a pentamer of thiophene functionalized in the central ring. The oxygen atoms enhance conjugation, shifting the electronic gap to lower energies, and yield higher electron affinity. The side chains improve solubility. The absorption band peaks at 450 nm, and it is structureless due to inhomogeneous broadening, probably caused by a distribution of conformers, which washes out the vibronic features. In transient experiments, we have shown that a coherent phonon of about 140 wavenumbers (period is 230 fs), with an assignment that is discussed later, is excited by sub-10-fs pulses, fulfilling the impulsive condition ($\tau_V/t_p = 23$).

Before describing vibrational dynamics, we need to briefly introduce the population dynamics that take place after excitation of T_5AO_2. The pump excitation

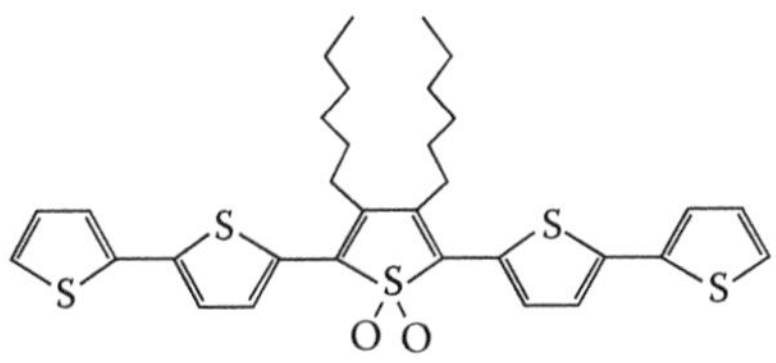

FIGURE 3.8 Chemical structure of quinquetiophene-S, S-dioxide (T_5AO_2).

redistributes population among a number of excited states with optical transitions that might be detected by the probe. Degenerate pump-probe experiments with sub-10-fs pulses show a broad SE band in the 500- to 700-nm spectral range, merging into photobleaching (PB) of HOMO-LUMO transition toward shorter wavelengths.

Time traces measured at several probe spectral components are plotted in Figure 3.9. The signal kinetics is nonexponential and wavelength dependent. It decays with an average time constant $\tau_{SE} \approx 350$ fs to a long-lived plateau, with larger amplitude of the fast component for a shorter wavelength. This indicates that the SE spectrum shifts to lower energy on a timescale of 1 ps. Similar behavior has been reported for several thiophene oligomers and assigned to a conformational readjustment in the excited state, which involves planarization [33, 34]. Superimposed onto such kinetics, strong oscillations are clearly visible at all wavelengths.

Since these are the objective of the present discussion, further analyses are done on the oscillating-only signal after subtraction of the slower-decaying background, as shown in Figure 3.10 for a sample wavelength. Fourier transform spectra are displayed in Figure 3.11 together with Raman spectra of solvent and solute molecule. All the spectra contain a peak at about 140 cm^{-1}, with some dispersion ranging between 136 and 144 cm^{-1} due to a distribution of conformers. The low wavenumber peak largely exceeds all other features, except for the 570-nm data, which contain peaks at 28 and 274 cm^{-1} (roughly double the previous one) and the relative height of the high-frequency modes is larger. The solvent lines are also more prominent at this wavelength, and there is a worse signal-to-noise ratio, suggesting it is not very reliable. Weaker structures, peaked at 824 and 1527 cm^{-1}, correspond to other T_5AO_2 modes that not discussed here.

We assigned the peak at 140 cm^{-1} to a vibration mode of T_5AO_2 in the electronic excited state because (i) it is not present in the Raman spectrum of the solvent, (ii) it does not correspond to any peak in the T_5AO_2 Raman spectrum in which a low-frequency mode is observed at 106 cm^{-1}, and (iii) the impulsive excitation implies that GS coherent modulation is negligibly small. Further assignment to a specific normal mode is difficult because no reliable calculation exists for ES molecular dynamics of such a molecule. Nevertheless, assignment to the inter-ring torsional mode driving planarization is tempting. Recent density functional calculations suggest that torsional motion in the ground state of T_5AO_2

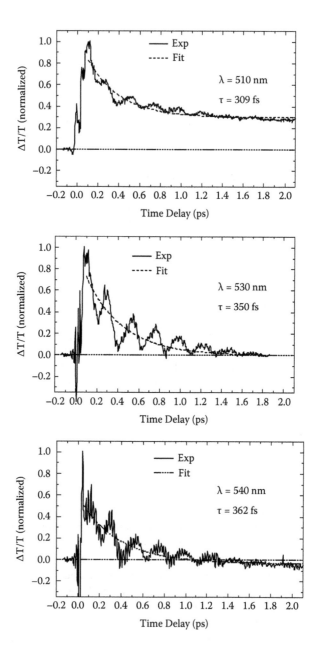

FIGURE 3.9 Difference in transmission time traces measured at several probe spectral components in T_5AO_2 in cyclohexane. Solid line, experiment; dashed line, exponential fit. Labels show probe wavelength component and time constant of the fit.

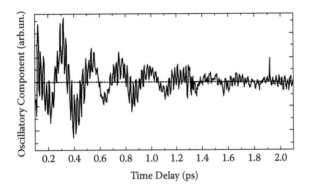

FIGURE 3.10 Oscillating-only signal, after subtraction of the slower-decaying background, from data at 540 nm (Figure 3.9).

occurs below 100 wavenumbers [35]. A detailed study of molecular dynamics in Stilbene [36], however, reports a sixfold increase of torsional frequency in the excited state. This is well understood by the formation of the quinoid structure, which tends to reverse the dimerization pattern by redistributing the electronic density. As a result, torsions occurring around the single bond are hardened in the quinoid state, and their frequency becomes higher. In conclusion, the observed wavenumber of about 140 cm^{-1} would be consistent with the assignment to the inter-ring torsional mode in the quinoid excited state. Note that electronic motion is much faster (adiabatic regime) than molecular readjustment, so that the change to the quinoid force constant is instantaneous, while planarization will take some time (estimated as a few picoseconds).

Full quantum mechanical calculations have been carried out to reproduce transient nonlinear coherent response in T_5AO_2 according to the model explained in the previous section. Note, however, that the example reported was considering a homogeneous lineshape, which is not the case in this experiment. Generally, the approach will be to work out all needed quantities (GS absorption and coherent amplitude) for a distribution of homogeneous responses and then build a convolution with a proper weight function, which is tested by reproducing the GS absorption lineshape. First, we show in Figure 3.12 the expected oscillation amplitude for three Raman modes of T_5AO_2, calculated for both ground and excited states. For the low-frequency mode, the ES amplitude is 150 times larger then the GS one. Higher frequencies have larger GS contribution, yet for them also the ES always exceeds the GS contribution. The result for the 140-cm^{-1} mode confirms the "delta-like" argument corroborating the assignment to ES motion. First, the amplitude in the ground state is negligibly small, Second, the relative amplitude with respect to higher frequencies is that expected for the ES. (In the GS spectrum, the low-frequency mode would be smaller.)

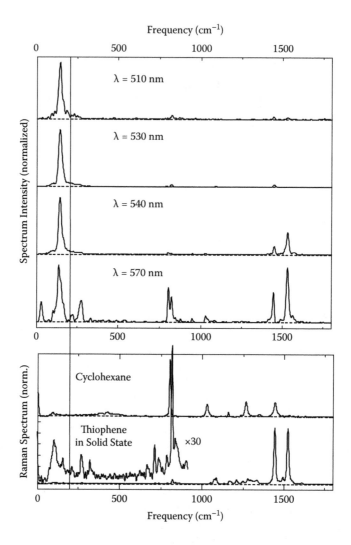

FIGURE 3.11 Coherent phonons spectra (upper panel) for T_5AO_2 and Raman spectra (off resonance) of solvent cyclohexane and T_5AO_2 thin film (lower panel).

In Figure 3.13 we compare experimental and calculated modulation depth at different spectral components. The experimental value is extracted from the measured trace around a pump-probe delay of about 0.5 ps; the theoretical value is estimated also considering amplitude decay according to the measured damping constant. Matching of the two curves is reasonable, indicating that the model can indeed predict the spectral dependence correctly. To evaluate the decay of the oscillating amplitude, we use the SWFT [37] of the oscillating signal (after subtraction of the slow background). This approach provides time-dependent

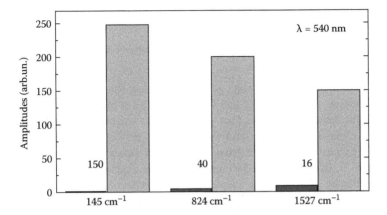

FIGURE 3.12 Calculated oscillation amplitudes for three Raman modes of T_5AO_2, calculated for both ground (black) and excited (gray) state.

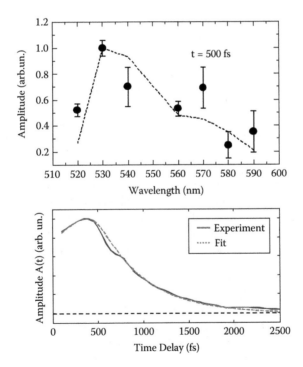

FIGURE 3.13 Comparison of experimental and calculated modulation depth at different spectral components for the low-frequency mode in T_5AO_2. The experimental value is extracted from the measured trace around pump-probe delay of about 0.5 ps; the theoretical value is estimated also considering amplitude decay, according to the measured damping constant.

Fourier spectra,

$$S(\omega, \tau) = \int_{-\infty}^{+\infty} s(t)g(t - \tau)\exp(-i\omega t)dt \qquad (3.8)$$

where τ is the pump-probe delay; $g(t)$ is the window function (in our case, g is a super-Gaussian of order $n = 6$ and 70-fs half width at half maximum (HWHM), enough for sampling three or four periods of the investigated oscillation); and $s(t)$ is the transient signal in the time domain. "Cuts" of the S function at fixed ω_0, $S(\omega_0\tau)$, provide amplitude decay at that chosen frequency. For instance, decay at 140 cm^{-1}, as measured at 540 nm, is shown in Figure 3.14 together with the exponential fit. Note that the shape of $S(\omega, \tau)$ reflects not only the coherence decay kinetics, but also the sliding-window algorithm. Thus, sweeping of the measuring temporal window causes the initial rise time across the time zero seen in Figure 3.14. In general, to fit the experimental $S(\omega, \tau)$, we build $f(t)$, the simulating function of $s(t)$, as

$$f(t) = \sum_{k=1}^{N} A_k(t)\cos(\omega_k t + \phi_k) \qquad (3.9)$$

where $A_k(t)$ is a decaying amplitude, ω_k is the frequency observed in the Fourier spectra, and ϕ_k is the phase constants to be determined. Then, the sliding-window transform of $f(t)$, namely, $G(\omega, \tau)$, is worked out according to the same procedure used for the experimental data and compared to $S(\omega, \tau)$.

For reproducing the experimental trace in this case, a single sinusoid in Eq. (3.8) is enough, with dephasing time T_2 of about 0.5 ps (see Table 3.1). Interestingly, T_2 roughly compares with half of τ_{SE}. This suggests that the mechanism responsible for SE decay is also responsible for vibrational coherence damping and that $T_2 = 2T_1 = 2\tau_{SE}$ (i.e., there is no pure phase loss). Here, T_1 is the vibrational lifetime, while τ refers to the spectral shift dynamics of the SE spectrum. We assign this phenomenon to the conformational readjustment in the excited state, probably planarization. We can then speculate that the observed coherent phonon dynamics directly maps the trajectory of motion of the wavepacket along the torsion angle coordinate, reproducing the damped oscillations of the molecular rings

TABLE 3.1
Damping Time Constant of the Oscillation (T_2), Decay Time Constant of Stimulated Emission (τ_{SE}) at Several Probe Spectral Components for the 140-cm^{-1} Phonon

λ (nm)	τ (fs)	T_2 (fs)
510	309	600
530	350	600
540	362	500
570	271	500

above and below the molecular plan until reequilibration in the flat conformation. Alternatively, it could be that planarization is still responsible for dephasing of the coherence, which is taking place along a different coordinate than the torsional ring. In this case, again the timescale of the process is determined by the conformational readjustment, yet the phase-space localization and trajectory do not describe it.

3.4 GROUND-STATE COHERENCE

Here, we report on real-time vibrational spectroscopy in polydiacetylene (PDA) chains isolated in benzene solution. PDAs [38] are long carbon-conjugated chains containing single, double, and triple bonds. The large π-electron delocalization corresponds to an optical gap in the green–red spectral region and to large polarizability. In general, PDAs possess high intrachain order and remarkable optical properties [39], which offer a unique opportunity for investigating one-dimensional, strongly correlated π-electron systems [40–43]. We studied a substituted PDA, poly[1,6-bis(3,6-dihexadecyl-N-carbazolyl)-2,4-hexadiyne] (polyDCHD-HS) [44]; PolyDCHD-HS is soluble in common organic solvents, and in benzene it displays an astonishingly sharp and intense room temperature excitonic resonance [45], indicative of a highly ordered material and negligible interchain interactions. The crucial point of the experiment is that in PDA ultrafast internal conversion takes place following optical excitation, to the lowest-lying dark state [46]. This process cancels out vibrational coherence in the excited state, allowing clear assignment of the long-lived coherent phonons to motion in the ground state [47].

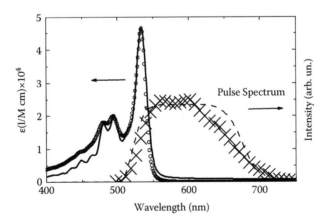

FIGURE 3.14 Ground-state absorption (circles) of poly(DCHD)-HS in C_6H_6 solution. Crosses are the measured pulse spectrum used for excitation and probing. The dashed line is the spectral band used for simulation; the dotted line is the simulated absorption (see text).

3.4.1 EXCITATION CONDITION

For these experiments, the pulses used for excitation and probing have a nearly transform limited duration of about 6 fs and a 500- to 680-nm bandwidth, as shown in Figure 3.14. The pump beam is focused at an energy density of about 0.2 mJ/cm^2 (spot diameter 120 μm) on solutions of polyDCHD-HS in benzene with a concentration of about 10^{-3} M and kept in a 0.2-mm-thick cuvette at room temperature. The probe beam is attenuated by about one order of magnitude and delayed in time with respect to the pump. Time-resolved measurements of transmission changes ($\Delta T/T$) at specific wavelengths are obtained by spectrally filtering the probe with 10-nm bandwidth interference filters positioned after the sample. The signal is recovered using lock-in amplifier-based detection. To control the phase of the pump pulse, which is the arrival time of its different spectral components, we add positive or negative chirp, inserting a 2-mm-thick plate of fused silica or adding two bounces onto the chirped mirrors, respectively. This results in both cases in slightly longer pump pulses (9–10 fs), while the probe ones are unchanged, allowing us to perform the experiments with high time resolution.

3.4.2 POPULATION RELAXATION IN POLYDCHD-HS

Figure 3.14 shows absorption and emission spectra of polyDCHD-HS in diluted benzene solution at room temperature. The purely electronic absorption peak at 534 nm is remarkably sharp, with FWHM of about 530 cm^{-1}. Vibronic replicas are assigned to the chain modes, mainly double and triple carbon bond stretching. The large ratio of the intensity of the 0–0 transition relative to the vibronic bands indicates a rather small electron–phonon coupling, with a Huang–Rhys factor of about 0.3 [45]. Both results suggest a small geometrical relaxation in the excited state. The fluorescence spectrum, also well structured, only slightly deviates from the absorption mirror image and shows an extremely small Stokes shift. The emission quantum yield is estimated to be 10^{-2} [45].

Based on previously reported transient transmission difference measurements on polyDCHD-HS [46] in C$_6$H$_6$ solution, the following picture, sketched in Figure 3.15, for photoexcitation dynamics in isolated polymer chains was proposed. Photoexcitation with sub-10-fs pulses creates a wavepacket in the singlet ionic manifold, which contains optically allowed states. The spectral signature is optical gain for a large portion of the pulse spectrum and PB of the GS transition. The initial wavepacket reaches a conical intersection, which provides an efficient nonadiabatic path to the covalent manifold, and undergoes a branching into two states, the covalent state (2^1A_g) and the relaxed ionic state (1^1B_u), responsible for emission. Fast PB recovery, observed on a timescale of about 100 fs, is tentatively assigned to this process, as caused by depletion of the optically coupled state, which contributes to the signal via SE. The energy located in the covalent state quickly funnels into the triplet manifold by singlet fission, thus hampering recovery to the original state. The buildup of an absorption band in the gain region is followed in time, providing the dynamics of the triplet formation, which

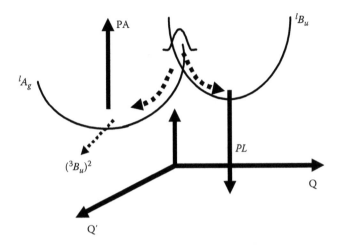

FIGURE 3.15 Proposed potential energy surface and relaxation path connecting the initially excited optical state to the lower dark state that undergoes fission. Q and Q' are different configurational coordinates along which the system is evolving. Those cannot be identified at this stage.

is completed in about 200 fs. At this point, the triplet–triplet absorption band is formed, peaking at 900 nm. The relaxed ionic state decays with lifetime in the 10- to 100-ps region, partially radiatively, while triplet pairs decay on a slightly longer timescale [48, 49].

3.4.3 COHERENT VIBRATIONAL DYNAMICS

Transmission difference data measured in polyDCHD-HS in solution after sub-10-fs optical excitation are reported in Figure 3.16. The time traces contain electronic coherence signatures at pump-probe delay $\tau < 0$, which have been discussed extensively in the literature (see, e.g., [50]). Briefly, at negative time delay oscillation appears in the ΔT signal due to pump perturbed probe free induction decay. Around $\tau = 0$, pump and probe fields can interfere to give a spike, known as *coherent coupling*, which does not reflect sample dynamics. We thus focus on the ΔT signal for positive delays once pulse overlap is over. At all wavelengths within the broad pulse band, periodic modulation of the signal is detected, even though with variable depth, superimposed on the slow PA buildup previously described. These oscillations are due to the time evolution of the nonstationary state induced by the impulsive excitation of the nuclear modes in polyDCHD-HS.

Before a detailed interpretation of the observed dynamics, we show in Figure 3.17 a number of selected Fourier power spectra of the corresponding traces in Figure 3.16. Observed frequencies have wavenumbers between 1000 and 2500 cm^{-1}, typical for nuclear motion in the conjugated backbone of polyDCHD-HS. The close correspondence with the continuous wave (CW) Raman spectrum

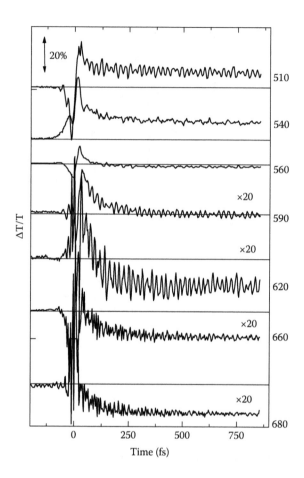

FIGURE 3.16 Transmission difference traces at several pulse spectral component as measured in poly(DCHD)-HS in C_6H_6 solution. Spectral component and enhancement factors (if used) are shown in the plots.

(excited by 1064-nm radiation) suggests, prima facie, that coherence is in the ground state. To this regard, remember that *fast state deactivation* of the optically prepared excited state in our sample is expected to destroy coherence in the excited state. The CW Raman spectrum contains strong lines due to the solvent, in particular the one at about 1000 cm^{-1}. This one also appears in the Fourier spectra due to the very large cross section, indicating a weak excitation of the solvent by impulsive SRS (i.e., the off-resonance case). Other solvent modes are much less evident in the Fourier spectra and can be disregarded.

Acceptable fitting of the experimental data (both Fourier spectra and amplitude decay) following the same procedure described above can be obtained for all wavelengths, assuming exponential amplitude decay. As an example, we report the results at 580 nm in Figure 3.18. At different wavelengths, we find slightly different frequencies, with variation on the order of 0.5%, which may be due to

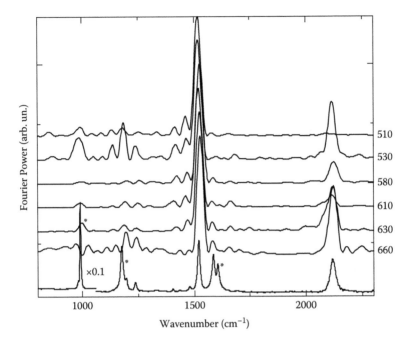

Fourier Power (arb. un.)

510
530
580
610
630
660

×0.1

1000   1500   2000

Wavenumber (cm^{-1})

FIGURE 3.17 Fourier power spectra of the oscillating part of the transient transmission difference traces at different spectral components measured in polyDCHD-HS after excitation with sub-10-fs pulses. Bottom line is the CW Raman spectrum for out-of-resonance excitation. The low-wavenumber region of the CW Raman spectrum has been rescaled.

inhomogeneous broadening or anharmonicity. No clear trend is found, however, and this phenomenon is not discussed further. The measured dephasing time, for the prominent mode at 1525 ± 7 cm^{-1}, displays a trend on changing the spectral component, shown in Figure 3.18d. T_2 is longer for shorter wavelengths. This weak dependence can be understood considering that long wavelengths probe the wavepacket when far out on the potential energy surface (PES), with large contribution of transitions from higher-lying vibrational levels. Higher quantum numbers are expected to be associated with faster dephasing, thus explaining the observation. The longer value found, about 750 fs, compares favorably with the one extracted from linewidth analysis of the CW Raman, of about 830 fs, supporting the assignment of the observed vibrational coherence to GS dynamics.

The periodic modulation of the vibrational amplitude that can be seen in these plots (Figs. 3.18b and 3.18c) comes generally from mode interference. Actually good fits of the data, reproducing the beating in the decaying amplitude, require introducing some minor modes too, which are deduced from the Fourier spectra.

In summary, the gathered information is as follows: Double-bond stretching is measured at 1525 ± 7 cm^{-1}, with variable dephasing as discussed above between 600 and 750 fs. Triple-bond stretching is detected at 2115 ± 13 cm^{-1} with a dephasing time of 400 fs. Minor modes at 596 and 994 cm^{-1} are sometimes

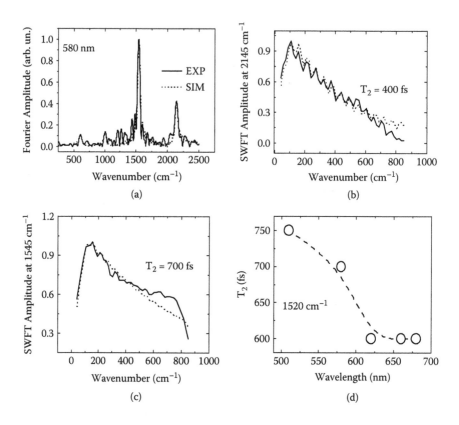

FIGURE 3.18 (a) Experimental and simulated Fourier amplitude spectrum; (b) and (c) amplitude decay for the double- and triple-bond stretching mode, experimental (thick), simulated (thin); (d) dephasing of the 1520-cm^{-1} mode measured at different spectral components.

included to better reproduce the oscillating pattern. From both frequency position and dephasing, it is conjectured that such modes also belong to the ground state.

3.4.4 GROUND-STATE VIBRATIONAL DYNAMICS: AN EXPERIMENTAL PROOF

To demonstrate that the dominant coherence observed is that of the ground state, we performed pump-probe experiments using variable-phase (chirped) pulses. The basic idea, pointed out in Ref. [51], is somewhat related to the concept of quantum control, that is, the possibility of controlling the radiation–matter interaction dynamics by changing the phase of the light pulse. A positively chirped pulse has red spectral components preceding blue ones. This implies that the initial resonance places amplitude onto lower excited vibronic levels, which fall out of resonance with the trailing edge of the pulse spectrum. The Raman-like interaction that generates GS coherence is thus inhibited. The other way around, when blue components

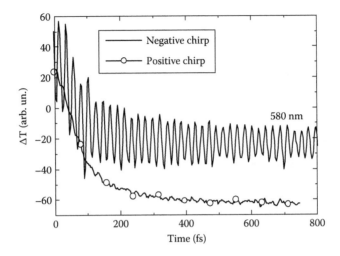

FIGURE 3.19 Transmission difference at 580-nm spectral component for negative (thick) and positive (thin) chirp pulses.

come first and the wavepacket relaxes during the interaction, the trailing edge of the spectrum finds resonance with the nonstationary state, bringing population back onto the ground state. The Raman-like process is now enhanced. Negative chirp then enhances GS coherence, while positive chirp suppresses it. Attention should be put onto the pulse duration because chirping increases it. Aware of this, we carefully prepared two pump pulses, with positive and negative chirp, of comparable time duration in order to highlight differences due to only the pulse phase. The result, in Figure 3.19, shows very clearly that larger oscillations are seen when negatively chirped pump pulses are used. Using positively chirped pulses almost washes out the coherent signal. This result confirms that the dominant oscillations come from motion in the ground state.

3.4.5 FULL QUANTUM MECHANICAL SIMULATION OF GROUND-STATE COHERENCE

Following the approach and the approximations presented above, we simulated our experimental observations. The lineshape of GS absorption, as worked out analytically, is plotted in Figure 3.14 assuming homogeneous broadening. The calculated population change, due to the pump excitation, is about 8% of the ground state, consistent with the experiment. The ability of reproducing absorption lineshape and population allows checking the validity of the model parameters used, such as field strength, dipole moment, and electron–phonon coupling. Once all those are set, we work out the initial displacement and momentum of the "one-dimensional" vibrational wavepackets for the two main modes, the double- and triple-bond stretching, around 1520 cm^{-1} ($\Delta = 0.8$) and 2120 cm^{-1} ($\Delta = 0.7$),

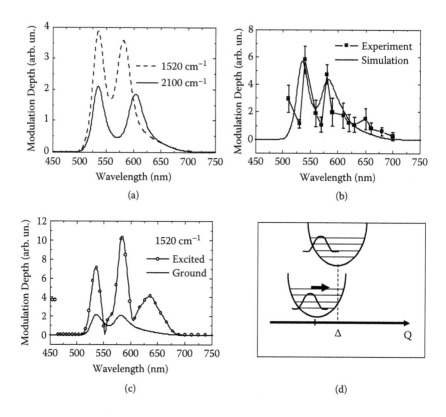

FIGURE 3.20 (a) Simulated modulation amplitude as a function of the probing wavelength (selected after the sample) due to coherent vibrational motion at 1520 cm^{-1} (dashed) and 2100 cm^{-1} (solid) in the ground state; (b) comparison of simulated and experimental amplitude modulation dependence on probe wavelength (see text); (c) simulated modulation amplitude as a function of the probing wavelength (selected after the sample) due to coherent vibrational motion at 1520 cm^{-1} in the ground (solid) and excited state (line plus circles); (d) schematic drawing of displaced harmonic potentials and initial wavepacket for the double stretching mode. The initial displacement in the excited state is $Q_{e0} = -0.17$, for ground state $Q_{g0} = -0.03$, and momentum $P_{g0} = 0.015$. Δ is the dimensionless displacement of the equilibrium position.

respectively (Δ is the dimensionless harmonic potential displacement expressing the strength of the coupling). Figure 3.20d is a sketch of the initial condition, as worked out by the model.

Ultimately the amplitude of the periodic modulation in ΔT, neglecting damping, was calculated using Eq. (3.3) (model section) properly expressed in terms of wavepacket attributes. In Figure 3.20a, we report the results for the GS contribution of the two modes, double- and triple-bond stretching. In Figure 3.20b, we show a comparison of the overall measured amplitude, evaluated at about 100-fs time delay by inspection of the experimental traces, with the calculated one. Here, the simulated trace contains both modes shown in (a), GS contribution, properly scaled for the measured dephasing (i.e., taking into account that after 100 fs the triple-bond stretching mode, with dephasing time $T_2 = 400$ fs, is about 80% smaller than the double-bond mode, with a relaxation time of $T_2 \sim 700$ fs). In Figure 3.20c, we plot the wavelength-dependent oscillation amplitude according to our simulation in both ground and excited state only for the double-bond stretching mode. The plot shows that the two contributions overlap completely, making their separation difficult (the same situation is found for the triple-bond stretching). It also points out that the ES contribution is expected to be large when dephasing is not included. The experimental observation that GS coherence dominates (see above) is thus a strong indication that ultrafast coherence loss is taking place in the excited state.

To complete the comparison between simulation and experiment, Figure 3.21 shows the "power" spectra worked out from the measured data and those obtained by the simulation assuming only GS contribution. All in all, the simulations, in

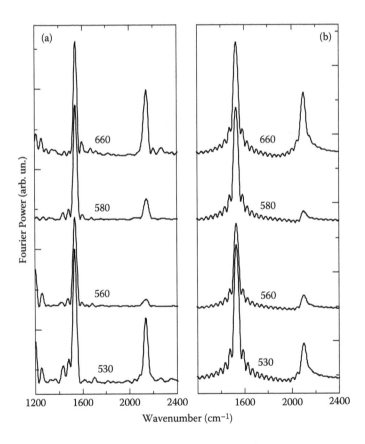

FIGURE 3.21 Fourier amplitude spectra at several probe spectral components according to the calculation described in the text. Experimental, solid line; simulated, dashed line.

spite of containing only linear coupling and no dephasing, reproduce satisfactorily the experimental data and corroborate the interpretation that GS coherence is the dominant observable. They also demonstrate that the strong wavelength dependence of the modulation depth comes from the vibronic structure, including electron–phonon coupling and pulse spectrum, while inhomogeneous broadening is small in our sample. The observed change in the frequency value at different spectral components may be due to the latter effect, and it is not reproduced by the simulation. To conclude this section, and as introduction for the next, we remark that in Figure 3.20b there is a deviation from the expected GS profile above 630 nm: the expected decay is not observed; on the contrary, the amplitude is rising. This might indicate that something different is occurring.

3.4.6 EXCITED-STATE SIGNATURES

So far, we have concentrated our attention on GS coherence, which is dominating the signal. In the Fourier amplitude spectra of long-wavelength transmission traces, however, broad features appear with *no correspondence to the CW Raman spectra* (see Figure 3.22). Consistently, SWFT cuts for the amplitude decay show extremely fast decay. Data can be reproduced only if we include new modes, with slightly different frequencies from those seen in the Raman spectrum, and ultrafast dephasing. The new modes are at 790 cm^{-1}, 1270 to 1300 cm^{-1}, and 1980 cm^{-1}. Fitting of experimental data is shown in Figure 3.22, adopting the same procedure discussed. Given their short lifetime, these frequencies cannot be determined with accuracy (the uncertainty can be 20–30 cm^{-1}). Their dephasing is in the range $T_2 = 25 - 35$ fs. We propose to assign these modes to the excited state, possibly the initial optically coupled (B_u) one reached by the pump transition. Assuming an average value $T_2 = 30$ fs, and if we conjecture dephasing is due to population relaxation ($T_2 = 2T_1$), then we get an estimate of the initial event, assigned

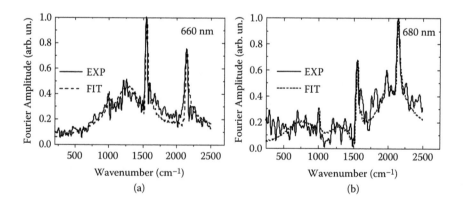

FIGURE 3.22 Fourier amplitude spectra of the oscillating signal at 660 nm (a) and 680 nm (b). Experimental, solid line; fitting, dashed line.

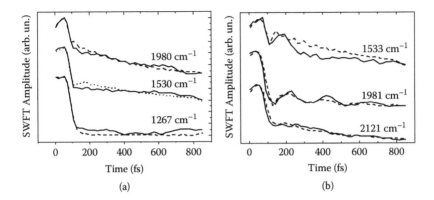

FIGURE 3.23 Cuts of sliding-window Fourier spectra at frequencies indicated in the frame and from data displayed in Figure 3.22, same labeling. Experiment, solid line; fitting, dashed line.

to crossing the conical intersection to reach the A_g state, of about 15 fs. This extremely short time could not be determined by our earlier experiments, even though ultrafast relaxation was expected. Measuring coherence loss can thus be an alternative tool for detecting ultrafast population dynamics when direct population decay cannot be detected. Our interpretation is that crossing to the dark state (A_g), with different geometry and modes, erases the coherence. The fast crossing could in principle initiate coherence in new modes of the A_g state, but with frequencies much lower because the transit timescale is around 50 fs. The appearance of ES signature at longer wavelength might be due to the lack of GS contribution there, as seen in Figure 3.20c, or to new electronic transitions.

The new modes found in our measurements suggest softening of the phonon frequencies of the carbon skeleton in the excited state, possibly still due to stretching of the single, double, and triple bonds. Note that in polyenes the vibrational frequency of the double-bond stretching in the S_2 excited state, deduced from the absorption envelope, is slightly smaller than that in S_1 for longer chains [52]. Yoshizawa et al. [6], using a three-beam technique, longer pulses, and SRS, were able to estimate a new mode at 1200 cm^{-1} in PDA-4BCMU, assigned to the excited state. Their results, however, can hardly be compared with the present ones because the observed timescales are quite different, the two studied PDAs are chemically different, and the photoexcitation dynamics seems also different. In addition, the region of the single carbon bond is particularly ambiguous due to a number of contributions from the backbone and the side groups. Given the lack of reliable theoretical prediction for PDA ES modes, we cannot further discuss their assignment.

In conclusion, we found that ultrafast deactivation substantially reduces ES coherence. The associated phase loss occurring with the 30-fs time constant yields an estimate for the fast evolution of the system from the initial optical state to the lower dark state. This timescale could hardly be determined by "population" spectroscopy. In addition, the weak "surviving" signature of ES coherence indicates frequencies are lower than those in the GS, even though with large uncertainty.

3.5 CONCLUSIONS

In this chapter, we provided a general introduction to the subject of coherent phonons, including quantum mechanics principles, experimental techniques, and some examples regarding their observation in linear conjugated chains. Quantum mechanical calculations were used for reproducing the wavelength-dependent signal due to coherent phonons. Essentially, these calculations allow bridging from the very intuitive and simple picture of wavepackets, as interference of vibrational states, to the third-order response function that was actually probed experimentally. By investigating a simple case, the connection between standard Raman spectra and CPS was shown, elucidating the differences and the added values of the latter. The localized nature of the vibrational wavepacket is the key issue in time domain experiments because it allows following the quasi-classical trajectory of motion in the phase space and then the local mapping of the potential energy surfaces. This reveals details on the dynamics of the geometrical evolution that cannot be achieved with other spectroscopy techniques. The experimental setup needed for such a type of investigation cannot be considered user friendly as could be a Raman spectrometer or similar, yet the rapid evolution of laser technology will soon overcome this limit. Growth in the number of laboratories detecting coherent phonons and their application in material science is thus expected in the near future.

The vibrational coherent state that we discussed is likely at the origin of quantum control, that is, the ability to shape light pulses to enhance or depress certain outcomes, as could be the yield of a photochemical reaction. The advantage of manipulating a coherent, nonthermal state is obvious; it enhances the probability of useful interaction over all possible outcomes because of the localization of the parameters (here, the molecular coordinates). In general, coherent phonons can thus be more than sophisticated diagnostic tools, and their study and application become independent research areas.

REFERENCES

1. B. S. Hudson, B. E. Kohler, and K. Schulten. Linear polyene electronic structure and potential surfaces. In *Excited States*, Vol. 6, pp. 1–95 (E. C. Lim, Ed.), Academic Press, New York, 1982.
2. L. X. Zheng, B. C. Hess, R. E. Benner, Z. V. Vardeny, and G. L. Baker. Resonant Raman-scattering spectroscopy of polydiacetylene films at high pressure. *Phys. Rev. B*, 47:3070–3077, 1993.
3. E. Ehrenfreund, Z. Vardeny, O. Brafman, and B. Horovitz. Amplitude and phase modes in *trans*-polyacetylene: resonant Raman scattering and induced infrared activity. *Phys. Rev. B*, 36:1535–1553, 1987.

4. G. Lanzani, S. Luzzati, R. Tubino, and G. Dellepiane. Polarized resonant Raman scattering of *cis*-polyacetylene. *J. Chem. Phys.*, 91:732–738, 1989.
5. T. Chen, A. Vierheiling, W. Kiefer, and A. Materny. Coherent vibrational dynamics of polydiacetylenes in their electronic ground state. *Phys. Chem. Chem. Phys.*, 3:5408–5415, 2001.
6. M. Yoshizawa, Y. Hattori, and T. Kobayashi. Femtosecond time-resolved resonance Raman gain spectroscopy in polydiacetylene. *Phys. Rev. B*, 49:13259–13262, 1994.
7. A. Vierheilig, T. Chen, P. Waltner, W. Kiefer, A. Materny, and A. H. Zewail. Femtosecond dynamics of ground-state vibrational motion and energy flow: polymers of diacetylene. *Chem. Phys. Lett.*, 312:349–356, 1999.
8. R. E. Peierls. *Quantum Theory of Solids*, Claredon, Oxford, 1955.
9. D. Baeriswyl, D. K. Campbell, and S. Mazumdar. *Conjugated Conducting Polymers* (H. Kiess, Ed.), Springer-Verlag, Berlin, 1992.
10. S. Mazumdar, and S. N. Dixit. Coulomb effects on one-dimensional Peierls instability: the Peierls-Hubbard model. *Phys. Rev. Lett.*, 51:292–296, 1983.
11. J. E. Hirsch. Effect of Coulomb interactions on the Peierls instability. *Phys. Rev. Lett.*, 51:296–299, 1983.
12. C. Rulliere, T. Amand, and X. Mrie. Spectroscopic methods for analysis of sample dynamics, In: *Femtosecond Laser Pulses*. (C. Rulliere, Ed.), Springer, Berlin, 1998.
13. G. Lanzani, L. X. Zheng, G. Figari, R. E. Benner, and Z. V. Vardeny. Photoexcition dynamics in polyacetylene probed bt transient photoinduced resonance Raman scattering. *Phys. Rev. Lett.*, 68:3104–3107, 1992.
14. H. Hashimoto and Y. Koyama. The C=C stretching Raman lines of β-carotene isomers in the S1 state as detetcted by pump-probe resonance Raman spectroscopy. *Chem. Phys. Lett.*, 154:321–325, 1989.
15. M. Yoshizawa, Y. Hattori, and T. Kobayashi. Femtosecond time-resolved resonance Raman gain spectroscopy in polydiacetylene. *Phys. Rev. B*, 49:13259–13262, 1994.
16. P. Kukura, D. W. McCamant, P. H. Davis, and R. A. Mathies. Vibrational structure of the S2(1Bu) excited state of diphenyloctatetraene observed by femtosecond stimulated Raman spectroscopy. *Chem. Phys. Lett.*, 382:81–86, 2003.
17. A. Laubereau and W. Kaiser. Vibrational dynamics of liquids and solids investigated by picosecond light pulses. *Rev. Mod. Phys.*, 50:607–664, 1978.
18. G. Lanzani, R. E. Benner, and Z. V. Vardeny. Picosecond dynamics of non-equilibrium phonons in trans-polyacetylene studied by transient photoinduced resonance Raman scattering. *Solid State Commun.*, 101:295–299, 1997.
19. T. Kobayashi, A. Shirakawa, H. Matsuzawa, and H. Nakanishi. Real-time vibrational mode-coupling associated with ultrafast geometrical relaxation in polydiacetylene induced by sub-5-fs pulses. *Chem. Phys. Lett.*, 321:385–393, 2000.
20. T. A. Pham, A. Daunois, J.-C. Merle, J. Le Moigne, and J.-Y. Bigot. Dephasing dynamics of the vibronic states of epitaxial polydiacetylene films. *Phys. Rev. Lett.*, 74:904–907, 1995.
21. A. Gambetta, C. Manzoni, E. Menna, M. Meneghetti, G. Cerullo, G. Lanzani, S. Tretiak, A. Piryatinski, A. Saxena, R. L. Martin, and A. R. Bishop. Real-time observation of nonlinear coherent phonon dynamics in single-walled carbon nanotubes. *Nat. Phys.*, 2:515–520, 2006.
22. A. E. Johnson and A. B. Myers. A comparison of time and frequency domain resonance Raman spectroscopy in tri-iodide. *J. Chem. Phys.*, 104:2497–2507, 1996.

23. A. B. Myers, R. A. Mathies, D. J. Tannor, and E. J. Heller. Excited state geometry changes from resonance intensities: isoprene and hexatriene. *J. Chem. Phys.*, 77:3857–3866, 1982.

24. S. Mukamel. *Principles of Nonlinear Optical Spectroscopy*, Oxford University Press, New York, 1995.

25. A. T. N. Kumar, F. Rosca, A. Widom, and P. M. Champion. Investigations of ultrafast nuclear response induced by resonant and nonresonant laser pulses. *J. Chem. Phys.*, 114:6795–6815, 2001.

26. A. T. N. Kumar, F. Rosca, A. Widom, and P. M. Champion. Investigations of amplitude and phase excitation profiles in femtosecond coherence spectroscopy. *J. Chem. Phys.*, 114:701–724, 2001.

27. M. Anni, G. Gigli, V. Paladini, et al. Color engineering by modified oligothiophene blends. *Appl. Phys. Lett.*, 77:2458–2460, 2000.

28. G. Gigli, O. Inganas, M. Anni, et al. Multicolor oligothiophene-based light-emitting diodes. *Appl. Phys. Lett.*, 78:1493–1495, 2001.

29. G. Barbarella, L. Favaretto, M. Zambianchi, et al. From easily oxidized to easily reduced thiophene-based materials. *Adv. Mater.*, 10:551–554, 1998.

30. G. Barbarella, L. Favaretto, G. Sotgiu, et al. Modified oligothiophenes with high photo- and electroluminescence efficiencies. *Adv. Mater.*, 11:1375–1379, 1999.

31. M. Anni, G. Gigli, R. Cingolani, et al. Amplified spontaneous emission from a soluble thiophene-based oligomer. *Appl. Phys. Lett.*, 78:2679–2681, 2001.

32. G. Lanzani, M. Nisoli, S. De Silvestri, et al. Femtosecond vibrational and torsional energy redistribution in photoexcited oligothiophenes. *Chem. Phys. Lett.*, 251:339–345, 1996.

33. G. Lanzani, M. Nisoli, S. De Silvestri, et al. Femtosecond vibrational and torsional energy redistribution in photoexcited oligothiophenes. *Chem. Phys. Lett.*, 251:339–345, 1996.

34. K. S. Wong, H. Wang, and G. Lanzani. Ultrafast excited-state planarization of the hexamethylsexithiophene oligomer studied by femtosecond time-resolved photoluminescence. *Chem. Phys. Lett.*, 288:59–64, 1998.

35. F. Della Sala, private communication.

36. J. Gierschner, H.-G. Mack, L. Luer, and D. Oelkrug. Fluorescence and absorption spectra of oligophenylenevinylenes: vibronic coupling, band shapes, and solvatochromism. *J. Chem. Phys.*, 116:8596–8603, 2002.

37. M. J. J. Vrakking, D. M. Villeneuve, and A. Stolow. Observation of fractional revivals of a molecular wave packet. *Phys. Rev. A*, 54:R37–R40, 1996.

38. D. N. Batchelder. *Polydiacetylene* (D. Bloor and R. R. Chance, Eds.), Nijhitt, Dordrecht, 1985.

39. T. Kobayashi. *Relaxation in Polymers*, World Scientific, Singapore, 1993.

40. B. I. Greene, J. F. Mueller, J. Orenstein, D. H. Rapkine, S. Schmitt-Rink, and M. Thakur. Phonon-mediated optical nonlinearity in polydiacetylene. *Phys. Rev. Lett.*, 61:325–328, 1988.

41. G. J. Blanchard, J. P. Heritage, A. C. Von Lehmen, M. K. Kelly, G. L. Baker, and S. Etemad. Excitonic and phonon-mediated optical Stark effects in a conjugated polymer. *Phys. Rev. Lett.*, 63:887–890, 1989.

42. R. Lécuiller, J. Berréhar, J. D. Ganière, C. Lapersonne-Meyer, P. Lavallard, and M. Schott. Fluorescence yield and lifetime of isolated polydiacetylene chains: evidence for a one-dimensional exciton band in a conjugated polymer. *Phys. Rev. B*, 66:125205, 2002.

43. F. Dubin, J. Berréhar, R. Grousson, T. Guillet, C. Lapersonne-Meyer, M. Schott, and V. Voliotis. Optical evidence of a purely one-dimensional exciton density of states in a single conjugated polymer chain. *Phys. Rev. B*, 66:113202, 2002.

44. B. Gallot, A. Cravino, I. Moggio, D. Comoretto, C. Cuniberti, and G. Delle Piane. Supramolecular organization in the solid state of a novel soluble polydiacetylene. *Liq. Cryst.*, 26:1437–1444, 1999.

45. M. Alloisio, A. Cravino, I. Moggio, et al. Solution spectroscopic properties of polydCHD-HS: a novel highly soluble polydiacetylene. *J. Chem. Soc., Perkin Trans.* 2:146–152, 2001.

46. G. Lanzani, G. Cerullo, M. Zavelani-Rossi, S. De Silvestri, D. Comoretto, G. Musso, and G. Delle Piane. Triplet-exciton generation mechanism in a new soluble (red-phase) polydiacetylene. *Phys. Rev. Lett.*, 87:187402, 2001.

47. G. Lanzani, M. Zavelani-Rossi, G. Cerullo, D. Comoretto, and G. Dellepiane. Real-time observation of coherent nuclear motion in polydiacetylene isolated chains. *Phys. Rev. B*, 69:134302, 2004.

48. B. Kraabel, D. Hulin, C. Aslangul, C. Lapersonne-Meyer, and M. Schott. Triplet exciton generation, transport and relaxation in isolated polydiacetylene chains: subpicosecond pump-probe experiments. *Chem Phys.*, 227:83–87, 1998.

49. G. Lanzani, S. Stagira, G. Cerullo, S. De Silvestri, D. Comoretto, I. Moggio, and G. Dellepiane. Triplet exciton generation and decay in a red polydiacetylene studied by femtosecond spectroscopy. *Chem. Phys. Lett.*, 313:525–529, 1999.

50. C. H. Brito Cruz, J. P. Gordon, P. C. Becker, R. L. Fork, and C. V. Shank. Dynamics of spectral hole burning. *IEEE J. Quantum Electron.*, 24:261–266, 1988.

51. C. J. Bardeen, Q. Wang, and C. V. Shank. Selective excitation of vibrational wave-packet motion using chirped pulses. *Phys. Rev. Lett.*, 75:3410–3413, 1995.

52. B. E. Kohler. *Conjugated Polymers: The Novel Science and Technology of Conducting and Non Linear Optically Active Materials.* (J. L. Bredas and R. Silbey, Eds.), Kluwer Press, Dordrecht, 1991.

4 Coherent Phonons in Bulk and Low-Dimensional Semiconductors

Michael Först and Thomas Dekorsy

CONTENTS

4.1 INTRODUCTION

The progress in the generation of ultrashort laser pulses has always been driven by the eagerness of scientists to watch physical phenomena *while they happen*. As the invention of the electron microscope and scanning probe techniques have led to atomically resolved pictures of the nanocosmos, the development of femtosecond lasers has allowed us to make movies of the dynamics of electrons and vibrational excitations on timescales shorter than typical scattering times. Especially in semiconductors and semiconductor nanostructures, ultrafast time-resolved optical techniques contributed to disentangle the time constants of the most important interactions [1].

With the first picosecond laser pulses available in the 1980s, first-time resolved observations of nonequilibrium phonon dynamics in semiconductors were performed by employing coherent anti-Stokes Raman scattering (CARS) [2–4]. CARS is based on time resolving the intensity of the anti-Stokes line; hence, the decay times of coherently excited phonons can be monitored in the time domain. This technique allowed the first determination of picosecond decay times of longitudinal optical (LO) phonons in semiconductors. However, this technique does not allow monitoring of the coherent atomic vibrations in amplitude and phase. This restriction is based on the fact that for time resolving a coherent vibration in the time domain, the laser pulse has to be shorter than half the vibrational period. Since the pulse duration is linked to the spectral width of the laser pulse via the time–bandwidth product [5], the bandwidth of the laser pulse necessarily exceeds the energy of the phonon. This is detrimental to CARS because, for the detection of the anti-Stokes Raman signal, the laser spectrum should not overlap the Raman-shifted scattered light. Nevertheless, important information on phonon decay has been gathered by this techniques [6–9].

The first observation of coherent vibrational excitation in amplitude and phase has been reported for coherent *acoustic* phonons in solid state [10,11]. Here, the interaction of the ultrashort laser pulse with the material is based on the thermoelastic effect. The absorption of the laser pulse generates a sound pulse (i.e., a pulse of acoustic phonons), which can be traced as it propagates back and forth in a given sample structure. This technique has given rise to the establishment of the field of picosecond ultrasonics, which has gained significant attention in recent years for the investigation of acoustic phonon propagation in thin films and nanostructures and moreover found applications in semiconductor metrology [12–16].

This book gives a broad overview of coherent vibrational dynamics in different material systems, so we omit a historical account of achievements not related to semiconductors. The first time-resolved detection of coherent *optical* phonons in semiconductors was reported in the early 1990s by two groups almost simultaneously. In the group of Ippen at the Massachusetts Institute of Technology (MIT), coherent phonons were observed in narrow-gap semiconductors and semimetals such as bismuth and antimony [17]. The coherent excitation of phonons in these materials is associated with a large transient modulation of the reflectivity with amplitudes up to 10^{-3} relative changes of the reflectivity. The coherent phonon excitation in these materials was interpreted within a model that explains the selective excitation of highly symmetric phonon modes, the so-called displacive excitation of coherent phonons (DECP) model [18]. In the group of Kurz at Aachen University, coherent LO phonons were observed in GaAs [19]. These modes appear in the time-resolved reflectivity via the linear electrooptic effect or Pockels effect due to an associated macroscopic polarization. The driving force for these modes is based on an ultrafast electric field change within the polar semiconductor on a timescale shorter than the phonon period. This model has often been refered to as the *current surge model.*

Experiments at both MIT and Aachen University were performed with the workhorse of femtosecond spectroscopy of those days: a colliding pulse mode-locked (CPM) dye laser that operated at 2-eV photon energy delivering pulses of about 50 fs duration and energies smaller than 1 nJ/pulse. The advent of Ti:sapphire Kerr-lens mode-locked oscillators and Ti:sapphire amplifiers gave rise to a large extension of the field of coherent phonon spectroscopy due to (i) the tunability of the laser wavelength, (ii) their higher pulse energies, and (iii) the higher power stability of these systems compared with femtosecond dye lasers.

This chapter summarizes the most recent achievements in the field of coherent phonon spectroscopy in semiconductors and semiconductor nanostructures. For reviews of coherent phonons, including other materials than semiconductors, refer to the reviews by Merlin [20], Dekorsy et al. [21], and Misochko [22]. The chapter is organized as follows: In Section 4.2, the most relevant excitation mechanisms for coherent phonons are distinguished. Because the excitation and detection process of coherent phonons need not neccessarily be based on the same physical process, the detection of coherent phonons in semiconductors is addressed separately in Section 4.3. The following sections refer to the relevance of these excitation and detection processes for the phonon dynamics in bulk semiconductors (Section 4.4), quantum wells (Section 4.5), and superlattices (Section 4.6).

4.2 EXCITATION MECHANISMS

The excitation of a coherent phonon displacement amplitude Q can be described phenomenologically as a harmonic oscillator driven by an external force. This external force is provided by the femtosecond laser pulse and represents an impulse shorter than the inverse of the phonon period under consideration. In a frequency domain picture, the same laser pulse is described by a broad spectrum that is linked to the pulse duration via the time–bandwidth product [5]. Thus, the requirement for the pulse duration in the time-domain translates to a pulse bandwidth that is larger than the frequency of the phonon.

The phenomenological equation of motion for the coherent phonon displacement amplitude Q, which can be derived from a quantum mechanical description [23], is expressed as [2, 24]

$$\mu^* \left(\frac{\partial^2 Q(t)}{\partial t^2} + 2\gamma_{phonon} \frac{\partial Q(t)}{\partial t} + \omega_{phonon}^2 Q(t) \right) = F^Q(t) \qquad (4.1)$$

where μ^* is the reduced lattice mass, γ_{phonon} is a damping constant, and $F^Q(t)$ is the driving force exerted by the laser pulse. The damping constant γ_{phonon} is related to the dephasing time T_2 of the coherent mode via $\gamma_{phonon} = 1/T_2$. The notion of a dephasing time has been established for coherent excitations; in a density matrix representation, T_2 describes the temporal evolution of the nondiagonal terms of the density matrix. This dephasing time T_2 is related to the population decay time T_1, which describes the decay of the diagonal terms of the density matrix via $2/T_2 = 1/T_1 + 1/T_p$, where T_p is the time related to truly phase-destroying processes [25]. The latter can usually be neglected for vibrational excitations in solids. Therefore, the observed dephasing time of a coherent phonon mainly reflects the population decay time; that is, the observed dephasing times in a time-resolved experiment should give the same values as derived from non-time-resolved inelastic light-scattering experiments. However, these experiments are often limited by their instrumental resolution, which becomes evident (e.g., in the case of long-lived acoustic phonons). For optical phonons, the population decay—neglecting electron–phonon interaction—is dominated by lattice anharmonicity, which gives rise to the decay into low-energy acoustic phonon modes.

4.2.1 COHERENT PHONON EXCITATION IN TRANSPARENT MEDIA

A common distinction for the driving force of coherent phonons is made into *impulsive* and *displacive* types of excitation. Especially in transparent materials in the absence of significant interband transitions, processes of impulsive nature have been identified as the dominant excitation mechanism. The driving force $F^Q(t)$ of a coherent phonon mode exerted by a Raman process is given in lowest order by the interaction of the electromagnetic field E of the laser pulse and the Raman tensor R_{jkl} of the medium,

$$F_j^Q(t) = R_{jkl} E_k E_l \qquad (4.2)$$

Here, the impulsive excitation is accomplished by the combined action of two field components, E_k and E_l, within the broad spectrum of the ultrashort laser pulse. According to wavevector and energy conservation rules, a lattice mode with frequency $\omega_l - \omega_k = \omega_{phonon}$ and wavevector $\mathbf{k}_l - \mathbf{k}_k = \mathbf{q}_{phonon}$ can be excited. This process is denoted *impulsive stimulated Raman scattering* (ISRS) [26–30]. ISRS is extensively discussed in a separate chapter of this book for the case of impulsive excitations in molecular systems. The initial phase of the coherent lattice displacement excited via ISRS should obey a $\sin(\omega_{phonon}t)$ dependence; however, ISRS may also exhibit a $\cos(\omega_{phonon}t)$ dependence under distinct resonant conditions [30]. By using two coincident excitation pulses, the phonon wavevector can be defined by the choice of their angles of incidence.

4.2.2 COHERENT PHONON EXCITATION IN OPAQUE MEDIA

In opaque materials such as low-bandgap semiconductors or semimetals, for example, the excitation of a coherent phonon mode is generally displacive as explained by the DECP model [17,18,31]. This model relates the coherent phonon excitation to a strong interband excitation from bonding to antibonding states. This process leads to a sudden change of the equilibrium position of the atoms and thereby to a coherent excitation of lattice vibrations maintaining the crystal symmetry, that is, A_1 modes for a large variety of materials. Displacively excited coherent phonons are characterized by a $\cos(\omega_{phonon}t)$ dependence of the coherent amplitude [18]. DECP has been treated in a more rigorous quantum-mechanical description by Kuznetsov and Stanton [23] and has often been referred to as a prominent example of a non-Raman-type excitation of coherent phonons. The group of Merlin has shown that the simple DECP description is compatible with a description based on Raman interaction when real and imaginary parts of the Raman tensor are considered. For the case of coherent A_{1g} symmetry phonons in antimony, they could prove that the spectral dependence of the coherently excited amplitude exactly follows the imaginary part of the dielectric tensor [32].

4.2.3 COHERENT PHONON EXCITATION IN POLAR SEMICONDUCTORS

In the case of polar semiconductors, the driving force in Eq. (4.1) can be described in the following way [24]:

$$F_j^Q(t) = R_{jkl}E_kE_l - \frac{e^*}{\epsilon_\infty\epsilon_0}P_j^{NL} \tag{4.3}$$

where R_{ijk} is again the Raman tensor, and P_j^{NL} represents a nonlinear longitudinal polarization along direction j. This nonlinear polarization P_j^{NL} can be divided into several different contributions:

$$P_j^{NL} = \chi_{jkl}^{(2)}E_kE_l + \chi_{jklm}^{(3)}E_kE_lE_m + \int_{-\infty}^t dt' J_j(t')$$

$$+ Ne\int_{-\infty}^t dt'\int_{-\infty}^\infty dx_j\langle\Psi(x_j,t')|x_j|\Psi(x_j,t')\rangle \tag{4.4}$$

where the first two terms represent the second- and third-order nonlinear susceptibilities, respectively. The third and fourth terms describe polarizations set up by ultrafast drift-diffusion currents or the polarization associated with coherent wavepacket dynamics, respectively. The ultrafast drift-diffusion currents can impulsively excite coherent LO phonons in a current surge model for the screening of surface fields of III–V semiconductors [19,33–35] (see Section 4.4) or via the ultrafast buildup of electric Dember fields [36–40]. The last term of Eq. (4.4) describes the case of coherent electronic wavefunctions, which may set up a macroscopic *intraband* polarization (e.g., quantum beats in semiconductor heterostructures) (see Sections 4.5.2 and 4.6.1).

For longitudinal phonons in polar semiconductors, the coupling of the lattice polarization given by a normalized displacement Q and an electronic polarization P is given by the coupled equations [35]

$$\frac{\partial^2}{\partial t^2} P_j + \gamma_{el} \frac{\partial}{\partial t} P_j + \omega_{el}^2 P_j = \frac{e^2 N}{\epsilon_\infty \mu^*} \left(E_j^{ext} - 4\pi \gamma_{12} Q_j \right) \quad (4.5)$$

$$\frac{\partial^2}{\partial t^2} Q_j + \gamma_{phonon} \frac{\partial}{\partial t} Q_j + \omega_{phonon}^2 Q = \frac{\gamma_{12}}{\epsilon_\infty} \left(E_j^{ext} - 4\pi P_j \right) \quad (4.6)$$

Here, E_j^{ext} is a macroscopic electric field that has to be determined via the Poisson equation, and γ_{12} is a coupling constant given by $\gamma_{12} = \omega_{TO}\sqrt{(\epsilon_0 - \epsilon_\infty)/(4\pi)}$. The generalized phenomenological damping constant γ_{el} describes the damping of the electronic polarization (e.g., the momentum scattering time of a dense carrier plasma or the intraband polarization dephasing time of a coherent electronic wavepacket). The frequency ω_{el} is either the plasma frequency or the frequency of the coherent wavepacket oscillation considered. The driving force in Eqs. (4.5) and (4.6) is set up by a combined action of the electric field dynamics and the polarization that is expressed in the different possible contributions in Eqs. (4.3) and (4.4). The determination of the relative strength of several possible contributions to the excitation process is one of the challenging tasks in the field of coherent phonon spectroscopy.

Equations (4.5) and (4.6) describe coupled electron–phonon dynamics in a macroscopic way. For the case of homogeneous densities and negligible damping, Eqs. (4.5) and (4.6) reproduce the well-known dispersion branches of coupled plasmon–phonon modes [41]

$$\omega_{PPM}^\pm = \frac{1}{2} \left(\omega_{LO}^2 + \omega_P^2 \pm \sqrt{\left(\omega_P^2 + \omega_{LO}^2\right)^2 - 4\omega_P^2 \omega_{TO}^2} \right)^{(1/2)} \quad (4.7)$$

with the plasma frequency $\omega_P^2 = (e_0^2 N)/(\epsilon_0 \epsilon_\infty m^*)$, N the electron density, and m^* the effective mass. The associated plasmon–phonon coupled modes were first observed in Raman experiments in doped GaAs [42, 43]. To temporally resolve coherent plasmon–phonon dynamics, we have to bear in mind that an optically excited plasma may be strongly inhomogeneous in real space due to the spatial profile of the laser beam and the absorption depth.

The wavevector of the excited phonon modes strongly depends on the excitation process. For experiments in transparent media where the coherent phonon excitation is based on first-order Raman processes, the wavevector is given by the wavelength λ of the pump laser; that is, $q = 2\pi n/\lambda$, where n is the refractive index. In two-beam excitation configurations based on $\chi^{(2)}$ processes, the coherent phonon wavevector can be selected via the angle between the two laser pulses (see above). In opaque materials, the excited phonon wavevector is mainly determined by the imaginary part of the refractive index κ; that is, $q = 2\pi\kappa/\lambda$. Since the dispersion of optical phonons is weak for small wavevectors, their wavevector cannot be determined from the observed frequency with high accuracy. However, for acoustic zone-folded phonons in semiconductor superlattices, this becomes possible due to backfolding of the acoustic phonon dispersion into the mini-Brioullin zone of the artificially periodic structure.

4.3 DETECTION OF COHERENT PHONONS

In a time-resolved experiment, the excitation and detection of a coherent lattice vibration can be well separated in contrast to the inelastic light-scattering process in Raman scattering. Many different possibilities exist for the detection of coherent phonons in the time domain. Most commonly employed is the detection of changes of the optical properties of the investigated material with a time-delayed probe pulse. This probe pulse may have the same energy (degenerate pump-probe) as the exciting laser pulse or may have a different energy (nondegenerate pump-probe). These experiments can be performed in reflection or transmission geometry. In addition, diffracted light may be detected as a function of time delay, which exhibits a periodic modulation due to coherently excited phonons in a four-wave mixing geometry. The modulation of the reflectivity ΔR of a sample with the frequency of the phonon can be explained on the base of the first-order Raman tensor $(\partial\chi)/(\partial Q)$, that is,

$$\Delta R = \frac{\partial R}{\partial n}\Delta n \sim \frac{\partial R}{\partial\chi}\frac{\partial\chi}{\partial Q}Q \qquad (4.8)$$

Since $(\partial\chi)/(\partial Q)$ is a tensor, different modes can be probed selectively via polarization analysis of the probe pulse [37,44,45]. In the case of fully symmetric phonons, the Raman tensor contains diagonal contributions of equal magnitude only. In this case, the reflectivity changes will be isotropic irrespective of the polarization of the probe pulse relative to the crystal orientation. For phonon modes with nondiagonal terms in the Raman tensor, $(\partial\chi)(\partial Q)_{jk} \neq 0$ for $j \neq k$, the reflectivity changes will depend on the angle of the light polarization with respect to the crystal orientation. Further detection schemes are based on Raman interaction with transmitted probe pulses, which results in a periodic shift of the spectral components of the probe pulse with the frequency of the lattice vibrations [20,46]. In crystals exhibiting a linear electrooptic effect (Pockels effect), coherent phonons may be detected sensitively via the associated longitudinal field [19]. In terms of Raman tensors, the so-called electrooptic contribution to the Raman tensor $(\partial\chi/\partial F)(\partial F/\partial Q)Q$ has to be added to Eq. (4.8). The electric field-induced phonon detection may be resonantly enhanced via third-order nonlinearities associated with the Franz–Keldysh effect close to interband resonances [47,48]. Coherently excited surface phonons may be selectively detected via the time-resolved detection of second-harmonic generation (SHG) [49,50].

Coherent infrared (IR)-active phonons, phonon–polaritons, or plasmon–phonon modes can also be detected via the emission of electromagnetic radiation in the terahertz frequency range [35, 37–40, 51–53]. The emission characteristics can be assumed to follow the emission of a dipole, which is set up within the volume in which coherent phonons form a macroscopic polarization. The pointing vector of the radiation is given by [35]

$$\mathbf{S}(t) = \frac{\omega_{phonon}^4 \mathbf{P}^2(t) \sin^2 \theta}{16\pi^2 \epsilon_0 r^2 c^3} \mathbf{s} \tag{4.9}$$

where θ is the angle between the direction of observation $\mathbf{s}$ and the polarization $\mathbf{P}$, and r is the distance to the detector. Experimentally, the time-resolved detection of terahertz electromagnetic fields has become possible by the application of ultrafast photoconductive switches [54] and the electrooptic detection scheme [51, 52, 55, 56]. In addition to the emission characteristics of the terahertz radiation, the outcoupling of the radiation through the semiconductor–air interface has to be considered. Due to the strong dispersion of the refractive index close to a phonon–polariton resonance, the Fresnel coefficient is strongly frequency dependent, thus leading itself to pronounced structures in the terahertz spectra [39, 51, 52, 57].

The most recent development in the detection of coherent phonons is based on ultrafast X-ray pulses produced from high-intensity femtosecond pulses [58–61]. Since these X-ray pulses are generated from an intense IR pulse, they are inherently synchronized and can be used to perform time-resolved Bragg diffraction on samples, which are excited by a portion of the optical pulse. The spatial variation of a Bragg scattered peak is the most direct observation of a coherent atomic displacement in the excited sample. The first experiment employing ultrafast X-ray pulses dealt with the detection of coherent acoustic motions [58, 59] in semiconductors. The improvement of this technique has led to the observation of coherent optical phonons in bismuth [60] and zone-folded acoustic phonons in semiconductor superlattices [61].

4.4   COHERENT PHONONS IN BULK SEMICONDUCTORS

The coherent phonon generation in bulk semiconductors was first observed by Cho et al. as minute oscillations in the time-resolved reflectivity changes in GaAs [19] and by Cheng et al. in Bi and Sb [31]. Since GaAs is a technologically important compound semiconductor, it has been studied extensively by means of Raman spectroscopy [62, 63]. The study of GaAs enlightened several pecularities in the excitation of coherent optical phonon modes in compound semiconductors.

4.4.1   COHERENT PHONON EXCITATION BY SURFACE-FIELD SCREENING

The laser system employed in the first subpicosecond time-resolved experiments on GaAs was a CPM dye laser that delivers light pulses at 2 eV with pulse durations of 50 fs. For the detection of coherent phonons, which often give small contributions on a large electronic background signal, a special fast-scanning detection system

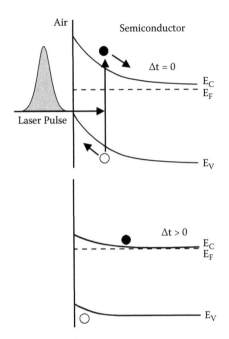

FIGURE 4.1 Schematic sketch of the field screening process at a bare surface of an n-doped semiconductor. The valence and conduction bands (E_V and E_C, respectively, E_F the Fermi level) are bent toward the surface, giving rise to an electric space charge field. Top: Excitation of free electron holes leads to rapid drift currents on the timescale of the exciting laser pulse. Bottom: After the excitation electron and hole distributions are spatially separated and the space charge field is screened.

has been developed [64]. This detection system is based on scanning the time delay between pump and probe pulses with a retroreflector mounted onto a vibrating membrane. This allows for delay scanning at frequencies of around 100 Hz. With a real-time data acquisition system, the data are accumulated without using further lock-in filtering techniques. This technique enabled the detection of time-resolved signal changes down to some 10^{-7} even with the CPM dye laser as laser source.

The excitation mechanism of coherent phonons in GaAs can be explained by taking into account electric fields at bare surfaces of III–V compounds. Charged surface states lead to a pinning of the Fermi level within the bandgap, hence bending the band structure toward the surface. The associated electric field is calculated within a Schottky barrier model and exhibits a square root dependence on the doping density [65]. An external manipulation of the surface field is possible via semitransparent Schottky contacts. Figure 4.1 sketches the underlying process for the generation of coherent LO phonons in a surface space charge field. The optical injection of carriers within the surface field region leads to an ultrafast current surge that rapidly screens the built-in electric field [34]. The ultrafast depolarization P^{NL} of the surface depletion field then leads to the coherent excitation of LO phonons according to Eqs. (4.3) and (4.4) [19,33]. The detection for wavelengths

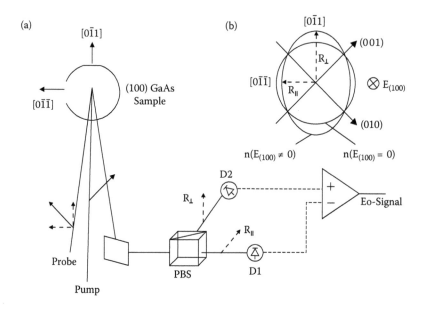

FIGURE 4.2 (a) Experimental setup for the time-resolved detection of anisotropic reflectivity changes from a (100) GaAs surface. PBS is a polarizing beam splitter; D1 and D2 are detectors. (b) Optical indicatrix on a (100) GaAs surface in the presence of an electric field E_{100} perpendicular to the surface.

far from resonances is accomplished via the electrooptic effect, also known as Pockels effect, in reflection geometry (reflective electrooptic sampling, REOS) [66]. On (100) GaAs surfaces, a longitudinal electric field along the [100] direction leads to induced birefringence in the (100) plane. By subtracting two polarization components of a linearly polarized probe pulse along the [011] and [01$\bar{1}$] direction, the anisotropic reflectivity change is given by [19,67]

$$\frac{\Delta R(t)}{R_0} = \frac{\Delta R_{[011]}(t) - \Delta R_{[01\bar{1}]}(t)}{R_0} = \frac{4\, r_{41} n_0^3}{n_0^2 - 1}\, \Delta E_{[100]}(t) \qquad (4.10)$$

where r_{41} is the electrooptic coefficient, n_0 is the unperturbed refractive index, and $\Delta E_{[100]}(t)$ is the time-dependent macroscopic electric surface field along the [100] direction. Figure 4.2 sketches the experimental setup and the electric field-induced anisotropy of the refractive index in the (100) plane of the crystal.

Figure 4.3 depicts the electrooptic reflectivity changes of an n-doped GaAs sample prepared with a transparent indium-tin oxide (ITO) Schottky contact on top recorded with a CPM dye laser. The experiments were performed at room temperature at an optical excitation density of $4 \times 10^{17} \mathrm{cm}^{-3}$. The electrooptic measurements in Figure 4.3 reveal the transient screening of the surface field $\Delta E_{\mathrm{surface}}(t)$ for different initial surface-field strengths. These strength can be varied by applying different voltages to the sample. The screening dynamics exhibit a fast component on the timescale of the exciting laser pulse (i.e., within one optical

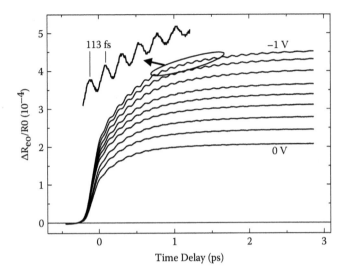

FIGURE 4.3 Time-resolved REOS signal from a (100) GaAs with a transparent Schottky contact on top for a variation of the electric surface field. The excitation density is $5 \times 10^{17} cm^{-3}$. The data are obtained for different reverse-bias voltages between 0 and 1 V at 300 K (from Ref. [34]). A part of the oscillatory signal is enlarged, revealing a single-frequency oscillation of 113 fs.

phonon period) and a slower component on a picosecond timescale. The associated electric field changes are in the order of 100 kV/cm [34]. The data are clearly modulated with oscillations of 8.75-THz frequency, matching the GaAs $q = 0$ LO phonon frequency at 300 K. The dephasing time T_2 of the oscillations, 4.0 ± 0.3 ps, does not depend on the electric field applied and is in agreement with the phonon lifetime derived from CARS experiments [7]. This agreement suggests that the observed decay of the coherent amplitude is determined by anharmonic decay into acoustic phonon modes and not by pure dephasing processes. Such an agreement between dephasing times derived from coherent phonon experiments and linewidths obtained in Raman scattering experiments has also been observed in a thorough study of single-crystal bismuth [68].

For the verification of surface-field screening as the responsible driving force of coherent LO phonons in GaAs, the phonon amplitude is plotted versus the reflectivity change at a time delay of 50 fs (Figure 4.4). Since this value represents the optically detected surface-field changes, that is, $\Delta R(50\,fs) \sim \Delta E_{surface}(50\,fs)$, it is proportional to the driving force for coherent LO phonons according to the third term in Eq. (4.4). The experimental data reveal a linear dependence, strongly confirming the proposed generation mechanism. It is, noteworthy that the phase of the oscillations is cosine-like $\sim \cos(\omega t)$. This behavior is expected for an oscillator starting at a displaced position (i.e., the atomic positions at negative time delays) driven to oscillate around a new equilibrium position that is given by the electric field at positive time delays.

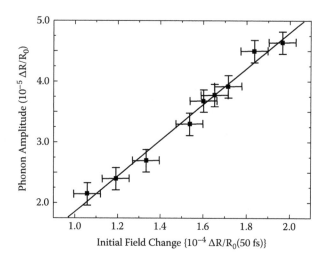

FIGURE 4.4 Amplitude of coherent LO phonons (solid squares) versus the electrooptic reflectivity changes in Figure 4.3 at a time delay of 50 fs. (Adapted from T. Pfeifer et al., *Appl. Phys. A*, 55:482, 1992).

Another verification of the field-screening mechanism is obtained from the variation of the excitation density. Figure 4.5 shows the electrooptic reflectivity changes of n-doped GaAs for excitation densities between 6×10^{16} cm^{-3} and 5×10^{18} cm^{-3} (i.e., a variation of nearly two orders of magnitude). For the lowest excitation densities, the screening of the surface field takes place over a picosecond timescale. These field dynamics do not have enough bandwidth to drive the LO phonon coherently. Only at densities above 6×10^{17}cm^{-3} do significant field changes occur on a timescale shorter than half the phonon period. With a further increase of the excitation density, the phonon amplitude increases. Simultaneously, the frequency changes due to plasmon–phonon coupling as described later.

The excitation process of coherent LO phonons observed in GaAs is *isotropic* in the sense that the coherent phonon amplitude is independent of the incident polarization of the pump pulse [33, 46]. This observation rules out a coherent phonon generation via $\chi^{(2)}$ processes [first right-hand term in Eq. (4.4)]. The last remaining Raman process that could in principle account for an isotropic generation process is a third-order nonlinear polarization with the third field being the surface field [second right-hand term in Eq. (4.4)]. This effect was observed in continuous wave (CW) Raman experiments close to electronic resonances and has been denoted as the inverse Franz–Keldysh effect [69]. However, it was shown to drop in scattering efficiency by orders of magnitude when the laser energy is detuned from interband resonances by only a few tens of millielectron volts. At the time when only femtosecond lasers with 2-eV photon energy were available, this effect could not be completely ruled out as a generation mechanism. Therefore, this question is addressed later in the context of resonant excitation with a tunable femtosecond Ti-sapphire laser.

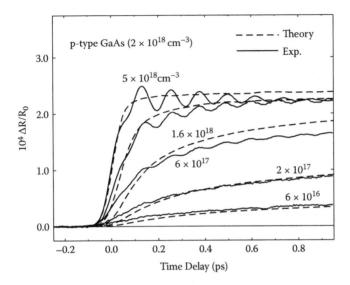

FIGURE 4.5 Electrooptic reflectivity changes from p-type GaAs for different excitation densities as given in the figure. The solid lines are experimental results and the dashed lines are simulations based on a drift-diffusion model. (Adapted from T. Dekorsy et al., *Phys. Rev. B*, 47:3842, 1993). The simulations do not incorporate the coherent optical phonon dynamics.

4.4.2 RESONANT COHERENT PHONON DETECTION VIA THE NONLINEAR FRANZ–KELDYSH EFFECT IN GaAs

The invention of the Kerr-lens mode-locked Ti:sapphire laser has allowed performing femtosecond time-resolved experiments over a wide wavelength range of roughly 700 to 1000 nm when using the fundamental frequency of the laser only. This range covers the electronic bandgap of GaAs (1.42 eV at room temperature), which has enabled the investigation of resonant and nonresonant contributions to the excitation and detection of coherent phonons.

Figure 4.6 depicts time-resolved reflectivity changes from a (100) GaAs surface for above-bandgap (2.0-eV) and near-resonant (1.47-eV) photon energies. In contrast to the electrooptic reflectivity changes shown in Figure 4.3, the probe pulses are detected separately for two different polarizations along principle axes of the electrooptic index ellipsoid (i.e., the [011] and [01$\bar{1}$] crystal directions). For above-bandgap energies, the observed reflectivity changes are composed of two contributions. The first is an *electronic* contribution due to changes of the refractive index or absorption that are associated with photogenerated free carriers. This contribution is *isotropic* in the sense that it does not depend on the probe pulse orientation and can be selectively detected in the [001] direction. The second is the *electrooptic* contribution due to electric field changes, which is of comparable magnitude as the isotropic reflectivity change. For a probe beam polarization along

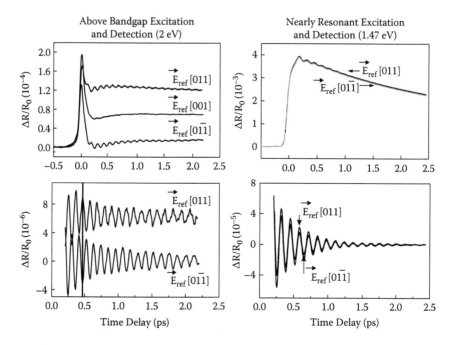

FIGURE 4.6 Comparison of reflectivity changes of (100) GaAs at excitation well above the band gap (2 eV, left) and close to the band gap (1.47 eV, right) for different polarizations of the reflected probe beam $\vec{E}_{ref}$ along the [001], [011], and [01$\bar{1}$] directions (top). In the lower part, the oscillatory contributions are compared for the [011] and the [01$\bar{1}$] polarization. Note the different scaling of the figures.

the [011] ([01$\bar{1}$]) crystal axis, this effect gives a positive (negative) contribution to the isotropic carrier-induced reflectivity changes. This sign reversal of the electrooptic contribution is also observed for the phonon-induced reflectivity changes, as can be seen in a π-phase shift between the oscillations. The same measurements performed close to the band edge at 1.47 eV exhibit a much larger isotropic reflectivity change and a phonon-induced contribution larger by more than an order of magnitude than the electrooptic contribution at above-bandgap excitation. In addition, the phase of the oscillations does not change with the probe polarization. Only a tiny difference exists between these two transients with a magnitude comparable with the difference between the two equivalent 2-eV curves. All these observations indicate that the phonon detection at the fundamental bandgap is not based on the electrooptic effect but on an isotropic optical nonlinearity. Cho et al. showed that this contribution arises from a third-order nonlinearity associated with the Franz–Keldysh effect [48]. The term in expansion of the nonlinear third-order polarization responsible for this detection process is

$$P(\omega) = \chi^{(3)} E_{surface}(0) \, E_{phonon}(\omega_{LO}) \, E(\omega) \qquad (4.11)$$

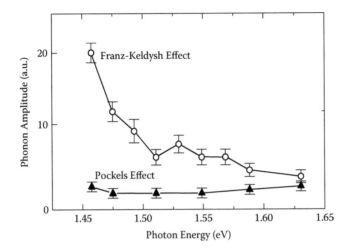

FIGURE 4.7 Amplitude of the coherent LO phonons for different detection energies close to the band gap of GaAs for the Franz–Keldysh effect (open dots) and electrooptic or Pockels effect contribution (solid triangles). (Adapted from G. C. Cho et al., *Phys. Rev. B*, 53:6904, 1996.)

where $E_{surface}(0)$ is the surface field, $E_{phonon}(\omega_{LO})$ is the field associated with the coherent LO phonons, and $E(\omega)$ is the laser field. The surface field $E_{surface}(0)$ is assumed to be nearly static with respect to the light frequency. However, via Eq. (4.11), the subpicosecond surface-field dynamics also influence the coherent phonon amplitude detected.

The detection of coherent LO phonons at different laser energies close to and above the bandgap enables the determination of the dispersion of electrooptic and Franz–Keldysh contributions, that is, the dispersion of $\chi^{(2)}$ and $\chi^{(3)}$ at the LO phonon frequency. Figure 4.7 reveals a nearly dispersion-free electrooptic effect, while the Franz–Keldysh-like detection is strongly enhanced close to the band edge. These data give clear evidence for the relevance of higher-order nonlinearities in the resonant detection of coherent optical phonons. A dispersion of the generation process can be ruled out since it would lead also to changes in the phonon amplitude detected in the electrooptic scheme. Consequently, these measurements rule out the inverse Franz–Keldysh effect discussed as a possible excitation mechanism.

4.4.3   COHERENT PHONON EXCITATION BY ULTRAFAST BUILDUP OF A DEMBER FIELD

The coherent phonon excitation so far described for the case of GaAs is based on the ultrafast screening of a built-in surface electric field. On the other hand, in some opaque materials, electric fields may buildup on the timescale of the exciting laser pulse. This effect is known as the *photo-Dember effect*, which is based on different mobilities of electrons and holes, respectively [36, 37, 70]. In a highly

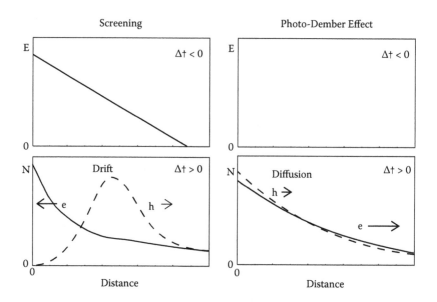

FIGURE 4.8 Comparison of the surface-field screening mechanism (left) and the photo-Dember effect (right). The top graph sketches the electric field distribution with (left) and without (right) a surface field before the optical excitation. The surface field is screened by drift currents of photogenerated electrons (solid line, bottom left) and holes (dashed line, bottom left). In the absence of a surface field, faster diffusion of electrons as compared with holes leads to the buildup of the Dember electric field (solid line, bottom right).

absorbing semiconductor with absorption coefficients on the order of 10^4 cm^{-1}, the photogenerated charge carrier density is characterized by a strong gradient perpendicular to the surface. The charge carrier species with the higher mobility μ (and larger diffusion constant D; both are connected via the Einstein relation $D/\mu = kT/e$ in thermal equilibrium)—usually the electrons—diffuse faster into the bulk than the other species. This charge carrier separation leads to the buildup of an electric field perpendicular to the surface. Once this electric field is established, the diffusion of electrons and holes is coupled, and the diffusion is dominated by the carrier species with the lower mobility. This is the reason for the description of diffusion in optically excited semiconductors through the ambipolar diffusion constant. Figure 4.8 schematically summarizes the difference between the screening and buildup of electric fields and the associated spatial carrier distributions. Note that, in contrast to the DECP mechanism introduced for the coherent excitation of symmetry-maintaining A_1 phonon modes in opaque materials in Section 4.2.2, the photo-Dember effect allows for only the coherent generation of IR-active phonon modes.

In femtosecond time-resolved experiments, the relevance of the Dember effect was first observed in the terahertz emission of single-crystal tellurium [37, 70]. Here, the absorption length of 2-eV photons is in the range of a few tens of nanometers, giving rise to extremely strong gradients of the carrier distribution close to the surface. By solving a drift-diffusion equation, it was shown that pulsed excitation of carrier densities in the range of 10^{18} cm^{-3} leads to the buildup of electric fields in the range of several tens of kilovolts per centimeter within 100 fs. These changes of the electric field are comparable to field changes observed in the field-screening model. Hence, polar phonons can be coherently excited via this mechanism (tellurium is a single-atom semiconductor but possesses low-frequency polar-optical phonons due to its particular crystal symmetry). The coherent excitation of plasmons via the photo-Dember effect is discussed in the next section.

4.4.4 COUPLED PLASMON–PHONON MODES IN DOPED SEMICONDUCTORS

Since the first observations by Mooradian and McWorther [43], coupled modes of plasmons and phonons in doped semiconductors have been an intensively investigated subject in CW Raman scattering experiments. Although the origin of this mode coupling is quite clear as described in Section 4.2, time-resolved experiments allow for the study of the coupled dynamics under nonequilibrium conditions. An intriguing question is the influence of a *cold* carrier plasma introduced via doping of the sample on plasma oscillation impulsively excited by optical carrier injection via femtosecond laser pulses. Such investigations are expected to provide insight into the mechanisms relevant for the dephasing of coherent plasmon oscillations [71].

Figure 4.9 depicts coherent plasmon–phonon signatures in time and frequency domain detected at constant optical excitation densities in n-doped GaAs crystals of different doping densities. The experiments were performed with 50-fs laser pulses tuned closely above GaAs bandgap (1.47-eV photon energy). The frequencies derived from these data in comparison with the theoretical plasmon–phonon dispersion curve are shown in Figure 4.10. Obviously, the data follow the expected dependence only if the sum over the optically excited density and the cold doping-related plasma is taken into account. This observation gives strong evidence that the background plasma initially thermalized participates in the coherent plasmon oscillation. Moreover, a closer analysis reveals that the dephasing time of the coherent plasmon oscillation decreases if the relative contribution of the optically excited density increases, even when the total carrier density is kept constant. This observation indicates that the plasmon dephasing is predominantly governed by electron-hole scattering [71]. Later, Hase and coworkors investigated the coherent plasmon–phonon modes in highly n-doped GaAs with 20-fs laser pulses [72]. The higher time resolution allowed for the observation of the upper branch of the plasmon–phonon coupled modes that had not been resolved before. At a doping density of 1×10^{18}cm^{-3}, a dephasing time of 130 ± 40 fs was obtained for the upper-branch mode at a frequency of 16 THz. In addition, the authors interestingly observed a linear increase of its dephasing rate with increasing excitation density. This result has led to the conclusion that the ultrafast decay of the upper branch of the coupled mode is mainly caused by loss of coherence in electron-hole plasmas when the photoexcited carrier density is higher than doping levels. It should be

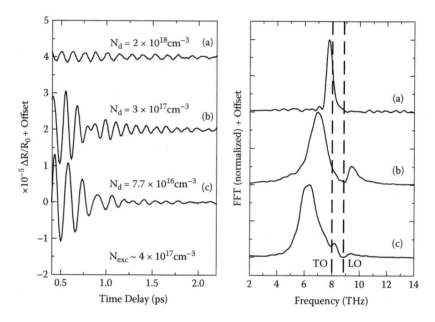

FIGURE 4.9 Oscillatory part of isotropic reflectivity changes of differently doped n-doped GaAs samples (doping density N_d given in the figure) at a constant excitation density (left). Fourier transforms of the time domain data (right). The experiments were performed with 50-fs pulses at 1.47-eV photon energy.

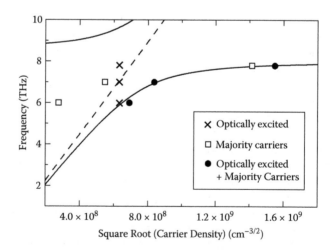

FIGURE 4.10 Plasmon–phonon dispersion curve (solid line) and plasma frequency (dashed line) for a given carrier density compared with the frequencies obtained from time-resolved data as a function of optically excited carrier density (crosses), doping densities (open squares), and the sum of both (closed circles). (Adapted from G. C. Cho et al., *Phys. Rev. Lett.*, 77:4062, 1996).

further noted that dephasing times in the 100-fs regime are not accessible by CARS techniques [8, 73].

The emission of terahertz radiation from coherent plasmon modes in p-i-n- and n-doped GaAs structures has been reported by Kersting et al. [74] and in biased GaAs photoconductors by Shen et al. [75]. These experiments demonstrated that tunable terahertz radiation can be simply obtained by varying the doping or excitation density. In time-resolved reflectivity experiments on GaAs at 77 K, plasmons far away from the phonon resonance were detected by Sha et al. [76]. Terahertz emission from coherently excited phonons and plasmon–phonon modes has been also reported for Te [37, 39, 57], PbTe, CdTe [39], InSb [40, 77], and InAs [78]. In all these materials, the excitation of the coherent longitudinal modes is based on either rapid surface-field screening or a rapid buildup of the Dember field (see Section 4.4.3). As an example, we mention the excitation of coherent plasmons in bulk (111)-InSb at low lattice temperatures, which was investigated by Hasselbeck and coworkers [40]. They proved that the excitation of cold plasma oscillations (i.e., where the carrier density is dominantly introduced by doping) takes place via the photo-Dember effect. A strong decrease of the emitted terahertz radiation at the plasmon frequency as a function of increasing temperature was observed. This effect could be explained by increased electron-hole scattering rates at higher temperatures, which in turn led to decreased carrier mobilities. Hasselbeck et al. furthermore demonstrated that, for the case of InSb, coherent LO phonons can also be excited via an impulsive stimulated Raman process while they are detected via their infrared activity through the emission of terahertz radiation [53]. This conclusion could be drawn from the dependence of the emitted terahertz radiation intensity on the pump polarization.

In adddition, coherent plasmon–phonon modes were observed in time-resolved SHG from bulk GaAs [79]. This detection is closely related to the electrooptic detection of plasmon–phonon modes discussed. Due to the possible selectivity of SHG toward the surface contribution, this technique opens an intriguing way to study the dynamics of coherently excited interface phonons [49, 80]. Chang and coworkers investigated the excitation of coherent plasmon–phonon modes in high-quality InN epitaxial layers by time-resolved SHG [81]. A coherent LO plasmon–phonon mode was observed at 13.54 THz. Surprisingly, the frequency of this mode did not show a dependence on the photoinjected carrier density up to $1.53 \times 10^{19} \text{cm}^{-3}$. This phenomenon was attributed to the hybridization of a coherent A_1 (LO) phonon mode with the intrinsic cold plasma accumulated in the near-surface region of InN, where the plasma density could reach values up to the order of 10^{20} cm^{-3}.

Recent progress in the generation and detection of large-bandwidth terahertz radiation has significantly increased the time resolution for the study of ultrafast excitations of semiconductors [55, 56]. Optical pump–terahertz probe experiments performed by Huber and coworkers allowed observation of changes on timescales much shorter than the LO phonon period in GaAs or the period of an optically excited carrier plasma [82]. The analysis of the dielectric function as a func-tion of time delay between IR and terahertz pulse reveals that it takes a finite time

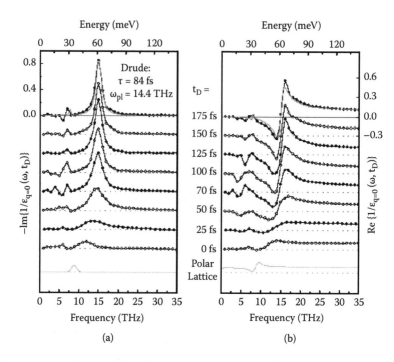

FIGURE 4.11 Imaginary (a) and real (b) part of the inverse dielectric function derived from terahertz transmission spectroscopy of a thin GaAs film. The spectra are depicted for different time delays after optical excitation. The lowest spectrum is the inverse dielectric function of the unexcited crystal showing only a resonance at the LO phonon. (Reprinted with permission from Macmillan Publishers Ltd., R. Huber et al., *Nature*, 414:286, 2001, copyright 2001.)

following the optical excitation until the excited charge carriers form a well-defined plasma with a corresponding plasma frequency. For the first time in solid state, these experiments enabled watching the formation of plasmons through many-particle interaction. Figure 4.11 shows the real and imaginary parts of the inverse dielectric function, (i.e., the electronic loss function) deduced from time domain data of the transmitted terahertz pulse for different time delays after optical excitation. The data clearly reveal the buildup of the plasmon pole on a timescale of the inverse plasma frequency. The experimental results confirmed previous quantum kinetic theories that predicted that on ultrashort timescales unscreened Coulomb collisions occur. Recently, the experimental results obtained in GaAs were confirmed in InP crystals [83]. Similar experiments on the very initial interaction of electrons and phonons were performed by Hase and coworkers via all-optical pump-probe experiments on Si surfaces [50]. With a time resolution of 10 fs, they could disentangle the initial forces exerted by the charge carriers on the lattice. A detailed analysis gave evidence of a Fano interaction based on the coherent superposition of phonons and the broad electronic continuum.

4.4.5 COHERENT PHONON EXCITATION IN WIDE-GAP SEMICONDUCTORS

Wide-gap semiconductors such as GaN and ZnO received significant attention in the last years due to their optoelectronic application for light emission in the blue and ultraviolet (UV) spectral range. Yee and coworkers first observed the excitation of coherent optical phonons in bulk GaN via femtosecond excitation at energies well below the bandgap where the generation of electron-hole pairs by linear absorption can be neglected [84]. In this case, ultrafast drift-diffusion currents or displacive lattice mechanisms on optical carrier injection cannot be expected to contribute to the coherent phonon generation.

The investigation of the polarization dependence of the excitation process revealed that ISRS is the driving force for phonons of A_1 and E_2 symmetry. These experiments are intriguing because the below-gap excitation allows the study of the phonon decay without the simultaneous excitation of free carriers as discussed previously. Distinctly different dephasing times could be measured for the A_1 and E_2 phonons at room temperature: the A_1 mode at 22.01 THz dephases with a T_2 of 1.5 ps, and two E_2 modes at 4.30 and 17.03 THz revealed dephasing times of 70 and 5.8 ps, respectively. The large difference in the dephasing times for the two E_2 modes was attributed to the difference in the final states for anharmonic decay, which is significantly larger for the high-frequency mode. The authors also investigated the effect of free carriers on the phonon decay by increasing the pump fluence and using small excitation spots (Figure 4.12). Via three-photon absorption, free carriers are generated, which reduces the dephasing time of the A_1 mode but not that of the dipole-forbidden E_2 mode. The reduced dephasing time of the A_1 mode is attributed to an increase of dephasing or population decay via carrier–phonon interaction.

Lee and coworkers observed coherent optical phonons in ZnO generated with excitation below the bandgap via ISRS [85]. They investigated the coherent amplitude as a function of the exciting laser bandwidth. Good agreement was found with the theory of ISRS that the exciting bandwidth has to be larger than the excited phonon frequency. In addition, the group of Merlin found an unusually long lifetime of the E_2-low mode in ZnO of 210 ps at 5 K. This has been attributed to the temperature-dependent anharmonic decay being determined by phonon difference processes [86].

4.5 COHERENT PHONONS IN QUANTUM WELLS

Electron–phonon interaction in low-dimensional semiconductor structures is of prime importance for the energy relaxation of charge carriers. Hence, this interaction has profound effects on the performance of optoelectronic devices such as heterostructure lasers, quantum cascade lasers, and optical modulators. The time-resolved study of electron–phonon interaction in low-dimensional semiconductor structures is the subject of numerous books and reviews [87,88]. The time-resolved optical analysis in quantum wells and semiconductor superlattices provides detailed insight into the relevant interaction mechanisms between coherent phonons

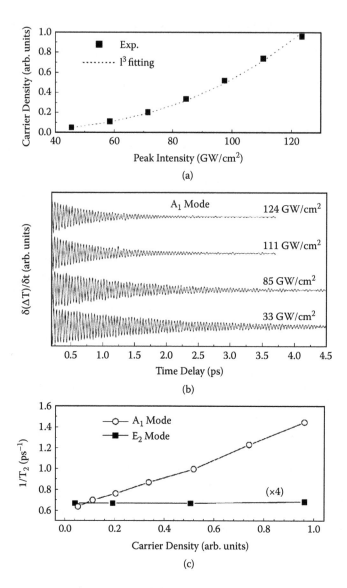

FIGURE 4.12 (a) Relative carrier density excited in GaN via three-photon excitation following an I^3 dependence. (b) Time-resolved coherent phonon traces of the A_1 mode at 22.01 THz for different laser intensities. (c) Dephasing rates of the A_1 mode (22.01 THz) and high-frequency E_2 mode (17.03 THz) versus the carrier density depicted in (a). (Reprinted with permission from the American Physical Society, K. J. Yee et al., *Phys. Rev. Lett.*, 88:105501, 2002. Copyright 2002.)

and coherently prepared electronic wavepackets (see Eqs. 4.5 and 4.6). When the electronic levels can be tuned by confinement parameters or electric and magnetic fields, the energy separation can be shifted into resonance with the optical phonon, allowing study of the electron–phonon resonances in the time domain.

4.5.1 EXCITATION MECHANISMS OF COHERENT OPTICAL PHONONS IN MULTIPLE QUANTUM WELLS

The generation of coherent optical phonons in quantum wells has been studied in detail by Yee et al. employing polarization-dependent excitation and an electrooptic detection scheme [89]. Distinct contributions from *allowed* and *forbidden* Raman scattering and electric field screening to the excitation of coherent LO phonons have been identified by varying laser intensity, energy detuning from the fundamental bandgap, lattice temperature, and barrier width. Most interesting to note is the fact that different processes simultaneously contribute to the phonon excitation, but the relative strength of their contribution can be separated by analyzing the amplitude and phase of the coherent lattice vibrations.

The GaAs/$Al_{0.36}Ga_{0.64}$As quantum wells investigated were grown on a [001] GaAs substrate. Figure 4.13 (left) depicts the pump polarization-dependent coherent phonon amplitudes for the case of low-intensity, resonant excitation of the first excitonic interband transition at 10 K lattice temperature. A strong dependence of the phonon amplitude on the polarization angle is observed exhibiting a maximum in the [110] and a minimum around the [100] direction. It is noteworthy that the angle for the observation of the minimum does not exactly coincide with the [100] direction, but slightly deviates from this as the result of an interference of

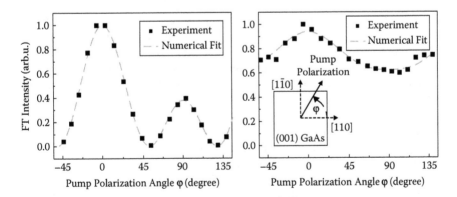

FIGURE 4.13 Polarization-dependent coherent LO phonon amplitudes detected in a GaAs/$Al_{0.36}Ga_{0.64}$As MQW for two different excitation energies: laser energy tuned to the first (left) and second (right) excitonic interband transitions. The dotted curves are numerical fits according to Eq. (4.12). (Adapted from K. J. Yee et al., *Phys. Rev. Lett.*, 86:1630, 2001.)

allowed and forbidden Raman scattering, both contributing to the coherent phonon excitation.

For an assessment of the relative strength of Raman scattering and electric field screening, the polarization-dependent amplitudes were fitted to the angle-dependent scattering intensity:

$$\frac{dS}{d\Omega} \propto |b + a \times \cos(2\varphi)|^2 + c^2 \tag{4.12}$$

which can be obtained from the associated Raman tensors and where φ is the polarization angle of the pump beam with respect to the [110] crystal axis (see inset of Figure 4.13). In this equation, a and b are the Raman polarizabilities for allowed scattering via deformation potential interaction and forbidden scattering via Fröhlich interaction, respectively, and c^2 represents the angle-independent contribution due to electric field screening. Numerically fitting the angular amplitude dependence with Eq. 4.12 yields relations for the fitting parameters of $b/a = 0.24$ and $c/a = 0$. From these values, the authors derived that, in the case of resonant excitation of the excitonic interband transition, the contribution of the dipole forbidden Raman scattering to the coherent phonon generation is stronger than the allowed Raman contribution. Moreover, electric field screening does not play a role under this excitation condition since the coherent phonon amplitude is close to zero at the minimum angle ($c/a = 0$). The absence of a significant contribution by field screening is intuitively clear because the transport perpendicular to the quantum wells layers is prohibited, and only polarization within these layers can contribute.

However, field screening comes into play if the laser intensity is increased or if the exciting laser wavelength is tuned above the second excitonic interband transition ($n = 2$) of the quantum well structure. In the case of doubling the laser intensity, the minimum phonon amplitude around the (100) direction is lifted to a finite value that cannot be explained without a contribution from electric field screening to the excitation process. Now, the fitting parameters are $b/a = 0.16$ and $c/a = 0.29$; that is, forbidden Raman scattering still is the dominant excitation mechanism, but the contribution of field screening is increased. Beyond this, field screening dominates the overall excitation process if the laser wavelength is tuned to the second excitonic interband transition. The corresponding polarization-dependent coherent phonon amplitudes shown on the right-hand side of Figure 4.13 exhibit a much smaller modulation compared with the measurements close to the fundamental bandgap. Here, fitting the angular dependence yields relations of $b/a = 1.4$ and $c/a = 3.4$, providing evidence for the predominance of the electric field screening and for the stronger contribution of allowed compared with forbidden Raman scattering. Finally, by varying the thickness of the $Al_{0.36}Ga_{0.64}As$ barriers between 5 and 1 nm, Yee and coworkers [89] could prove the transition from Raman scattering to electric field screening as the dominant excitation mechanism with decreasing barrier thickness (i.e., with the transition of a multiple quantum well (MQW) to the more "bulk-like" superlattice).

At higher excitation densities on the order of 10^{18} cm^{-3}, coupled intersubband plasmon–phonon modes can be observed in quantum wells on the femtosecond timescale [90]. These are collective oscillations based on the coupling between optical phonons and intersubband transitions in a two-dimensional confined system. In contrast to coupled plasmon–phonon modes in bulk semiconductors, the electric field screening is strongly suppressed in such systems due to the charge carrier confinement in the quantum well layers, which restricts transport along the growth direction (see above). Consequently, the observation of the screened LO phonon mode at the TO frequency is strongly suppressed compared with the three-dimensional system [71]. Furthermore, the nonthermal population of different subbands after pulsed optical excitation strongly alters the dielectric reponse of the carrier–phonon system.

4.5.2 COUPLED MODE OF COHERENT LO PHONONS AND EXCITONIC QUANTUM BEATS IN MULTIPLE QUANTUM WELLS

In the theoretical description of different excitation mechanisms in Section 4.2, we took into account that—beyond Raman scattering and electric field screening—the nonlinear longitudinal polarization P_j^{NL} driving the phonons may also result from a coherent *intraband* polarization (see Eq. 4.4). Important examples are not only quantum beats of heavy hole (HH) and light hole (LH) excitons in MQWs that we refer to in this section, but also Bloch oscillations in semiconductor superlattices, which are extensively discussed in Section 4.6.1. (A common requirement for effective coherent phonon excitation is in both cases that the frequency of the quantum beat is in resonance with the optical phonon frequency. In quantum wells, this beat frequency is determined by the energy splitting between HH and LH excitons, which can be easily varied by tailoring the height and width of the barriers and wells during growth (bandgap engineering). Excitonic quantum beats driven by femtosecond laser pulses have particularly drawn attention in the sense of the emission of coherent terahertz radiation [91]. Beyond, optical pump-probe experiments devoted to the quantum coherence of HH and LH continuum states in the valence band of MQWs gained further insight into fundamental aspects of this coherent excitation [92]. Via the longitudinal polarization along the growth direction that arises from band mixing between the HH and LH states, excitonic quantum beats are expected to couple to LO phonons and thus to provide a driving force for coherent lattice oscillations under resonance conditions.

Mizoguchi and coworkers carefully studied the coupled dynamics in GaAs/AlAs quantum wells in which the HH–LH beat frequency was tuned by shifting the energy spacing $\Delta_{LH\text{-}HH}$ via the quantum confinement parameters of the sample structure [93]. Time-resolved REOS measurements were carried out at low excitation densities with the wavelength of a Ti:sapphire laser tuned to the center of the HH and LH exciton transition energies at 10 K (Figure 4.14, left). Each transient is composed of a strong oscillatory contribution on short timescales below 0.5 ps, which shifts in frequency for the different samples. For longer time delays, coherent GaAs LO phonons are observed with amplitudes that strongly

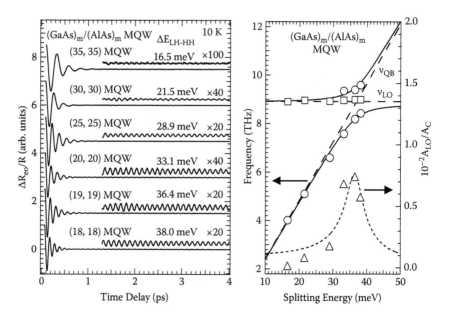

FIGURE 4.14 Time-resolved REOS signals detected in GaAs/AlAs MQWs of different splitting energies $\Delta_{LH\text{-}HH}$ of the heavy hole and light hole excitons (left). Peak frequencies and amplitude relation of the coupled and LO phonon modes are plotted on the right. The solid curve is the dispersion relation of the coupled quantum beat–phonon mode. (Figure adapted from K. Mizoguchi et al., *J. Phys.: Condens. Matter*, 14:L103, 2002.)

depend on the splitting energy $\Delta_{LH\text{-}HH}$ derived from photoluminescence measurements. Obviously, the phonon amplitude is largest for resonant conditions, that is, when the frequency of the strong initial oscillation matches the LO phonon frequency illustrated by the triangles in the right-hand plot of Figure 4.14. Moreover, Fourier spectra calculated for different time slices reveal that the frequency spectrum of the initial contribution splits into two branches when its center frequency becomes close to the the LO phonon frequency (open circles in Figure 4.14, right). This anticrossing, similar to the characteristics of coupled plasmon–phonon modes introduced in Section 4.4.4, clearly manifests the *linear* coupling of the excitonic quantum beat to the LO phonon. As predicted by Eq. (4.7), the linear coupling results in the splitting of the dispersion relation into an upper and a lower branch (solid lines in Figure 4.14, right) in best agreement with the experimental data.

Most recently, coherent interface optical phonon modes were observed by the time-resolved detection of SHG in GaInP/GaAs/GaInP single quantum wells [94]. This detection scheme is favored due to its intrinsic sensitivity to carrier and phonon dynamics at surfaces and interfaces. The excitation of a coherent interfacial phonon mode at a frequency of 9.4 THz localized in the GaInP/GaAs interface has been shown to be based on both electric field screening and Raman scattering. This study has gained special attention because the analysis of the dephasing time of this coherent phonon mode allows gaining information on the quality of buried interfaces in semiconductor heterostructures.

4.6 COHERENT PHONONS IN SUPERLATTICES

In semiconductor superlattices, discrete electronic states that appear in single or uncoupled MQWs broaden into minibands with energy widths that depend on the coupling strength between neighboring wells. However, the coupling strength can be easily reduced by applying static electric fields along the growth direction, which leads to the localization of the electronic wavefunctions and thus to the splitting of corresponding minibands into Wannier–Stark states [95, 96]. Concerning the excitation of coherent phonons, the quantum beat of adjacent Wannier–Stark states (i.e., coherent Bloch oscillations) provides an effective driving force if they are tuned into resonance with optical phonons by the applied electric field. Furthermore, the artificial periodicity of a superlattice leads to a backfolding of acoustic and optic phonon dispersion curves of bulk semiconductors into the mini-Brillouin zone of the superlattice. In the case of the acoustic branch, this effect enables the excitation of coherent acoustic phonons in the center of the Brillouin zone.

4.6.1 COHERENTLY COUPLED BLOCH–PHONON OSCILLATIONS IN BIASED SUPERLATTICES

An intriguing example for the resonant excitation of coherent phonons by coherent charge carrier dynamics is the coupled dynamics of Bloch oscillations and LO phonons in semiconductor superlattices. Bloch oscillations—theoretically described by Bloch and Zener around the 1930s [97, 98] and proposed for semiconductor superlattices by Esaki and Tsu [99]—are a coherent frequency-tunable electronic excitation. In a superlattice of period d, they can be excited via the superposition of adjacent Wannier–Stark states by ultrashort laser pulses [100, 101]. The Bloch frequency ν_{BO} is determined by the energy spacing between the excited Wannier–Stark states, and thus $\nu_{BO} = eFd/h$ is tunable by an applied electric field F. For a detailed description, refer to a review by Leo [102].

Bloch–phonon coupling can be observed in superlattices with large electronic miniband widths for which Bloch oscillations can be tuned across the LO phonon frequency. In resonance, the coherent carrier oscillations are able to effectively launch coherent lattice vibrations even at low excitation densities [103]. A first theoretical description of this effect was given by Ghosh and coworkers for the case of a GaAs/AlGaAs superlattice [104]. In contrast to the HH–LH quantum beat–phonon coupling introduced above, the authors showed that a Bloch oscillating electronic wavepacket acts like a *nonlinear* pendulum linearly coupled to the LO phonon. Contrary to the general expectations of mode coupling in the linear regime (i.e., an anticrossing of the linear relation between Bloch frequency and electric field strength) coupled Bloch–phonon modes were predicted not to have a gap in their frequency spectra. Nonetheless, the coherent amplitude of the GaAs LO phonon should be strongly enhanced if the Bloch oscillations are in resonance with the LO phonon.

The first experimental observation of coupled Bloch–phonon oscillations was accomplished in a GaAs/Al$_{0.3}$Ga$_{0.7}$As superlattice [103]. The 35-period super-lattice (composed of 51-Å-thick GaAs wells and 17-Å-thick Al$_{0.3}$Ga$_{0.7}$As barriers) is embedded in a p-i-n diode to enable a Bloch frequency tuning by varying the applied DC (direct current) reverse bias. For the excitation of Bloch oscillations, the wavelength of the Ti:sapphire laser used was chosen to coherently superpose Wannier–Stark states in the first electronic miniband. With a miniband width of 60 meV, the frequency of the Bloch oscillations can be tuned beyond the LO phonon at 8.8 THz. Polarization dynamics arising from the coherently coupled Bloch–phonon dynamics were recorded in time-resolved REOS measurements at a low excitation density of approximately 4.4×10^{14} electron-hole pairs per cm^3 and 10 K lattice temperature. The corresponding Fourier transforms for different reverse-bias voltages applied to the superlattice are depicted in Figure 4.15. Small and sharp peaks contribute at the GaAs TO and LO frequencies of 8.0 and 8.8 THz, respectively. Over a wide voltage range, their amplitudes are independent of the applied voltage, which is attributed to the excitation of coupled plasmon–phonon modes in the highly doped cladding layers of the sample (see Section 4.4.4). In addition, spectrally broad peaks appear that are identified with Bloch oscillations due to their voltage-dependent linear frequency shift (see inset of Figure 4.15a).

Focusing on the voltage range in which the Bloch frequency is tuned into resonance with the LO phonon provides insight into the coupled dynamics.

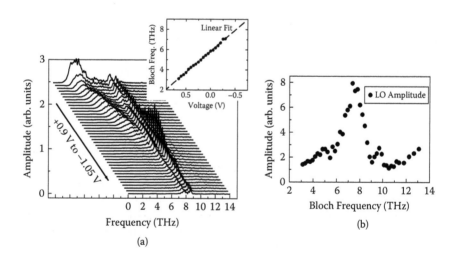

FIGURE 4.15 (a) Fourier transforms of the oscillatory contributions to time-resolved REOS signals recorded for different reverse biases in the GaAs/AlGaAs superlattice. The inset shows the Bloch frequency versus the bias voltage applied. (b) Amplitude of the LO phonons as a function of the Bloch oscillation frequency. (Figure adapted from T. Dekorsy et al., *Phys. Rev. Lett.*, 85:1080, 2000.)

In agreement with the theory of Ghosh et al. [104], a bending of neither the LO phonon frequency nor the Bloch oscillation frequency close to the resonance is observed. The second and even more intriguing feature is the pronounced enhancement of the LO phonon amplitude for the case of resonant Bloch and LO phonon frequencies. This becomes obvious in Figure 4.15b, in which the amplitude of the LO phonon (as extracted from the Fourier spectra) is plotted versus the Bloch frequency. Surprisingly, the absolute position of the maximum phonon amplitude at 7.5 THz deviates from the eigenfrequency of the bare GaAs LO phonon of 8.8 THz. Starting from the background level, the LO phonon amplitude passes through its absolute maximum and rapidly drops toward a minimum at the LO phonon frequency, before finally passing a second local maximum. This behavior is attributed to result from a reduced driving force for the coherent LO phonons in resonance with the Bloch oscillations. The resonant enhancement of electron–phonon scattering at the LO phonon frequency leads to increased dephasing rates of Bloch oscillations in resonance [105] and thus to a resonantly reduced driving force for the coherent optical phonons. In the theoretical treatment of Bloch–phonon coupling, this effect can be considered by the introduction of a frequency-dependent dephasing rate of the Bloch oscillations [104].

Further insight into the coupled dynamics was achieved in a narrow-well superlattice that was specifically designed in the $In_{0.53}Ga_{0.47}As/In_{0.52}Al_{0.48}As$ material system for the observation of Bloch–phonon coupling in the presence of Zener tunneling [106]. At high electric fields, Zener tunneling is an efficient dephasing mechanism of coherent Bloch oscillations in superlattices in which the first electronic miniband is energetically shifted close to the confining barrier potential [107]. One can expect that this dephasing process modifies the shape of the Bloch–phonon coupling resonance. In addition, a selective coupling of Bloch oscillations to optical phonons of different frequencies should become feasible in the ternary system.

The superlattice investigated here consists of $In_{0.53}Ga_{0.47}As$ wells and $In_{0.52}Al_{0.48}As$ barriers with thicknesses 8.4 Å and 51.6 Å, respectively [108]. These parameters lead to the formation of a 60 meV wide first electronic miniband that is weakly bound below the confining barrier potential. Figure 4.16 (a) depicts the Fourier transforms of Bloch–phonon coupling-induced polarization dynamics detected for different reverse biases at 10 K. Again, spectrally broad Bloch oscillations linearly shifting in their frequency with increasing reverse bias (see inset) and sharp peaks arising from optical phonons (here at frequencies of the bare LO phonons at 7.1, 8.0, and 11.0 THz) contribute to the spectra. The LO phonon amplitudes are plotted versus the Bloch oscillation frequency in part (b) of the figure. They reach maximum values in resonance of the BO frequencies, with the corresponding bare phonon eigenfrequencies marked by dashed vertical lines [109, 110]. Again, this enhancement clearly demonstrates the resonant coupling of coherent Bloch oscillations and optical phonons. Moreover, the selective coupling to phonons of both, well and barrier, materials proves their excitation by long-range Coulomb forces that arise from the macroscopic polarization of the Bloch oscillating electronic wavepackets.

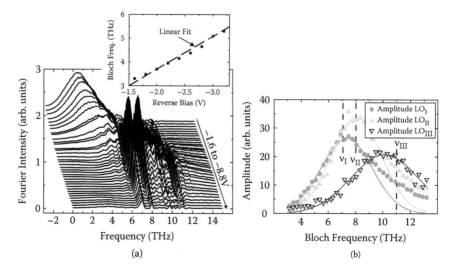

FIGURE 4.16 (a) Fourier spectra of oscillatory REOS contributions recorded for different reverse biases in the InGaAs/InAlAs superlattice. The inset shows the Bloch frequency versus the bias voltage applied. (b) Amplitude of the LO phonons as a function of the Bloch oscillation frequency. (Reprinted with permission from the American Physical Society, M. Först et al., *Phys. Rev. B*, 67:085305, 2003. Copyright 2003.)

Most remarkably, the spectral shape of the Bloch–phonon resonance differs from the observation in the GaAs/AlGaAs superlattice; that is, the phonon amplitudes are asymmetrically enhanced on the high-frequency wing of the resonance. This effect is the result of Zener tunneling processes, which are favored in this narrow-well superlattice due to the energetic position of the electronic miniband [111]. The tunneling of optically excited carriers into above-barrier continuum states leads to a steady increase of the Bloch oscillation dephasing rate with increasing electric field strength and thus to an additional current $J(t)$ (see Eq. 4.4), which is able to launch coherent phonons if the bandwidth provided by this process is large enough [106]. This excitation mechanism superimposed on the resonant Bloch–phonon coupling leads to the asymmetric increase of the coherent phonon amplitude above the resonance. This interpretation was recently manifested in a study of Bloch–phonon coupling in $In_{0.53}Ga_{0.47}As/In_{0.52}Al_{0.48}As$ superlattices in which the energetic position of the electronic miniband in the conduction band was tailored by varying the superlattices' well widths [112].

4.6.2 COHERENT EXCITATION OF ZONE-FOLDED ACOUSTIC PHONONS BY IMPULSIVE STIMULATED RAMAN SCATTERING

Besides the investigation of coherent optical phonons, periodic low-dimensional semiconductors allow for the generation and detection of coherent *acoustic* lattice modes. A prerequisite for the excitation of these phonon modes in a superlattice is the backfolding effect of the bulk phonon dispersion into the mini-Brillouin zone of the artificially periodic heterostructure [113]. Within the Rytov model, the dispersion relation of the superlattice zone-folded modes can be calculated on the basis of layered elastic continuum model [114]. Here, zone folding results in a series of acoustic phonon branches with finite frequencies $\omega \neq 0$ at vanishing wavevectors $\mathbf{q} = 0$. For this reason, light scattering from acoustic modes in superlattices is commonly referred to as Raman scattering in contrast to Brillouin scattering of acoustic phonons in bulk semiconductors. Intensive studies on folded acoustic phonons in semiconductor heterostructures have been performed by CW Raman scattering experiments (for a review, see, e.g., Ref. [115]). Beyond these experiments, insight into phonon-scattering processes can be gained on ultrashort timescales for which phonons can be driven coherently. The dephasing times of coherent acoustic phonons are on the order of several tens of picoseconds so that coherent phonon spectroscopy becomes feasible with an extremely high-frequency resolution of up to 0.3 GHz, which corresponds to linewidths of 0.01 cm^{-1} in CW Raman experiments.

The coherent excitation of terahertz acoustic phonons by femtosecond laser pulses was first observed in 1994 by Yamamoto et al. in GaAs/AlAs superlattices [116]. In measurements of the time derivative of phonon oscillation–induced reflectivity changes, a single mode of the first-order backfolded acoustic phonon branch was detected. The selective generation of this B$_2$ symmetry mode (symmetric mode with respect to a midplane of a superlattice layer) was explained to result from the selective excitation of charge carriers in the superlattice wells, (i.e., the preparation of a periodic carrier distribution with a wavevector determined by the superlattice period). Stress that is instantaneously induced in the wells by carrier absorption was held responsible for launching the coherent phonons due to symmetry considerations. A contribution of Raman processes to the coherent excitation, however, was not considered because Raman selection rules prohibit the excitation of the B$_2$ symmetry mode observed in the investigated superlattices.

The identification of ISRS as the excitation mechanism of coherent longitudinal acoustic (LA) phonons in GaAs/AlAs superlattice was given later by parallel investigations of three independent groups [117–119]. Here, we focus on the work of Bartels et al., who observed the femtosecond laser-induced generation of two phonon triplets in the first- and second-order backfolded acoustic branch of several superlattices [119]. Figure 4.17 (top) displays time-resolved reflectivity changes detected in a 40-period superlattice consisting of 19-monolayer-thick GaAs wells and 19-monolayer-thick AlAs barriers, that is, a (19/19) GaAs/AlAs superlattice. The data are traced after resonant excitation of the lowest excitonic interband transitions with 50-fs laser pulses of a Ti:sapphire laser. The signal is dominated by the excitation of charge carriers at zero time delay and their subsequent dynamics on the timescale of several hundred picoseconds (shown in the inset). These reflectivity changes are modulated by rather small oscillations with an amplitude of $\Delta R_{osc}/R \approx 10^{-5}$ that arise from coherently excited acoustic phonons. The oscillations are numerically extracted from the background signal and plotted in Figure 4.17.

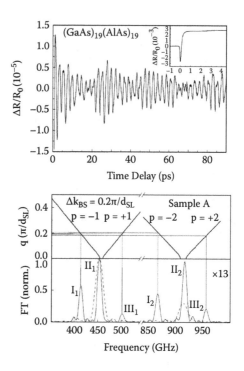

FIGURE 4.17 Top: Time-resolved reflectivity changes of a (19/19) GaAs/AlAs superlattice induced by resonant excitation at room temperature. Bottom: Acoustic phonon dispersion relation of the superlattice and Fourier spectrum of the detected phonon oscillations. (Reprinted with permission from the American Physical Society, A. Bartels et al., *Phys. Rev. Lett.*, 82:1044, 1999. Copyright 1999.)

The Fourier spectrum of the oscillatory signal components and the LA phonon dispersion of the superlattice calculated by the above-mentioned Rytov model [114] are shown on the bottom of Figure 4.17. Two triplets of phonon modes are observed that—by comparison with the dispersion relation—can be assigned to first- and second-order zone-folded acoustic phonons. Both phonon triplets are composed of an intense mode with wavevector $q = 0$ and two satellite modes with wavevectors $q = 2 \times k_{laser}$, where $k_{laser} = 2\pi n/\lambda_{laser}$ is the laser's wavevector (n is the averaged refractive index of the superlattice). In agreement with the symmetry properties of acoustic vibrations in the superlattice [120], the center frequencies can be identified as A_1 symmetry modes excited by ISRS in the forward-scattering direction. The weaker satellite peaks are mixed-symmetry modes excited in the backward scattering direction. Further evidence for identifying ISRS as the excitation mechanism of the coherent acoustic phonons has been provided by comparing the experimentally obtained relative amplitudes of phonons within the first-order triplet with a calculation of normalized Raman scattering intensities. In additional measurements with spectrally resolved probe pulses, the detection of the coherent LA phonons was proved to be based on the modulation of the interband transition energy via the acoustic deformation potential [119].

An intriguing concept in solid-state physics is the coherent control of dynamic processes by time sequential double- and multipulse excitation. In semiconductors, this technique allows for detailed investigation of electron–phonon and phonon–phonon interactions, for example. Coherent control experiments on zone-folded acoustic phonons were also performed by Bartels et al. [121]. By using adequate time spacing between two excitation pulses, first-order acoustic modes were efficiently silenced while second-order modes were simultaneously enhanced due to destructive and constructive interference of the coherently prepared acoustic sound waves, respectively. By this means, the phonon system could be driven into a nonequilibrium state that is not achievable by either CW or single-pulse excitation. Although coherent control experiments of single optical phonons had been carried out previously [45, 122], this study demonstrated the feasibility of an undisturbed investigation of any desired lattice mode within a phonon multiplet.

We stated above that the B_2 symmetry mode of the first-order acoustic phonon branch cannot be excited via ISRS due to Raman scattering selection rules. It is noteworthy, that Mizoguchi and coworkers showed that this argument does not hold for finite-size superlattices with a limited number of periods [123]. In a striking experiment, the authors compared the excitation of coherent LA phonons in superlattices with period numbers between 20 and 195. The B_2, mode, which is Raman forbidden in superlattices with infinite periodicity, appeared in the finite-size superlattice due to the breakdown of wavevector selection rules in the symmetry-reduced structure. The time-resolved data and corresponding Fourier spectra are shown in Figure 4.18 and depict a phonon triplet for the 195-period superlattice, but a phonon quartet for the 20-period superlattice in the single-color measurements (pump and probe wavelength = 720 nm). Moreover, it was found that basically for the same reasons the bandwidth of coherent acoustic modes increases with decreasing number of superlattice periods.

4.6.3 PROPAGATION OF COHERENT ACOUSTIC PHONONS IN SUPERLATTICES

Finite-size superlattices also provide experimental access to the investigation of acoustic phonon propagation because the observation of phonon wavepackets is shortened in time due to propagation out of these structures. This effect was convincingly proved in two-color pump-probe experiments performed by the same research group [123]. In this study, Mizoguchi et al. used frequency-doubled Ti:sapphire laser pulses in the UV range for the excitation of narrow phonon packets near the superlattice surface and infrared pulses of the fundamental wavelength for their detection throughout the entire structure (Figure 4.18). The escape of such a narrow acoustic phonon wavepacket from a 20-period superlattice was observed as a sudden disappearance of the corresponding time-resolved reflectivity changes after 22 ps. On the other hand, even after 100 ps the time-resolved signal hardly decays in the 195-period superlattice because the escape time of the acoustic phonon wave is calculated to be a factor of two longer. The short escape

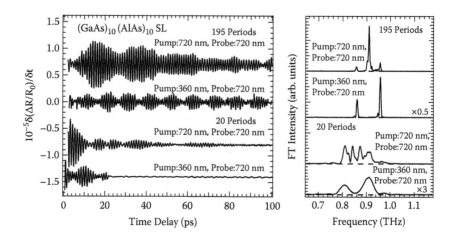

FIGURE 4.18 Single- and two-color pump-probe experiments of (10/10) GaAs/AlAs superlattices with 195 and 20 periods, respectively. Oscillatory parts of the time-resolved reflectivity changes (left) and corresponding Fourier spectra (right). (Figures adapted from K. Mizoguchi et al., *J. Phys.: Condens. Matter*, 14:L103, 2002.)

time of the coherent phonons out of the finite-size superlattice is further reflected by the increased bandwidth of the corresponding Fourier spectrum as compared with that of the quasi-infinite 195-period superlattice. The difference in the center frequencies of the two samples was ascribed to slight fluctuations of the constituent layer thicknesses [123].

In optical pump-probe experiments, coherent acoustic phonons are detected as time-dependent modulations of the sample reflectivity or transmissivity. A direct measure of the acoustic phonon propagation has become possible by time-of-flight techniques that make use of incoherent phonon detection via thin-film superconducting bolometers attached to the backside of a superlattice sample.

A first experimental study of a (19/19) GaAs/AlAs superlattice was carried out by Hawker et al. at low temperatures. After excitation with femtosecond laser pulses, an increase of the bolometer temperature at time delays of approximately 150 ns was observed, a value that coincides with the calculated propagation time of LA phonons across the GaAs substrate [124]. An analysis of the detector response at various laser excitation energies below and above the bandgap enabled a separation of the coherently excited propagating LA phonons from incoherent TA phonons for which the latter are emitted within the relaxation process of charge carriers photogenerated in the superlattice and the substrate. Nonetheless, the bolometer detection is not able to provide information about the spectral composition of the acoustic phonons, so an unambigious identification with coherently excited phonons in the superlattice was not possible. Therefore, the same research group made use of frequency-dependent phonon scattering in the GaAs substrate as a spectral filter [125]. By this means, they could prove that the acoustic phonon spectrum detected at the backside of the sample substrate has a strong monochromatic component at the frequency $\nu = \upsilon_{LA}/d_{SL}$ (d_{SL} superlattice period and υ_{LA} phonon speed), which agrees with the center frequency of the first-order zone-folded phonon triplet at the Brillouin zone center (see above).

To further illustrate the propagation and the monochromacity of propagating acoustic phonons, the authors inserted a second superlattice into the sample structure [126]. Coherent first-order zone-folded LA phonons were excited by femtosecond Ti:sapphire pulses in the first *generator* superlattice, which is separated from the bolometer detector by a second *filter* superlattice. According to an early publication by Tamura et al., the spectral phonon transmission through such a superlattice can be tailored to form some tens of gigahertz wide stop bands for propagating LA phonons [127]. Acoustic phonon-induced time-delayed bolometer signals were recorded for two samples comprising different generator superlattices (first-order LA phonons excited at frequencies of 0.54 and 0.66 THz, respectively) but identical filter superlattices with an 80-GHz-wide stop band centered at 0.67 THz. Tuning the photon energy of the exciting laser above the first excitonic interband transition of the generator superlattices revealed an increase of the LA signal contribution only when the filter stop band is off-resonant with the impulsively generated acoustic phonon. In contrast, a 100% attenuation of the propagating phonons is noticed when the filter stop band matches the frequency of the generator superlattice. This observation indicates that the linewidth of the propagating acoustic phonons impulsively excited in a superlattice is less than 80 GHz, that is, the spectral width of the filter stop band.

4.6.4 COHERENT ACOUSTIC PHONONS IN WIDE-BANDGAP NITRIDE-BASED STRUCTURES

The variety of coherent phonon excitation mechanisms is exemplified in the generation of zone-folded acoustic lattice modes in wide-bandgap nitride-based heterostructures. Here, large strain-induced piezoelectric fields not only modify the static optical properties [128], but also affect the dynamics of photogenerated charge carriers, which in principle are able to launch coherent phonons (see Eq. 4.6).

Sun et al. demonstrated the generation of coherent acoustic phonons in InGaN/GaN MQWs by the subpicosecond carrier screening of the built-in piezoelectric field [129, 130]. Within this study, several samples with InGaN wells of varying width between 12 and 62 Å and GaN barriers of constant thickness (43 Å) were investigated. The electronic bandgap of the MQWs is in the range between 3.18 and 3.39 eV (dependent on the well width), thus requiring frequency doubling of Ti:sapphire laser pulses to resonantly excite electron-hole pairs. Coherent phonon oscillations were traced in time-resolved transmission changes at excitation densities on the order of 10^{13} cm^{-2} per well. Comparable to experiments in GaAs/AlAs superlattices (see Figure 4.17), acoustic phonon-induced signal contributions were observed as periodic modulations of the slowly varying carrier relaxation dynamics [129, 130]. In contrast, however, the phonon oscillations on the order of $\Delta T_{osc}/T_0 \approx 10^{-2}$ are by three orders of magnitude larger than previously reported for the III–V semiconductor system that has been ascribed to different excitation and detection mechanisms in the nitride-based heterostructure. The excitation of the coherent LA phonons was identified as the result of the rapid screening of the strain-induced piezoelectric field by optically excited electron-hole pairs. This field screening—comparable to the surface-field screening in bulk GaAs (see Section 4.4)—quasi-instantaneously results in a new equilibrium position of the crystal lattice. Consequently, coherent phonon oscillations are launched via displacive excitation that is manifested by a cosine-like modulation of the sample transmissivity with respect to zero time delay. The frequency of the excited phonons is determined by the period d_{MQW} of the MQW structure via the acoustic phonon dispersion. Since carriers are only excited in the InGaN wells, the periodic carrier distribution selectively couples to acoustic phonon modes with wavevector $q = 2\pi/d_{MQW}$ of the bulk dispersion, which corresponds to the first-order zone-folded ($q = 0$) acoustic phonon of the heterostructure. In the samples investigated, the phonon frequency varied between 0.67 and 1.23 THz depending on the MQW period. The detection of the coherent phonon oscillations is based on the quantum-confined Stark effect. According to this, the optical transmission of the MQW is periodically changed in time due to coherent phonon-induced modulations of the built-in piezoelectric field [129,130]. As reported for the GaAs/AlAs system, coherent control experiments of acoustic phonons were also carried out in InGaN-based MQWs [131].

4.6.5 COHERENT ACOUSTIC PHONON DETECTION BY TIME-RESOLVED X-RAY DIFFRACTION

The availability of subpicosecond, kilohertz repetition rate X-ray sources (see Section 4.3) most recently enabled the spatially resolved observation of zone-folded acoustic phonons in a time-resolved X-ray diffraction experiment [61]. Coherent oscillations were launched by intense 50-fs laser pulses (2-mJ/cm^2 fluence) in a GaAs/Al$_{0.4}$Ga$_{0.6}$As superlattice with a period of 16 nm, and structural changes were traced by the diffraction pattern of time-delayed X-ray pulses. Particularly, these experiments are noteworthy since a quantitative determination of lattice displacements from changes of the X-ray diffraction angle is rendered possible. However, high excitation densities above the saturation threshold of the first excitonic interband transition and thick superlattice periods were required in this study due to the limited signal-to-noise ratio and limited time resolution, respectively. However, at the high excitation densities used, the coherent phonon excitation is rather displacive, which is in contrast to the identification of ISRS as the excitation mechanism of coherent LA phonons described above under different excitation conditions. Further experiments with higher spatial and temporal resolution will be necessary to enable the X-ray diffraction-based detection of acoustic phonon triplets at lower excitation densities. It is expected that the further development of time-resolved X-ray diffraction will strongly contribute to the deeper understanding of structural dynamics on femtosecond timescales.

4.7 CONCLUSION

In this chapter, we summarized the fundamental aspects and recent developments in the field of coherently excited phonons in semiconductors and semiconductor heterostructures. The physics of coherent phonon states in various condensed matter systems underlies a multitude of excitation and detection mechanisms that have been shown to be distinguishable by specific experimental techniques. It is again worth noting here that, in contrast to CW Raman scattering, combined carrier and phonon dynamics can be studied under nonequilibrium conditions, thus providing profound insight into their interaction processes. Due to the strong growth of the field in the past years, we probably forgot or overlooked some important publications and apologize for doing so. Other developments, such as the study of phonon squeezing [132–135], phonon noise [22, 136], and anharmonic phonon dynamics at high excitation fluences [60, 137–142], have been omitted because their discussion would have gone beyond the scope of this chapter. However, future research in the field of coherent phonon spectroscopy will truly benefit from the availability of novel detection schemes employing ultrabroadband terahertz radiation sources and femtosecond X-ray sources for sub-10-fs temporal resolution or nanometer spatial resolution, respectively.

ACKNOWLEDGMENT

We thank A. Bartels, G. C. Cho, A. Ghosh, O. V. Misochko, K. Mizoguchi, and D. S. Kim for valuable discussions. Deutsche Forschungsgemeinsschaft is gratefully acknowledged for their support (KU 540/42-1.2 and DE 567/9-1).

REFERENCES

1. J. Shah. *Ultrafast Spectroscopy of Semiconductors and Semiconductor Nanostructures*, 2nd ed., Vol. 115, *Springer Series in Solid-State Sciences*, Springer, Berlin, 1999.
2. W. E. Bron, J. Kuhl, and B. K. Rhee. Picosecond-laser-induced transient dynamics of phonons in GaP and ZnSe. *Phys. Rev. B*, 34:6961, 1986.
3. D. von der Linde. *Ultrashort Laser Pulses*, Vol. 60, *Topics in Applied Physics*, Springer, Berlin, 1988.
4. J. A. Kash and J. C. Tang. *Spectroscopy of Nonequilibrium Electrons and Phonons*, Elsevier, Amsterdam, 1992, p. 113.
5. J. C. Diels, W. Rudolph, P. F. Liao, and P. Kelley. *Ultrashort Laser Pulse Phenomena*, Academic Press, New York, 1996.
6. W. E. Bron. *Coherent Optical Interactions in Semiconductors*, Vol. 330, *NATO ASI B*, Plenum, New York, 1994, p. 199.
7. F. Vallée and F. Bogani. Coherent time-resolved investigation of LO-phonon dynamics in GaAs. *Phys. Rev. B*, 43:12049, 1991.
8. F. Vallée. Time-resolved investigation of coherent LO-phonon relaxation in III–V semiconductors. *Phys. Rev. B*, 49:2460, 1994.

9. F. Ganikhanov and F. Vallée. Coherent to phonon relaxation in GaAs and InP. *Phys. Rev. B*, 55:15614, 1997.
10. C. Thomsen, H. T. Grahn, H. J. Maris, and J. Tauc. Surface generation and detection of phonons by picosecond light pulses. *Phys. Rev. B*, 34:4129, 1986.
11. H. T. Grahn, H. J. Maris, and J. Tauc. Picosecond ultrasonics. *IEEE J. Quant. Electr.*, 25:2562, 1989.
12. O. B. Wright. Ultrafast nonequilibrium stress generation in gold and silver. *Phys. Rev. B*, 49:9985, 1994.
13. G. A. Antonelli, H. J. Maris, and S. G. Malhotra. Picosecond ultrasonics study of the vibrational modes of a nanostructure. *J. Appl. Phys.*, 91:3261, 2001.
14. T. Saito, O. Matsuda, and O. B. Wright. Picosecond acoustic phonon pulse generation in nickel and chromium. *Phys. Rev. B*, 67:205421, 2003.
15. J. Bokor and N. W. Pu. Study of surface and bulk acoustic phonon excitations in superlattices using picosecond ultrasonics. *Phys. Rev. Lett.*, 91:076101, 2003.
16. O. Matsuda, O. B. Wright, D. H. Hurley, V. E. Gusev, and K. Shimizu. Coherent shear phonon generation and detection with ultrashort optical pulses. *Phys. Rev. Lett.*, 93:095501, 2004.
17. T. K. Cheng, J. Vidal, M. J. Zeiger, G. Dresselhaus, M. S. Dresselhaus, and E. P. Ippen. Mechanism for displacive excitation of coherent phonons in Sb, Bi, Te, and Ti_2O_3. *Appl. Phys. Lett.*, 59:1923, 1991.
18. H. J. Zeiger, J. Vidal, T. K. Cheng, E. P. Ippen, G. Dresselhaus, and M. S. Dresselhaus. Theory for displacive excitation of coherent phonons. *Phys. Rev. B*, 45:768, 1992.
19. G. C. Cho, W. Kütt, and H. Kurz. Subpicosecond time-resolved observation of coherent phonons in GaAs. *Phys. Rev. Lett.*, 65:764, 1990.
20. R. Merlin. Generating coherent terahertz phonons with light pulses. *Solid State Commun.*, 102:207, 1997.
21. T. Dekorsy, G. C. Cho, and H. Kurz. In: *Light Scattering in Solids*, Vol. 8, Coherent phonons in condensed media, Springer, Berlin, 2000, p. 169.
22. O. V. Misochko. Coherent phonons and their properties. *J. Exp. Theor. Phys. (JETP)*, 92:246, 2001.
23. A. V. Kuznetsov and C. J. Stanton. Theory of coherent phonon oscillations in semiconductors. *Phys. Rev. Lett.*, 73:3243, 1994.
24. W. Kütt. *Coherent phonons in III–V compounds*, Vol. 32, Vieweg, Braunschweig, 1992, p. 113.
25. A. Laubereau and W. Kaiser. Vibrational dynamics of liquids and solids investigated by picosecond light pulses. *Rev. Mod. Phys.*, 50:607, 1978.
26. Y.-X. Yan, E. B. Gamble, and K. A. Nelson. Impulsive stimulated scattering: general importance in femtosecond laser pulse interactions with matter and spectroscopic applications. *J. Chem. Phys.*, 83:5391, 1985.
27. S. DeSilvestri, J. G. Fujimoto, E. B. Gamble, L. R. Williams, and K. A. Nelson. Femtosecond time-resolved measurements of optic phonon dephasing by impulsive stimulated Raman scattering in small α-perylene crystal from 20 to 300 K. *Chem. Phys. Lett.*, 116:146, 1985.
28. Y.-X. Yan and K. A. Nelson. Impulsive stimulated light scattering. I. General theory. *J. Chem. Phys.*, 87:6240, 1987.
29. Y.-X. Yan and K. A. Nelson. Impulsive stimulated light scattering. II. Comparison to frequency-domain light-scattering spectroscopy. *J. Chem. Phys.*, 87:6257, 1987.

30. J. Chesnoy and A. Mokthari. Resonant impulsive-stimulated Raman scattering on malachite green. *Phys. Rev. A*, 38:3566, 1988.

31. T. K. Cheng, S. D. Brorson, A. S. Kazeroonian, J. S. Moodera, G. Dresselhaus, M. S. Dresselhaus, and E. P. Ippen. Impulsive excitation of coherent phonons observed in reflection in bismuth and antimony. *Appl. Phys. Lett.*, 57:1004, 1990.

32. T. E. Stevens, J. Kuhl, and R. Merlin. Coherent phonon generation and the two stimulated Raman tensors. *Phys. Rev. B*, 65:144304, 2002.

33. T. Pfeifer, T. Dekorsy, W. Kütt, and H. Kurz. Generation mechanism of coherent LO phonons in surface-space-charge fields of III–V compounds. *Appl. Phys. A*, 55:482, 1992.

34. T. Dekorsy, T. Pfeifer, W. Kütt, and H. Kurz. Subpicosecond carrier transport in GaAs surface space charge fields. *Phys. Rev. B*, 47:3842, 1993.

35. A. V. Kuznetsov and C. J. Stanton. Coherent phonon oscillations in GaAs. *Phys. Rev. B*, 51:7555, 1995.

36. H. Dember. *Phys. Z.*, 32:554, 1931.

37. T. Dekorsy, H. Auer, C. Waschke, H. J. Bakker, H. G. Roskos, H. Kurz, V. Wagner, and P. Grosse. Emission of submillimeter electro-magnetic waves by coherent phonons. *Phys. Rev. Lett.*, 74:738, 1995.

38. T. Dekorsy, H. Auer, H. J. Bakker, H. G. Roskos, and H. Kurz. Terahertz emission by coherent infrared active phonons. *Phys. Rev. B*, 53:4005, 1996.

39. M. Tani, R. Fukasawa, H. Abe, S. Matsuura, K. Sakai, and S. Nakashima. Terahertz radiation from coherent phonons excited in semiconductors. *J. Appl. Phys.*, 83:2473, 1998.

40. M. P. Hasselbeck, L. A. Schlie, and D. Stalnaker. Coherent plasmons in InSb. *Appl. Phys. Lett.*, 85:6116, 2004.

41. B. B. Varga. Coupling of plasmons to polar phonons in degenerate semiconductors. *Phys. Rev.*, 137:A1896, 1965.

42. A. Mooradian and G. B. Wright. Observation of the interaction of plasmons with longitudinal optical phonons in GaAs. *Phys. Rev. Lett.*, 16:999, 1966.

43. A. Mooradian and A. L. McWorther. Polarization and intensitiy of Raman scattering from plasmons and phonons in gallium arsenide. *Phys. Rev. Lett.*, 19:849, 1967.

44. G. A. Garrett, T. F. Albrecht, J. F. Whitaker, and R. Merlin. Coherent terahertz phonons driven by light pulses and the Sb problem: what is the mechanism? *Phys. Rev. Lett.*, 77:3661, 1996.

45. M. Hase, K. Mizoguchi, H. Harima, S. Nakashima, M. Tani, K. Sakai, and M. Hangyo. Optical control of coherent optical phonons in bismuth films. *Appl. Phys. Lett.*, 69:2474, 1996.

46. W. A. Kütt, W. Albrecht, and H. Kurz. Generation of coherent phonons in condensed media. *IEEE J. Quantum Electron*, QE 28:2434, 1992.

47. M. Kuball, N. Esser, T. Ruf, C. Ullrich, M. Cardona, K. Eberl, A. Garcia-Cristobal, and A. Cantarero. Electric-field induced Raman scattering in GaAs: Franz-Keldysh oscillations. *Phys. Rev. B*, 51:7353, 1995.

48. G. C. Cho, H. J. Bakker, T. Dekorsy, and H. Kurz. Time-resolved observation of coherent phonons by the Franz-Keldysh effect. *Phys. Rev. B*, 53:6904, 1996.

49. Y. M. Chang, L. Xu, and H. W. K. Tom. Observation of coherent surface optical phonon oscillations by time-resolved second harmonic generation. *Phys. Rev. Lett.*, 78:4649, 1997.

50. M. Hase, M. Kitajima, A. M. Constantinescu, and H. Petek. The birth of a quasi-particle in silicon observed in time–frequency space. *Nature*, 426:51, 2003.

51. A. Leitenstorfer, S. Hunsche, J. Shah, M. C. Nuss, and W. H. Knox. Femtosecond charge transport in polar semiconductors. *Phys. Rev. Lett.*, 82:5140, 1999.
52. A. Leitenstorfer, S. Hunsche, J. Shah, M. C. Nuss, and W. H. Knox. Femtosecond high-field transport in compound semiconductors. *Phys. Rev. B*, 61:16642, 2000.
53. M. P. Hasselbeck, L. A. Schlie, and D. Stalnaker. Emission of electromagnetic radiation by coherent vibrational waves in stimulated Raman scattering. *Appl. Phys. Lett.*, 85:173, 2004.
54. X.-C. Zhang and D. H. Auston. Optoelectronic measurement of semiconductor surfaces and interfaces with femtosecond optics. *J. Appl. Phys.*, 71:326, 1992.
55. Q. Wu and X. C. Zhang. Seven terahertz broadband GaP electro-optic sensor. *Appl. Phys. Lett.*, 70:1784, 1997.
56. R. Huber, A. Brodschelm, F. Tauser, and A. Leitenstorfer. Generation and field-resolved detection of femtosecond electromagnetic pulses tunable up to 41 THz. *Appl. Phys. Lett.*, 76:3191, 2000.
57. T. Dekorsy, H. Auer, H. J. Bakker, H. G. Roskos, and H. Kurz. Terahertz, electromagnetic emission by coherent infrared-active phonons. *Phys. Rev. B*, 53:4005, 1996.
58. A. Cavalleri, C. W. Siders, F. L. H. Brown, D. M. Leitner, C. Toth, J. A. Squier, C. P. J. Barty, K. R. Wilson, K. Sokolowski-Tinten, M. Horn von Hoegen, D. von der Linde, and M. Kammler. Anharmonic lattice dynamics in germanium measured with ultrafast X-ray diffraction. *Phys. Rev. Lett.*, 85:586, 2000.
59. D. A. Reis, M. F. DeCamp, P. H. Bucksbaum, R. Clarke, E. Dufresne, M. Hertlein, R. Merlin, R. Falcone, H. Kapteyn, M. M. Murnane, J. Larsson, Th. Missalla, and J. S. Wark. Probing impulsive strain propagation with X-ray pulses. *Phys. Rev. Lett.*, 86:3072, 2001.
60. K. Sokolowski-Tinten, C. Blome, J. Blums, A. Cavalleri, C. Dietrich, A. Tarasevitch, I. Uschmann, E. Förster, M. Kammler, M. Horn von Hoegen, and D. von der Linde. Femtosecond X-ray measurement of coherent lattice vibrations near the Lindemann stability limit. *Nature*, 422:287, 2003.
61. M. Bargheer, N. Zhavoronkov, Y. Gritsai, J. C. Woo, D. S. Kim, M. Woerner, and T. Elsaesser. Coherent atomic motions in a nanostructure studied by femtosecond X-ray diffraction. *Science*, 306:1771, 2004.
62. A. Cantarero, C. Trallero-Giner, and M. Cardona. Excitons in one-phonon resonant Raman scattering: deformation-potential interaction. *Phys. Rev. B*, 39:8388, 1989.
63. A. Cantarero, C. Trallero-Giner, and M. Cardona. Excitons in one-phonon resonant Raman scattering: Fröhlich and interference effects. *Phys. Rev. B*, 40:12290, 1989.
64. M. J. Rosker, F. W. Wise, and C. L. Tang. Femtosecond relaxation dynamics of large molecules. *Phys. Rev. Lett.*, 57:321, 1986.
65. S. M. Sze. *Physics of Semiconductor Devices*, 2nd ed., Wiley, New York, 1981.
66. A. Yarif. *Introduction to Optical Electronics*, Holt, Rinehart and Winston, New York, 1971.
67. L. Min and R. J. D. Miller. Subpicosecond reflective electro-optic sampling of electron-hole vertical transport in surface-space-charge fields. *Appl. Phys. Lett.*, 56:524, 1990.
68. M. Hase, K. Mizoguchi, H. Harima, S. Nakashima, and K. Sakai. Dynamics of coherent phonons in bismuth generated by ultrashort laser pulses. *Phys. Rev. B*, 58:5448, 1998.

69. J. G. Gay, J. D. Dow, E. Burstein, and A. Pinzcuk. *Light Scattering in Solids*, Flammarion, Paris, 1971, p. 33.
70. T. Dekorsy, R. Ott, H. Kurz, and K. Köhler. Bloch oscillations at room temperature. *Phys. Rev. B*, 51:17275, 1995.
71. G. C. Cho, T. Dekorsy, H. J. Bakker, R. Hövel, and H. Kurz. Generation and relaxation of coherent majority plasmons. *Phys. Rev. Lett.*, 77:4062, 1996.
72. K. Mizoguchi, M. Hase, S. Nakashima, H. Harima, and K. Sakai. Ultrafast decay of coherent plasmon–phonon coupled modes in highly doped GaAs. *Phys. Rev. B*, 60:16526, 1999.
73. F. Vallée, F. Ganikhanov, and F. Bogani. Dephasing of LO-phonon-plasmon hybrid modes in n-type GaAs. *Phys. Rev. B*, 56:13141, 1997.
74. R. Kersting, K. Unterrainer, G. Strasser, H. F. Kaufmann, and E. Gornik. Few-cycle terahertz, emission from cold plasma oscillations. *Phys. Rev. Lett.*, 79:3038, 1997.
75. Y. C. Shen, P. C. Upadhya, E. H. Linfield, H. E. Beere, and A. G. Davies. Terahertz generation from coherent optical phonons in a biased GaAs photoconductive emitter. *Phys. Rev. B*, 69:235325, 2004.
76. W. Sha, A. L. Smirl, and W. F. Tseng. Coherent plasma oscillations in bulk semiconductors. *Phys. Rev. Lett.*, 74:4273, 1995.
77. P. Gu, M. Tani, K. Sakai, and T. R, Yang. Detection of terahertz radiation from longitudinal optical phonon–plasmon coupling modes in InSb film using an ultrabroadband photoconductive antenna. *Appl. Phys. Lett.*, 77:1798, 2000.
78. M. P. Hasselbeck, D. Stalnaker, L. A. Schlie, T. J. Rotter, A. Stintz, and M. Sheik-Bahae. Emission of terahertz radiation from coupled plasmon–phonon modes in InAs. *Phys. Rev. B*, 65:233203, 2002.
79. W. de Jong, A.F. van Etteger, P. J. van Hall, and T. Rasing. Coherent plasmon surface phonon oscillations in a GaAs schottky barrier. *Surface Science*, 377:355, 1997.
80. Y. M. Chang, L. Xu, and H. W. K. Tom. *Ultrafast Phenomena X*, Vol. 62, *Springer Series in Chemical Physics*, Springer, Berlin, 1996, p. 391.
81. Y. M. Chang, C. T. Chuang, C. T. Chia, K. T. Tsen, H. Lu, and W. J. Schaff. Coherent longitudinal optical phonon and plasmon coupling in the near-surface region of InN. *Appl. Phys. Lett.*, 85:5224, 2004.
82. R. Huber, F. Tauser, A. Brodschelm, M. Bichler, G. Abstreiter, and A. Leitenstorfer. How many-particle interactions develop after ultrafast excitation of an electron–hole plasma. *Nature*, 414:286, 2001.
83. R. Huber, C. Kübler, S. Tübel, A. Leitenstorfer, Q. T. Vu, H. Haug, F. Köhler, and M.-C. Amann. Femtosecond formation of coupled phonon-plasmon modes in InP: ultrabroadband terahertz experiment and quantum kinetic theory. *Phys. Rev. Lett.*, 94:027401, 2005.
84. K. J. Yee, K. G. Lee, E. Oh, D. S. Kim, and Y. S. Lim. Coherent optical phonon oscillations in bulk GaN excited by far below the band gap photons. *Phys. Rev. Lett.*, 88:105501, 2002.
85. I. H. Lee, K. J. Yee, K. G. Lee, E. Oh, D. S. Kim, and Y. S. Lim. Coherent optical phonon mode oscillations in wurtzite ZnO excited by femtosecond pulses. *J. Appl. Phys.*, 93:4939, 2003.
86. C. Aku-Leh, J. Zhao, R. Merlin, and J. Menendez. Coherent optical phonons with very large quality factors: the E_2-low mode in ZnO. In: *CLEO/IQEC and PHAST Technical Digest on CD-ROM*, Optical Society of America, Washington, DC, 2004, p. IMJ 2.

87. J. Shah. *Ultrafast Spectroscopy of Semiconductors and Semiconductor Nanostructure*, Vol. 115, *Solid State Science*, Springer, Berlin, 1996.
88. K. T. Tsen. *Ultrafast Dynamical Processes in Semiconductors*, Vol. 92, *Topics in Applied Physics*, Springer, Berlin, 2004.
89. K. J. Yee, Y. S. Lim, T. Dekorsy, and D. S. Kim. Mechanism for the generation of coherent longitudinal-optical phonons in GaAs/AlGaAs multiple quantum wells. *Phys. Rev. Lett.*, 86:1630, 2001.
90. T. Dekorsy, A. M. T. Kim, G. C. Cho, H. Kurz, A. V. Kuznetsov, and A. Förster. Subpicosecond coherent carrier-phonon dynamics in semiconductor heterostructures. *Phys. Rev. B*, 53:1531, 1996.
91. P. C. M. Planken, M. C. Nuss, I. Brenner, K. W. Goosen, M. S. C. Luo, S. L. Chuang, and L. Pfeifer. Terahertz emission in single quantum wells after coherent optical excitation of light hole and heavy hole excitons. *Phys. Rev. Lett.*, 69:3800, 1992.
92. T. Dekorsy, A. M. T. Kim, G. C. Cho, H. J. Bakker, S. Hunsche, H. Kurz, K. Köhler, and S. L. Chuang. Quantum coherence of continuum states in the valenceband of GaAs quantum wells. *Phys. Rev. Lett.*, 77:3045, 1996.
93. K. Mizoguchi, O. Kojima, T. Furuichi, M. Nakayama, K. Akahane, N. Yamamoto, and N. Ohtani. Coupled mode of the coherent optical phonon and excitonic quantum beat in GaAs/AlAs multiple quantum wells. *J. Phys.: Condens. Matter*, 14:L103, 2002.
94. Y.-M. Chang, H. H. Lin, C. T. Chia, and Y. F. Chen. Observation of coherent interfacial optical phonons in GaInP/GaAs/GaInP single quantum wells. *Appl. Phys. Lett.*, 84:2548, 2004.
95. E. E. Mendez, F. Agulló-Rueda, and J. M. Hong. Stark localization in GaAs/GaAlAs superlattices under an electric field. *Phys. Rev. Lett.*, 60:2426, 1988.
96. P. Voisin, J. Bleuse, C. Bouche, S. Gaillard, C. Alibert, and A. Regreny. Observation of the Wannier-Stark ladder quantization in a semiconductor superlattice. *Phys. Rev. Lett.*, 61:1639, 1988.
97. F. Bloch. Über die Quantenmechanik der Elektronen in Kristallgittern. *Z. Phys.*, 52:555, 1928.
98. C. Zener. The theory of the electrical breakdown of solid dielectrics. *Proc. R. Soc. London*, A 145:523, 1934.
99. L. Esaki and R. Tsu. Superlattice and negative differential conductivity in semiconductors. *IBM J. Res. Dev.*, 14:61, 1970.
100. J. Feldmann, K. Leo, J. Shah, D. A. B. Miller, J. E. Cunningham, T. Meier, G. von Plessen, A. Schulze, P. Thomas, and S. Schmitt-Rink. Optical investigation of Bloch oscillations in a semiconductor superlattice. *Phys. Rev. B*, 46:7252, 1992.
101. K. Leo, P. Haring-Bolivar, F. Brüggemann, R. Schwedler, and K. Köhler. Observation of Bloch oscillations in a semiconductor superlattice. *Solid State Commun.*, 84:943, 1992.
102. K. Leo. *High Field Transport in Semiconductor Superlattices*, Vol. 187, *Springer Tracts in Modern Physics*, Springer-Verlag, Berlin, 2003.
103. T. Dekorsy, A. Bartels, H. Kurz, K. Köhler, R. Hey, and K. Ploog. Coupled Bloch–phonon oscillations in semiconductor superlattices. *Phys. Rev. Lett.*, 85:1080, 2000.
104. A. W. Ghosh, L. Jönsson, and J. W. Wilkins. Bloch oscillations in the presence of plasmons and phonons. *Phys. Rev. Lett.*, 85:1084, 2000.
105. J. Hader, T. Meier, S. W. Koch, F. Rossi, and N. Linder. Microscopic theory of the intracollisional field effect in semiconductor superlattices. *Phys. Rev. B*, 55:13799, 1997.

106. M. Först, T. Dekorsy, H. Kurz, and R. P. Leavitt. Bloch–phonon coupling and tunneling-induced coherent phonon excitation in semiconductor superlattices. *Phys. Rev. B*, 67:085305, 2003.

107. B. Rosam, D. Meinhold, F. Löser, V. G. Lyssenko, S. Glutsch, F. Bechstedt, F. Rossi, K. Köhler, and K. Leo. Field-induced delocalization and Zener breakdown in semiconductor superlattices. *Phys. Rev. Lett.*, 86:1307, 2001.

108. J. L. Bradshaw and R. P. Leavitt. Observation of an electron-wave-function coherence length approaching the theoretical limit in a nearly ideal semiconductor-alloy superlattice. *Phys. Rev. B*, 50:17666, 1994.

109. P. Bhattacharya, Ed. *Properties of Lattice-Matched and Strained InGaAs*, INSPEC, London, 1993.

110. L. Pavesi, R. Houdré, and P. Giannozzi. Strain and alloying effects on the electronic and vibrational properties of $In_yAl_{1-y}As$ on InP. *J. Appl. Phys.*, 78:470, 1995.

111. M. Helm, W. Hilber, G. Strasser, R. De Meester, F. M. Peeters, and A. Wacker. Continuum Wannier-Stark ladders strongly coupled by Zener resonances in semiconductor superlattices. *Phys. Rev. Lett.*, 82:3120, 1999.

112. M. Först, T. Dekorsy, H. Kurz, and R. P. Leavitt. Well-width dependence of coupled Bloch phonon oscillations in biased InGaAs/InAlAs superlattices. *Phys. Status Solids (c)*, 1:2702, 2004.

113. M. Cardona and P. Y. Yu. *Fundamentals of Semiconductors*, 3rd ed., Springer-Verlag, Berlin, 2003.

114. S. M. Rytov. *Acoust. Zh.*, 2:71, 1956.

115. T. Ruf, *Phonon Raman Scattering in Semiconductors, Quantum Wells and Superlattices*, Vol. 142, *Springer Tracts in Modern Physics*, Springer, Berlin, 1998.

116. A. Yamamoto, T. Mishina, Y. Masumoto, and M. Nakayama. Coherent oscillations of zone-folded phonon modes in GaAs-AlAs superlattices. *Phys. Rev. Lett.*, 73:740, 1994.

117. K. Mizoguchi, K. Matsutani, M. Hase, S. Nakachima, and M. Nakayama. Resonance effect of coherent folded acoustic phonons generated by ultrashort light pulses in GaAs/AlAs superlattices. *Physica B*, 249:887, 1998.

118. T. Mishina, Y. Iwazaki, Y. Masumoto, and M. Nakayama. Real time-space dynamics of zone-folded phonons in GaAs/AlAs superlattices. *Solid State Commun.*, 107:281, 1998.

119. A. Bartels, T. Dekorsy, H. Kurz, and K. Köhler. Coherent zone-folded longitudinal acoustic phonons in semiconductor superlattices: excitation and detection. *Phys. Rev. Lett.*, 82:1044, 1999.

120. B. Jusserand, D. Paquet, F. Mollot, F. Alexandre, and G. Le Roux. Influence of the supercell structure on the folded acoustical Raman line intensities in superlattices. *Phys. Rev. B*, 35:2808, 1987.

121. A. Bartels, T. Dekorsy, H. Kurz, and K. Köhler. Coherent control of acoustic phonons in superlattices. *Appl. Phys. Lett.*, 72:2844, 1998.

122. T. Dekorsy, W. Kütt, T. Pfeifer, and H. Kurz. Coherent control of LO phonon dynamics in opaque semiconductors by femtosecond laser pulses. *Europhys. Lett.*, 23:223, 1993.

123. K. Mizoguchi, H. Takeuchi, T. Hino, and M. Nakayama. Finite-size effects on coherent folded acoustic phonons in GaAs/AlAs superlattices. *J. Phys.: Condens. Matter*, 14:L103, 2002.

124. P. Hawker, A. J. Kent, L. J. Challis, A. Bartels, T. Dekorsy, H. Kurz, and K. Köhler. Observation of coherent zone-folded acoustic phonons generated by Raman scattering in a superlattice. *Appl. Phys. Lett.*, 77:3209, 2000.

125. A. J. Kent, N. M. Stanton, L. J. Challis, and M. Henini. Generation and propagation of monochromatic acoustic phonons in gallium arsenide. *Appl. Phys. Lett.*, 81:3497, 2002.

126. N. M. Stanton, R. N. Kini, A. J. Kent, M. Henini, and D. Lehmann. Terahetz phonon optics in GaAs/AlAs superlattice structures. *Phys. Rev. B*, 68:113302, 2003.

127. S. Tamura, D. C. Hurley, and J. P. Wolfe. Acoustic-phonon propagation in superlattices. *Phys. Rev. B*, 38:1427, 1988.

128. T. Takeuchi, S. Sota, M. Katsuragawa, M. Komori, H. Takeuchi, H. Amano, and I. Akasaki. Quantum-confined Stark effect due to piezoelectric fields in GaInN strained quantum wells. *Jpn. J. Appl. Phys.*, 36:L382, 1997.

129. C. K. Sun, J. C. Liang, C. J. Stanton, A. Abare, L. Coldren, and S. P. DenBaars. Large coherent acoustic phonon oscillation observed in InGaN/GaN multiple-quantum-wells. *Appl. Phys. Lett.*, 75:1249, 1999.

130. H. K. Sun, J. C. Liang, and X. Y. Yu. Coherent acoustic phonon oscillations in semiconductor multiple quantum wells with piezoelectric fields. *Phys. Rev. Lett.*, 84:179, 2000.

131. C. K. Sun, J. C. Liang, C. J. Stanton, A. Abare, L. Coldren, and S. P. DenBaars. Coherent optical control of acoustic phonon oscillations in InGaN/GaN multiple quantum wells. *Appl. Phys. Lett.*, 78:1201, 2001.

132. X. Hu and F. Nori. Squeezed phonon states: modulating quantum fluctuations of atomic displacements. *Phys. Rev. Lett.*, 76:2294, 1996.

133. G. A. Garrett, A. G. Rojo, A. K. Sood, J. F. Whitaker, and R. Merlin. Vacuum squeezing of solids: macroscopic quantum states driven by light pulses. *Science*, 275:1638, 1997.

134. M. Artoni. Detecting phonon vacuum squeezing. *J. Nonl. Opt. Phys. Mat.*, 7:241, 1998.

135. A. Bartels, T. Dekorsy, and H. Kurz. Impulsive excitation of phonon-pair combination states by second-order Raman scattering. *Phys. Rev. Lett.*, 84:2981, 2000.

136. O. V. Misochko, K. Sakai, and S. Nakashima. Phase-dependent noise in femtosecond pump-probe experiments on Bi and GaAs. *Phys. Rev. B*, 61:11225, 2000.

137. S. Hunsche, K. Wienecke, T. Dekorsy, and H. Kurz. Impulsive mode softening of coherent phonons. *Phys. Rev. Lett.*, 75:1815, 1995.

138. M. F. DeCamp, D. A. Reis, P. H. Bucksbaum, and R. Merlin. Dynamics and coherent control of high amplitude optical phonons in bismuth. *Phys. Rev. B.*, 64:092301, 2001.

139. M. Hase, M. Kitajima, S. Nakashima, and K. Mizoguchi. Dynamics of coherent anharmonic phonons in bismuth using high density photoexcitation. *Phys. Rev. Lett.*, 88:067401, 2002.

140. P. Tangney and S. Fahy. Density-functional theory approach to ultrafast laser excitation of semiconductors: application to the A_1 phonon in tellurium. *Phys. Rev. Lett.*, 65:054302, 2002.

141. A. M. T. Kim, C. A. D. Roeser, and E. Mazur. Modulation of the bonding-antibonding splitting in Te by coherent phonons. *Phys. Rev. B*, 68:012301, 2003.

142. O. V. Misochko, M. Hase, K. Ishioka, and M. Kitajima. Observation of an amplitude collapse and revival of chirped coherent phonons in bismuth. *Phys. Rev. Lett.*, 92:197401, 2004.

5 Coherent Dynamics of Halogen Molecules in Rare Gas Solids

Markus Gühr

CONTENTS

5.1 INTRODUCTION

A molecular vibrational wavepacket consists of coherently coupled vibrational eigenfunctions in a molecular electronic potential. *Coherence* in this context means that the phase relation among the corresponding eigenfunctions is well defined and does not change in an arbitrary way. The coherent superposition of those eigenfunctions leads to constructive and destructive interference phenomena. They are a prerequisite for many coherent control schemes (e.g., the Tannor–Rice method [126, 127]) and determine the temporal propagation of the wavepacket.

Since most chemical reactions take place in condensed solvents, it is important to study the influence of such environments on the molecular coherence. In general, any environment has the potential to destroy the well-defined relative phase among molecular eigenstates, and the wavepacket can decay into an incoherent statistical superposition of eigenstates. Furthermore, energy will be transferred from the molecular chromophore to the solvent after the excitation of the chromophore. These processes are crucial on all levels of complexity: from the simple diatomic molecule in solution to the most complicated biological molecules in their functional environment. This work aims at establishing molecule–solvent interactions (energy relaxation, coherence decay) for the conceptually simple system of diatomic halogens in a rare gas solid (RGS), that can also be simulated in great detail [13, 14, 17, 19, 46, 80, 95–97, 99, 100, 138, 139].

The small molecules I_2 and Br_2 in the gas phase have been studied via excitation with ultrashort pulses (see Refs. [49, 134]) and they show long lasting vibrational and rotational wave packets indicating a vibrational and rotational coherence in the several hundred ps range. The anharmonicity effects of electronic potentials on vibrational wavepackets have been studied, and a dispersion (anharmonicity-induced broadening) followed by wavepacket revivals has been found [15, 62, 112, 134]. The phenomena are used in this study to gain detailed insight into vibrational and electronic coherence decay for small molecules in RGS. The RGS has a large ionization energy, allowing use of high-intensity laser pulses. The rare gas atoms are chemically inert; thus, molecules do not react after photoexcitation, and adopted gas-phase potentials can be used for the molecular dopants. Since the vibrational-level spacing of the halogens is close to the Debye frequency of the RGS, the molecules undergo efficient ultrafast relaxation and decoherence according to the energy gap law [94].

The molecules I_2 and Br_2 replace two nearest-neighbor atoms in a Kr respective Ar crystal, according to their van der Waals radii. Because of the asymmetry of this "double-substitutional site," the molecular rotation is blocked in contrast to the case of smaller molecules residing in "single-substitutional sites" such as ClF in solid argon [7]. Thus, only vibrational wavepackets (and no rotational ones) will form after ultrafast excitation, and the molecular motion will be guided by the cage. The surrounding keeps its face centered cubic structure and because of this high environment symmetry, the molecular vibrational coherence survives even strong collisions of the molecule with its cage. Collisions of the molecule with the matrix can lead to the creation of coherent wavepackets built from vibrational states not populated before the collision.

The vibrational and electronic coherences of molecular iodine in RGSs have been studied extensively by pump-probe spectroscopy [124, 140, 141, 144, 145] and coherent anti-Stokes Raman spectroscopy (CARS) in the group of Apkarian [18, 20, 68–70, 121]. The method employed in this study is the femtosecond pump-probe spectroscopy. A vibrational wavepacket is created on a bound covalent state by a first ultrashort laser pulse acting on the electronic ground-state population. The vibrational wavepacket is probed by a second time-delayed laser pulse. This probe pulse transfers population from the covalent state to a charge transfer state. In turn, we will detect the resulting charge transfer fluorescence in the ultraviolet (UV) range. By chirping the pump laser pulse, the phase relation in the eigenstates and the resulting wavepacket time evolution can be changed allowing for wavepacket focusing and the coherent control of the wavepacket propagation.

The environment is usually considered a source for decoherence, and its internal coherent motions are neglected. We show that specific RGS host atoms in the vicinity of the molecule move coherently, which is confirmed via pump-probe spectra of the molecular guest. We attribute the observed host motion to a coherent zone boundary phonon (ZBP) of the Ar solid for the Br_2:Ar case. To support the ZBP character, spectra of I_2 in solid krypton are also presented, revealing coherent ZBP of the solid Kr host.

Usually, optically excited coherent phonons in crystals have a zone center character. Due to a special excitation mechanism, via the change in the molecular electronic wavefunction, ZBP creation becomes possible. The pump pulse induces a transition from the electronic ground state X to the covalent states. The molecular electronic cloud is blown up in this transition, the matrix atoms are impulsively repelled from the molecule, and phonons are created in the environment. The zone center phonons travel away from the excited molecule with their finite-group velocity v_g. The ZBP remains near the molecule due to its vanishing v_g. The phonons shift the solvation energy of the molecular charge transfer states and thereby modulate in the pump-probe spectrum of the molecule with their characteristic frequency.

The experimental setup with the vacuum apparatus and the laser system is given in Section 5.2. The results for intramolecular coherences are presented in Section 5.3. The part discusses the Br_2:Ar system only. The pump-probe spectra with energy relaxation and coherence transfer among vibrational levels are explained in Section 5.3.1, followed by the method of wavepacket focusing (Section 5.3.2). The focusing can be used to disentangle reversible wavepacket broadening in a Morse oscillator (dispersion) from irreversible decoherence, as described in Section 5.3.3. In contrast to other articles in the field, the words *decoherence* and *dephasing* are used as synonyms in this contribution. They describe an irreversible loss of phase information. To get more detailed insight into intramolecular vibrational coherence, the fractional revivals of wave packet s are introduced in Section 5.3.4. They are used to deduce the coherence time of several coupled vibrational levels and electronic coherence (Sections 5.3.5 and 5.3.6) implementing the coherent control of fractional revivals.

The detection of coherent phonons is described in Section 5.4. Besides the results on Br_2:Ar, experiments on I_2:Kr are added to support the assignment.

The excitation mechanism supported by simple model calculations is presented in Section 5.4.2, and the probe mechanism is described in Section 5.4.3. A comparison with related experiments and with theoretical approaches finishes this chapter.

5.2 EXPERIMENTAL SETUP

5.2.1 VACUUM SYSTEM

Briefly, the halogen and the rare gas are mixed and afterwards sprayed on a cooled substrate in the cryostat. On this substrate, the gas mixture freezes, and a polycrystalline film of the halogen-doped rare gas grows.

The ultrahigh vacuum system (UHV) used for all experiments documented in this chapter is divided into two parts (see Figure 5.1). Figure 5.1(a) represents the bromine mixing system used to mix the Br_2 with argon; Figure 5.1(b) represents the cryostat chamber. The gray-shaded parts in Figure 5.1 consist of stainless steel, whereas the transparent parts consist of glass (if not stated otherwise). A third part, the I_2:Kr mixing system, is not described here. It is similar to Figure 5.1a but is essentially made from stainless steel components only.

We tried to prepare the bromine–rare gas mixture in a stainless steel vessel, but we did not succeed in growing a bromine-doped crystal because the bromine reacts strongly with stainless steel walls. Thus, a mixing apparatus made from glass had to be constructed. The setup is shown in Figure 5.1a. A central mixing volume of about one liter is connected to a bromine reservoir and an argon vessel. The seals in this case are made of Teflon and Viton. All components are chemically inert concerning bromine.

Before preparing the Br_2:Ar mixture, freeze-thaw cycles are repeated several times to purify the bromine. Afterwards, the desired concentration ratio of Br_2:Ar is prepared in the mixing volume. The mixture is introduced in the cryostat exclusively by glass and teflon transfer lines to avoid any metal contact before the gas reaches the cooled substrate.

The cryostat chamber (Figure 5.1b) consists of a UHV stainless steel vessel evacuated by the turbo pump. A closed cycle refrigerator or alternately a liquid He flow cryostat cools a 1 mm thick CaF_2 substrate via a copper rod. A heater and a temperature diode (Lakeshore) are connected to the copper rod to control and measure the actual temperature. In the case of the closed-cycle He refrigerator, a temperature of about 17 K can be reached at the substrate, whereas 5 K can be achieved with the liquid He flow cryostat. The cooling acts as a cryopump and reduces the pressure to 10^{-9} mbar together with the turbo pump.

For crystal preparation, the substrate is positioned 5 cm in front of the glass tube ends. The valves connecting the mixing vessel and the cryostat are opened, and the gas mixture flows on the substrate. In a systematic study, the optimal gas flow parameters for highest optical sample quality were searched using the pressure in the cryostat chamber measured by the ionization gauge as an indicator. The crystals are crack free at the end of the growing process and develop some cracks in the course of time, most probably due to temperature fluctuations and resulting

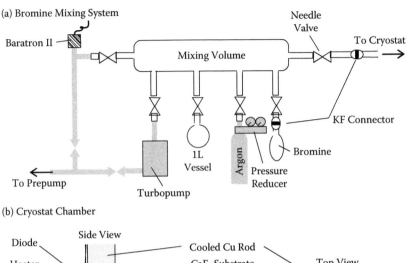

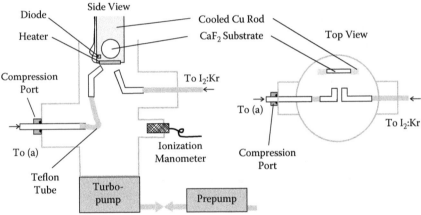

FIGURE 5.1 Vacuum setup used in the experiments. (a) Mixing system for Br_2 and Ar gas. The transparent parts are made of glass, the gray shaded of stainless steel. The setup for I_2 and Kr is similar to this but built completely from stainless steel. (b) Side and top view of the cryostat chamber. A substrate is mounted on a cooled copper rod. The halogen:rare gas mixtures are sprayed on the substrate through tubes under UHV conditions (see text).

strain. In case of the I_2-doped Kr, the cryocrystals are purple (as gas phase I_2), whereas the color of Br_2-doped Ar crystals is brown-orange (as Br_2 gas). Cryostat and copper rods are placed on a UHV manipulator, and thus their position can be controlled. The manipulator allows for placing the cryocrystals in the laser beams.

5.2.2 LASER SETUP

A commercially available CPA (chirped pulse amplifier) with 1-kHz repetition rate pumps to home-built noncollinear optical parametric amplifiers (NOPAs) as shown in Figure 5.2. A general introduction to short-pulse parametric amplifiers

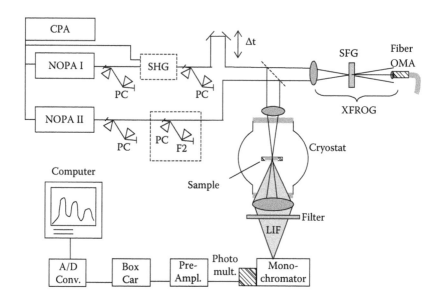

FIGURE 5.2 Experimental setup. The laser pulses are generated in a commercial chirped pulse amplifier (CPA). NOPAs generate the pump and the probe pulses. The dispersion of vacuum windows and lenses in the beam is compensated by prism compressors (PC), allowing for a variation of the pump-pulse chirp. The pulses are delayed by Δt on a computer-controlled delay stage. An SFG FROG (often called XFROG) or PG FROG allows for pulse characterization. The pump and probe pulses induce a sample fluorescence in the vacuum chamber, which is collected and imaged on a photomultiplier after spectral selection with filters and a monochromator. The photomultiplier signal is amplified and integrated in a boxcar. A computer records the integrated signal as a function of time delay Δt.

is given in Ref. [29]. The NOPAs in our lab are based on the design of the Riedle group [9, 137]. They produce spectrally tunable pulses (from 480 to 750 nm) with durations in the sub-30-fs range. Prism compressors (PCs) made of fused silica are used to compensate the dispersion of the cryostat windows and focusing lenses. A further prism compressor (PC F2) in the pump beam from NOPA II is used to impose a strong negative chirp. The probe pulse from NOPA I can be frequency doubled (second-harmonic generation) to probe by UV light. Alternatively, the pulses of NOPA I can be used directly to probe via two-photon transition (for Br_2:Ar) or a one-photon transition (for I_2:Kr). A time delay Δt between pump and probe pulse is introduced via a computer-controlled delay stage. Sum-frequency generation (SFG) and polarization gating (PG) frequency-resolved optical gating (FROG) are used to characterize the pump and the probe pulse. In addition, an autocorrelator is used for a quick pulse diagnosis.

Since in the pump-probe scheme used here the charge transfer states are finally populated, their fluorescence ($\sim$ 320 nm for Br_2:Ar, $\sim$ 420 nm for I_2:Kr) is collected and separated from the laser pulses by filters and a monochromator. A photomultiplier (PM) detects the fluorescence signal, which is then preamplified and further processed in a boxcar integrator to reduce noise. The computer records the fluorescence signal as a function of pump-probe delay. The resulting data set is called a *pump-probe spectrum*.

5.3 INTRAMOLECULAR COHERENCES

The pump-probe method is described in this section. Based on the pump-probe spectra, an effective potential is constructed for the B state of Br_2 in solid argon. Close to the free molecule dissociation limit, the potential reflects the influence of the argon cage. Near the potential minimum, the effective potential is similar to that of the gas-phase Br_2 molecule. Therefore, we directly compare simulations for gas-phase Br_2 (without decoherence) with experimental results for Br_2:Ar (with decoherence) to learn about the decoherence. To remove the dispersive broadening of wavepackets from the experimental spectra, we spatially focus the wavepackets. A general decoherence time of 3 ps is deduced from those data as shown.

To get further insight in the coherence of a distinct group of vibrational eigenstates, the concept of fractional revivals is used. The revival structures would appear after 3 ps in the gas phase, which is after loss of coherence in the rare gas solid. Therefore, a coherent control scheme is worked out to shift the revival structures forward in time into the coherent regime. The 1/6 revival is observed, and the corresponding four-level coherence is deduced. Finally, a lower limit for the electronic coherence between X and B state is derived in the last part of the section.

5.3.1 POTENTIAL CONSTRUCTION AND VIBRATIONAL RELAXATION

The solid curves in Figure 5.3a show three electronic potentials of free Br_2 (X, B and E). Because the molecule is embedded in a RGS at 20 K, all population in the electronic ground state will be in the $v = 0$ level (as indicated by the Gaussian curve). The pump pulse acts on X $v = 0$ and populates *several* vibrational levels in the electronic B state coherently because of its broad spectral width. Thereby, an intramolecular vibrational wavepacket is formed. The wavepacket is "born" in the Franck–Condon range at the inner potential limb and moves to the outer part of the potential. A second, time-delayed laser pulse (probe pulse) is used to detect the vibrational wavepacket. The probe pulse photon energy is in resonance with the E-B difference energy at the internuclear separation R_{win}. The corresponding B-state energy is called E_{win}. When the wavepacket has reached R_{win}, the probe pulse can promote population from the B state to the charge transfer E state, which will emit its typical fluorescence. Figure 5.3b shows the charge transfer fluorescence signal of Br_2 in argon as a function of pump-probe delay Δt for different central wavelengths of the pump pulse λ_{pump}. The probe wavelength λ_{pump} is kept constant at 348 nm.

The lower, dotted spectrum is recorded for $\lambda_{pump} = 560$ nm. At zero time delay, the vibrational wavepacket has not yet reached the probe window R_{win}, and thus no population is promoted to the E state, and no charge transfer fluorescence occurs.

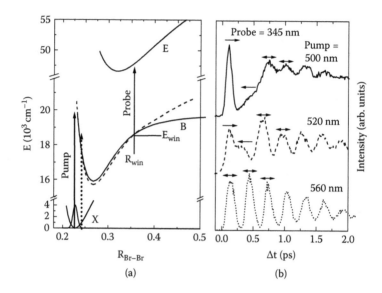

FIGURE 5.3 (a) Electronic potentials of free Br_2. Shown are the $X(^1\Sigma_{0g})$, $B(^3\Pi_{0u})$, and $E(0_g)$ states. The pump pulse induces a transition from X to B, thereby creating a vibrational wavepacket. The wavepacket is probed to the E state. The probe window is located at R_{win}, where the probe photon energy overcomes the potential difference E-B. The charge transfer state E fluoresces to lower covalent states. This fluorescence is recorded as a function of the pump-probe delay Δt in the time-resolved spectra. The dashed line gives the constructed RKR potential of the B state in the Ar matrix. (b) Three pump-probe spectra with $\lambda_{pump} = 560$, 520, and 500 nm and $\lambda_{probe} = 345$ nm. The arrows mark the wavepacket's direction of motion in the potential when probed at R_{win}. The double arrow indicates that R_{win} sits at the wavepacket turning point.

After half an oscillation period T, the wavepacket passes R_{win}, which is exactly at the outer (right) turning point of the wavepacket excited at $\lambda_{pump} = 560$ nm. One oscillation period later, the wavepacket has returned to the outer turning point, and a further peak is visible in the pump-probe spectrum. The time difference between first and second peak thus gives the oscillation period T of the wavepacket. Since the NOPA sources are tunable over a large spectral range, the pump as well as the probe pulse wavelength can be changed. Figure 5.3b shows two more spectra with different $\lambda_{pump} = 520$ and 500 nm. When going to $\lambda_{pump} = 520$ nm (dashed spectrum in Figure 5.3b), the first peak splits into two. Since we excited above E_{win}, the wavepacket crosses R_{win} twice per round-trip, on the outward motion ($\rightarrow$ in Figure 5.3b) and once more on the inward motion ($\leftarrow$ in Figure 5.3b). Similar features are observed for a λ_{pump} of 500 nm; however, the splitting between outward and inward passage increases. The 500-nm excitation (20,000 cm^{-1}) of Br_2:Ar is located above the dissociation limit of the free Br_2 molecule. If we would perform the experiment in the gas phase, then only the first peak ($\rightarrow$) on the way to dissociation should be visible in the pump-probe spectrum. However, the solid curve in Figure 5.3b shows a recombining signal of the wavepacket, reflecting the stabilizing influence of the rare gas cage.

The cage effect was first explored in the work of Franck and Rabinowitch [39,106,107]. With the help of ultrafast pump-probe spectroscopy, we are now able to construct an effective potential of the chromophore Br_2 in its environment. This can be accomplished in the following way. For a series of pump wavelengths, pump-probe spectra are recorded, and the vibrational period T can be read from the wave packet features. The vibrational period is given by the time separation between the first peak and a later peak, marked by ($\leftrightarrow$) in the pump-probe spectrum. The method has been repeated for 12 different pump wavelengths. Thus, we obtain the vibrational period T as a function of the pump photon energy. Using the Rydberg–Klein–Rees (RKR) algorithm [75,82,108,113], the distance ΔR between inner and outer turning point of a trajectory can be deduced from the oscillation period. With all different excitation wavelengths, ΔR as a function of pump photon energy can be constructed. The inner potential limb follows the gas-phase values, as absorption spectroscopy proves [51]. This allows for the construction of the full potential $V(R)$ from the ΔR data. The details of the method can be found in Refs. [6,51,56].

Now, we determine the vibrational energy relaxation of the Br_2 molecules in the argon crystal. It will be exemplified for the spectrum excited with $\lambda_{pump} = 500$ nm (solid line in Figure 5.3b). The first period from $\rightarrow$ to $\leftrightarrow$ is much longer than the following vibrational periods (from $\leftrightarrow$ to $\leftrightarrow$). The period of $T = 300$ fs visible beyond 1 ps matches exactly the period of the spectrum with $\lambda_{pump} = 560$ nm. The vibrational period T indicates the actual energy of the wavepacket in the anharmonic B state potential. We can therefore derive the energy loss of the wavepacket in the first collision with the cage. It is given by $(hc/500 \text{ nm}) - (hc/560 \text{ nm}) = 2140 \text{ cm}^{-1}$, which is about half of the maximal kinetic energy at $\lambda_{pump} = 500$ nm. Because the excitation occurred above the gas-phase dissociation limit, the vibrational wavepacket reaches far out in the intranuclear coordinate, and the atoms can effectively interact with the matrix. For smaller elongation, the matrix interaction decreases, and the energy loss in the first collision falls off exponentially while lowering the excitation photon energy [56]. This trend has been proved for many molecules [6,8].

Despite this strong vibrational relaxation, a coherent wavepacket structure remains visible. This is in strong contradiction to the standard statistical models using the usual time constants T_1 (energy relaxation time) and T_2 (phase relaxation time). In those approaches [76,77], the dephasing rate has to be at least half the energy relaxation rate or $T_2 \leq 2T_1$. Here, T_1 is less than one period; however, coherent wavepackets are observed for several cycles, indicating a T_2 of several oscillation periods T. The result is unique and is due to the high environmental symmetry. All molecules excited and probed "see" the same environment, and due to the "guided" motion in the double substitutional site, the scattering events are very similar for all molecules in the ensemble.

It is worth mentioning here that a high-resolution excitation spectrum exists for Br_2 in solid argon [22]. The spectrum shows a vibrational progression consisting of sharp zero phonon lines (ZPLs) and broader phonon sidebands from the bottom of the potential to about 1500 cm^{-1} above (corresponding to 570-nm excitation wavelength). Our RKR and the excitation spectroscopy data agree in this range. Beyond 1500 cm^{-1}, the excitation spectrum is unstructured, not allowing for potential construction. However, the RKR potential can be constructed from the pump-probe spectroscopy even beyond the gas-phase dissociation limit. The difference between frequency-resolved spectroscopy and time-resolved coherent spectroscopy lies in the sensitivity to inhomogeneous and homogeneous line broadening [56]. The strong energy relaxation presented above contributes to homogeneous line broadening. After the first collision with the cage, the wavepacket does not return to the inner turning point where it was born at $\Delta t = 0$. In the language of correlation functions [60], the energy relaxation leads to an unstructured autocorrelation function. This in turn will result in a broad absorption or excitation spectrum. A quickly decaying electronic coherence between X and B state enhances the broadening effect. Pump-probe spectroscopy, however, is only sensitive to the vibrational coherence, and the wavepacket oscillation period T delivers information about the potential. Even in case of strong energy relaxation, the probe window can be systematically shifted in the potential, allowing the relaxed wavepacket to be caught. Furthermore, the pump-probe spectrum is less sensitive to inhomogeneous line broadening since it only measures the distance between vibrational lines [56].

The lower part of the B state RKR potential (dashed line in Figure 5.3a) can be expressed with a simple Morse potential. The features of vibrational wavepackets in a Morse potential are elaborated in the next section.

5.3.2 DISPERSION OF WAVEPACKETS AND WAVEPACKET FOCUSING

The Morse potential is a standard way to parameterize the molecular potential of a diatomic [61,63]. The potential V as a function of the internuclear distance R is given as

$$V(R) = D_e(1 - e^{-\alpha(R-R_e)})^2 \tag{5.1}$$

with

$$\alpha = \sqrt{\frac{\mu}{2D_e}}\omega_e \tag{5.2}$$

$$D_e = \hbar\omega_e/4x_e \tag{5.3}$$

where D_e is the dissociation energy, R_e the equilibrium distance, ω_e the harmonic eigenfrequency, and x_e the anharmonicity.

The classical frequency ν of a trajectory started with energy E in a Morse oscillator is given as

$$\nu(E) = \frac{\omega_e}{2\pi}\sqrt{1 - \frac{E}{D_e}} \tag{5.4}$$

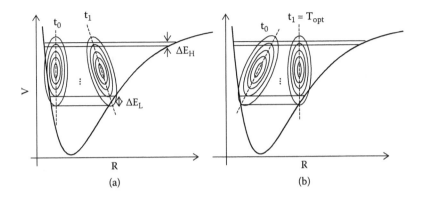

FIGURE 5.4 Explanation for wavepacket dispersion and focusing. (a) The wavepacket is excited at t_0 by an unchirped laser pulse. The high vibrational energy and low vibrational energy parts are therefore excited at the same time. The vibrational spacing in the high-energy range ΔE_H is smaller than the vibrational spacing in the low-energy range ΔE_L. The succeeding oscillation times T_H are therefore longer than T_L, and the low-energy parts of the wavepacket advance the high-energy parts after some oscillations at a time $t_1 > t_0$. This is called *wavepacket dispersion*. The wavepacket is always plotted when moving from left to right. (b) The dispersion can be suppressed by starting the "slow" blue components earlier than the "fast" red ones, as indicated by the dashed line at t_0. At a time T_{opt}, the red ones will have caught up with the blue wavepacket components.

Considering this expression, it is immediately clear that an ensemble of classical trajectories with energy spread representing a classical wavepacket will broaden with time (*disperse*). The trajectories in the ensemble having a high vibrational energy E ("blue" ones) have a lower oscillation frequency than the low-energy ("red") parts. Therefore, the trajectories excited in the red will travel ahead of the trajectories excited in the blue spectral range. Figure 5.4a gives a pictorial interpretation of the dispersion with the help of quantum mechanics. In panel a, the energy-R distribution of a wavepacket excited by an unchirped laser pulse is shown. At excitation time t_0, the spectral components start at the same internuclear distance. As depicted in Figure 5.4a, the vibrational spacing ΔE is much larger in the low-energy range of the Morse oscillator than in the high-energy range ($\Delta E_L > \Delta E_H$). Therefore, parts of the wavepacket excited at low energies will oscillate faster than those excited at high energies (as in the classical trajectory analogue). Again, this leads to the spreading of the wavepacket at later times t_1. The wavepacket shows a sort of "tilt" at time t_1 in the representation chosen in Figure 5.4, indicated by the dashed line. The dispersion time T_{disp}, which is the time until the wavepacket has completely spread out on all possible values of R, fulfills the following equation (see appendix in Ref. [51] for a detailed calculation):

$$T_{\text{disp}} = v/\omega_e x_e \delta E \qquad (5.5)$$

where δE is the full spectral width of the wavepacket, and ν is the central vibrational frequency. For reasons of convenience, all units are given in cm^{-1}. In case of vanishing anharmonicity $\omega_e x_e = 0$ (harmonic oscillator), no dispersion occurs; thus, T_{disp} goes to infinity. The larger the anharmonicity, the faster the wavepacket broadens due to dispersion. The $1/\delta E$ proportionality in Eq. (5.5) can be explained via the extremes: in case of a monochromatic excitation, all trajectories have absolutely the same energy and will not show any broadening due to dispersion. In a broadband excitation, the oscillation frequency difference between the upper (blue) and lower (red) trajectories scales with the width δE of the excitation source, and the dispersion time decreases. It is possible, however, to compensate for the dispersion by special excitation pulses.

A compensation scheme is depicted in Figure 5.4b. As shown above, the red components of a wavepacket or classical ensemble oscillate faster than the blue ones. Now, one starts the slower oscillating blue components earlier than the red ones (see Figure 5.4b at t_0) by a specially tailored laser pulse. The instantaneous optical frequency of the pulse is described by a linear function in time. Such a pulse is negatively linear chirped. The red components will have caught up with the blue components after a time T_{opt}, which is called the *focusing time*.

A detailed analysis in Refs. [5, 25, 51] leads to

$$T_{opt} = -\frac{\beta' \nu^2}{4\pi \omega_e x_e} \tag{5.6}$$

where $\beta' = \frac{\beta(\nu)}{c}$ is the linear chirp parameter (fs cm), ν is the oscillation frequency at the wavepackets' center (cm^{-1}), and $\omega_e x_e$ are the respective molecular constants of the Morse oscillator [see Eq. (5.2)]. A short example is given here. A laser pulse has a chirp of $\beta' = -1.78$ fs cm. The molecular vibration ν at an excitation energy of 570 nm in the Br_2 B state is 118 cm^{-1}, and $\omega_e x_e = 1.6361$ cm^{-1}. This leads to a T_{opt} of 1.2 ps.

The focusing scheme was theoretically worked out in Refs. [25,79] and applied successfully to free I_2 molecules [78].

The dispersion of a wavepacket destroys the modulation contrast in a pump-probe spectrum, as can be seen in the simulated solid spectrum presented in Figure 5.5b. The simulation was carried out without including the decoherence induced by the crystalline environment. If the decoherence is of the same time order as the dispersion, one has to find a way how to disentangle the two processes, and wave packet focusing will prove to be an ideal tool for this purpose.

5.3.3 DISENTANGLING DISPERSION AND DECOHERENCE

The probe window in the simulation in Figure 5.5b is always located at the outer limb of the B state potential. The wavepacket narrowing for negative chirped excitation leads to the highest modulation contrast in a pump-probe spectrum at the focusing time T_{opt} (see Figure 5.5b, dashed line). If the wavepacket is located in the probe window R_{win}, then a pronounced maximum is visible; if it is located at the inner turning point, then the signal decays to zero. Comparing the two simulated

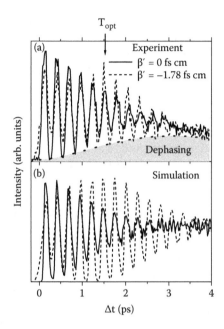

FIGURE 5.5 (a) Pump-probe spectra of Br_2 in Ar for chirps $\beta' = 0$ (solid) and -1.78 (dashed) fs cm excited at $\lambda_{pump} = 567$ nm and probed at 600 nm. The dashed gray shaded line gives the experimentally determined vibrational dephasing (decoherence) background. (b) Simulations using the same laser parameters for a free Br_2 molecule. (Reproduced from M. Gühr et al., *Phys. Chem. Chem. Phys.*, 6:5353–5361, 2004, by permission of the PCCP owner societies.)

pump-probe spectra for a free Br_2 gives the expected results. An unchirped excitation leads to a good modulation contrast in the beginning of the spectrum (see
Figure 5.5b, solid line), and the contrast is then decaying due to dispersion. The
spectrum excited with a negative chirp of $\beta' = -1.78$ fs cm begins with meager
modulation in the beginning and evolves to an optimal contrast at $T_{opt} \approx 1.4$ ps, as
predicted by Eq. (5.6). All eigenstates contributing to the wavepacket "remember"
their initial phase given by the negatively chirped laser pulse since no decoherence
is included in the calculation.

The experimental curves of the Br_2 in solid argon are given in Figure 5.5a
for unchirped excitation (solid line) and for negatively chirped excitation (dashed
line) with $\beta' = 1.78$ fs cm. The two curves show a general decay of the overall
signal due to vibrational relaxation of the wavepacket below the probe window.
This reduces the wavepacket detection efficiency. The solid curve for unchirped
excitation shows a loss of the modulation contrast, which is due to *dispersion and
decoherence* induced by the environment. To separate them, the wavepacket is
focused, thereby suppressing the dispersion at T_{opt}.

The qualitative behavior of the experiment (Br_2:Ar) is similar to the free Br_2 simulation. Even the focusing time T_{opt} has the correct value, which is not surprising since the Br_2:Ar RKR potential is similar to the gas-phase potential in the selected range (see Figure 5.3a). However, the experimental minima at T_{opt} for the dashed curve in Figure 5.3a are not at a signal level of zero, in contrast to the simulation. This background at T_{opt} is caused by vibrational decoherence (irreversible dephasing). Full focusing contrast only occurs if all eigenstates making up the vibrational wavepacket remember their initial phase. Vibrational decoherence destroys the phase relation and produces a diffuse background of statistical population distributed all over the possible R values, which gives rise to the background signal at T_{opt}.

We now change the focusing time T_{opt} by shifting the negative chirp parameter of the pump pulse. The different focusing times read from the experiment are shown as solid squares in Figure 5.6. The error bars reflect an uncertainty of one oscillation period T. The solid circles in Figure 5.6 represent the calculated values for T_{opt} according to Eq. (5.6). A good agreement is realized for theoretical values of free Br_2 and the data for Br_2:Ar.

The background contribution for all different T_{opt} is given as the gray shaded area in Figure 5.5a. This background cannot be suppressed by wavepacket focusing and reflects the decoherence of the vibrational wavepacket. Removing the energy relaxation and fitting the background by an exponential rise yields a vibrational decoherence (dephasing) time of $T_{deph}^{vib} = 3$ ps. Thus, no coherent control schemes or quantum interference phenomena will work beyond this time.

We now quantify the coherence properties of a distinct group of vibrational levels. Therefore, the concept of fractional revivals is used.

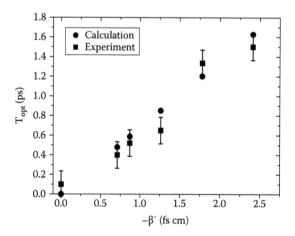

FIGURE 5.6 Focusing times T_{opt} versus negative chirp β'. The experimental values (solid squares) are compared with the ones calculated using $T_{opt} = -\frac{\beta' v^2}{4\pi \omega_e x_e}$ (solid circles). (Reproduced from M. Gühr et al., *Phys. Chem. Chem. Phys.*, 6:5353–5361, 2004, by permission of the PCCP owner societies.)

5.3.4 Fractional Revivals of Wavepacket in a Morse Oscillator

Up to now, the simulated pump-probe spectra reached only out to 4 ps (see Figure 5.5b). However, in case of full vibrational coherence, interesting features occur after 4 ps, as can be seen in Figure 5.7a. For example, after about 11 ps, the dispersed wavepacket re-forms, and the full modulation contrast is visible in the pump-probe spectrum. This phenomenon is called a *wavepacket revival*; the feature at 11 ps is called a *half revival* in the common terminology [2]. The theory of revivals has been established in theoretical literature [1–3, 84, 87, 109, 118, 125, 129, 130]. Revivals occur not only in Morse oscillators but also in all quantum state superpositions for which the potential is not harmonic. Accordingly, revivals have been found in vibrational [3, 62, 84, 87, 112, 125, 129, 130, 134], rotational [83, 89, 105, 111], and electronic Rydberg systems [104].

In a general theory [2, 118, 125], one expands energy eigenvalues E_n in a Taylor series around $E_{\bar{n}}$: $E_n = E_{\bar{n}} + \frac{\partial E}{\partial \bar{n}}(n - \bar{n}) + \frac{1}{2}\frac{\partial^2 E}{\partial \bar{n}^2}(n - \bar{n})^2$, neglecting the higher order. Setting $k = (n - \bar{n})$, one can write the wavepacket as

$$\Psi(R, t) = \sum_k |c_k|^2 \psi_k(R) \exp\left(-2\pi i \left(k\frac{t}{T} + k^2\frac{t}{T_{rev}}\right)\right) \quad (5.7)$$

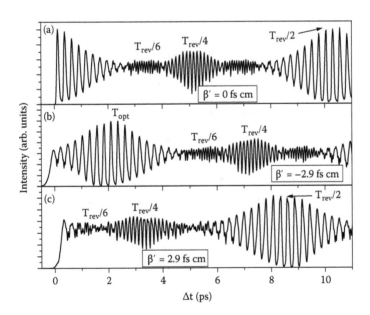

FIGURE 5.7 Simulated revival pattern of Br_2 in pump-probe spectra for different chirps. (a) Revival pattern for an unchirped pump pulse described in the text probed in the range of the outer turning point. (b) Same conditions as in (a) but with a *negative* excitation chirp of -2.9 fs cm. The revival structures are shifted *forward* in time. (c) Same conditions as (a) but with a *positive* excitation chirp of $+2.9$ fs cm. The revivals are shifted *backward* in time.

with the vibrational eigenfunctions $\psi_k(R)$, the local oscillation period $T = 1/\nu_{cl} = h(\frac{dE}{dn})^{-1}$, and the revival time $T_{rev} = 2h(\frac{d^2E}{dn^2})^{-1} = \frac{2}{\hbar}(\nu_{cl}\frac{d\nu_{cl}}{dE})^{-1} = \frac{2\pi}{\omega_e x_e}$ for the case of a Morse oscillator. For the B state of Br_2, the revival time T_{rev} is equal to 21.08 ps.

The appearance of the fundamental frequency at $T_{rev}/2$ and T_{rev} in the pump-probe spectrum is easy to prove. Therefore, we turn to the 1/4 and 1/6 fractional revival at $T_{rev}/4$ and $T_{rev}/6$ (see Figure 5.7a). Suppose the pump pulse is so long and thus narrow in frequency space that it couples only two adjacent vibrational levels n and $n+1$ with energy E_n and E_{n+1} coherently. The energy of level n in the Morse oscillator is given by

$$E_n = \hbar\omega_e(n + 1/2) - \hbar\omega_e x_e(n + 1/2) \tag{5.8}$$

The vibrational angular frequency follows from

$$(E_{n+1} - E_n)/\hbar = \omega_e(1 - 2nx_e - 2x_e) \tag{5.9}$$

and the time pattern is equivalent to a harmonic oscillator with this resonance frequency. Now, we couple three vibrational levels n, $n+1$, and $n+2$ by a shorter and broader pulse. We obtain a period T from $(E_{n+2} - E_{n+1})$, which is close to that from $(E_{n+1} - E_n)$. Furthermore, a doubled frequency or half fundamental period arises from $(E_{n+2} - E_n)$.

Figure 5.7a demonstrates that this new $T/2$ period dominates the time course around $T_{rev}/4$, and it leads to a beating pattern with enhanced amplitude.

With an even shorter pulse, we may couple four levels from n to $n+3$. Now, a third period $T/3$ from $(E_{n+3} - E_n)$ contributes besides T and $T/2$. This again causes a beating pattern, now at $T_{rev}/6 = 3.5$ ps, and the period $T/3$ dominates in this beating region (see Figure 5.7a). The pattern at $T_{rev}/6$ is bound to two conditions, which have to be fulfilled simultaneously. We define $2\pi/T$ in this special case as

$$\frac{2\pi}{T} = \frac{E_{n+2} - E_{n+1}}{\hbar} = \omega_e(1 - 2nx_e - 4x_e) \tag{5.10}$$

First, the phase differences of the periods from the successive energy differences have to acquire $(2\pi)/3$. With Eq. (5.10), this condition reads as

$$2\pi/3 = \left(\frac{2\pi}{T} - \frac{E_{n+3} - E_{n+2}}{\hbar}\right)t = \left(\frac{E_{n+1} - E_n}{\hbar} - \frac{2\pi}{T}\right)t = 2\omega_e x_e t \tag{5.11}$$

and leads to the appearance of the 1/6 revival at

$$t = 2\pi/6\omega_e x_e = T_{rev}/6 \tag{5.12}$$

Second, the fastest component from $(E_{n+3} - E_n)$ has to add up constructively with the other components. The energy difference

$$\frac{E_{n+3} - E_n}{\hbar} = 3\omega_e(1 - 2nx_e - 4x_e) = 3\frac{2\pi}{T} \tag{5.13}$$

indeed yields a constructive interference at $T_{rev}/6$, and a characteristic oscillation period of $T/3$ for the pump-probe spectra is observable. Higher fractional revivals can be explained in an analogous way, with more coupled Morse vibrational eigenstates.

A good pictorial representation of the wavepacket behavior during factional revivals can be given with the help of the Wigner representation $f_W(R, p)$. The Wigner function represents the quantum mechanical analogue of a phase space density with the nuclear distance R and momentum coordinate p:

$$f_W(R, p) = \frac{1}{2\pi\hbar} \int_{-\infty}^{\infty} e^{\frac{i}{\hbar} p(x-x')} \langle x'|\Psi\rangle \langle\Psi|x\rangle ds \qquad (5.14)$$

where $x = R + s/2$, and $x' = R - s/2$. Thus, $f_W(R, p)$ is the Fourier transform of $\langle x'|\Psi\rangle\langle\Psi|x\rangle = \Psi(x')\Psi^*(x)$ along the difference coordinate $s = x - x'$. The projections of the Wigner functions on R or p show the absolute square of the wavepacket on the respective coordinate. Nevertheless, $f_W(R, p)$ itself might become negative. This is in conflict with its assignment to a classical phase space probability distribution, which is discussed in the literature [125].

The Wigner function of a wavepacket in a harmonic potential without dissipation would travel on isoenergetic circles in phase space. In an anharmonic oscillator, it travels on egg-shape isoenergetic surfaces, as indicated by the white arrows in Figure 5.8a. The revival structure can be perfectly observed in the Wigner representation, as Figure 5.8 shows. In Figure 5.8a, the almost undispersed wavepacket is shown at the outer turning point of the anharmonic potential (the B state of free Br_2 is chosen here). In Figure 5.8b, the Wigner function at $T_{rev}/8$ is shown. One observes four subwave packets (A, B, C, D) that interfere with each other. The substructures are visible better in Fig. 5.8c and 5.8d, where the 1/6 and 1/4 revival are shown, respectively. Especially in Figure 5.8d, the doubling of the frequency for the 1/4 revival is easy to understand: two wavepackets exist that travel on isoenergetic lines in phase space. The interference structure in the middle shows their coherence.

The 1/6 fractional revival, consisting of three wavepackets and their interferences, indicates a four-level coherence, as shown above. Thus, if we could measure the 1/6 revival, we would know that at least four levels are coherently coupled. The 1/6 revival occurs at $T_{rev}/6 = 3.5$ ps (see Figures 5.7a and 5.8c), which is beyond the coherence range of 3 ps. Because the full vibrational coherence is needed for the development fractional revivals, the 1/6 revival will *not* be visible. However, there is a possibility to shift the revival forward into the coherence range between 0 and 3 ps, as demonstrated in the next section.

5.3.5   COHERENT CONTROL OF FRACTIONAL REVIVALS AND FOUR-LEVEL COHERENCE

The 1/6, 1/4, and 1/2 revivals are marked in Figure 5.7a at their appearance. The pump-probe spectrum simulation in Figure 5.7a was executed with unchirped excitation. By excitation with a negatively chirped pulse, as done in Figure 5.7b,

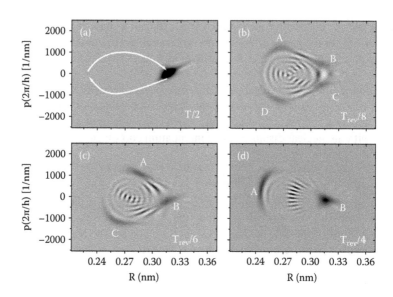

FIGURE 5.8 Wigner functions of fractional revivals; positive values are plotted dark, negative values light. Panel (a) shows the wavepacket $T/2 = 130$ fs after excitation. The wavepacket shows no interference structure. In the course of time, the packet broadens and splits into subwavepackets that travel on isoenergetic lines. (b) At $T_{rev}/8$, the splitting into four wavepackets with interferences among them is shown. The wavepacket maxima are denoted A, B, C, and D. (c) At $T_{rev}/6$, the splitting into three wavepackets A, B, and C with interferences can be observed. (d) Two substructures A and B with interference pattern in the middle can be observed at the quarter revival occurring at $T_{rev}/4$.

the vibrational wavepacket is focused at T_{opt}. The time zero of Figure 5.7a is now located at T_{opt} in Figure 5.7b, and all fractional revivals of order q are shifted to $T_{opt} + T_{rev}/q$. The classical interpretation of the phenomenon was given in context with Figure 5.4: the wavepacket tilting due to the molecular anharmonicity is eliminated at T_{opt}. An untilted wavepacket corresponds to time zero when exciting with an unchirped pump pulse.

Our aim now is to bring the revival structure forward in time. Clearly, a negatively chirped pump pulse does the opposite, as can be seen in Figure 5.7b. Therefore, a positively chirped pulse is tested. Indeed, the revival structures are shifted forward in time, as shown in Figure 5.7c. The classical interpretation of the shift is once more given by the wavepacket's tilt. This tilt (dashed line in Figure 5.9a) gets stronger with longer propagation time. Thus, one can spool forward in the history of the vibrational wavepacket by starting it with the appropriate tilt. A positively chirped pump starts the red components earlier than the blue wavepacket components and thus mimics the right tilt. The wavepacket revivals are now shifted forward in time. The focusing time T_{opt} is now negative by definition in Eq. (5.6). The chirp of the pump laser pulse was chosen to be $\beta' = 2.9$ fs cm, leading to a T_{opt} of -2.2 ps. The chirp elongates the pump pulse by a factor of ten from

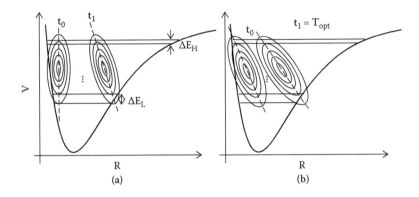

FIGURE 5.9 (a) Similar to Figure 5.4. The wavepacket is excited unchirped at t_0 and broadens due to dispersion. The wavepacket has evolved a "tilt" at time t_1. This kind of tilt can be accomplished at excitation time t_0 using a positively chirped pump pulse. This pulse excites the red components of the wavepacket earlier than the blue ones. In the course of time, the wavepacket develops an even stronger tilt.

about 30 to 300 fs. Nevertheless, well-structured revivals appear in the simulated spectrum. The 1/6 revival, which would normally occur at 3.5 ps, is now shifted to $(3.5 - 2.2)$ ps $= 1.3$ ps.

One can consider this scheme in terms of compensation and addition of molecular dispersion by laser pulse dispersion (chirp). The *positive dispersion* of the wavepacket on a molecular state can be compensated by a *negatively dispersed* laser pulse with negative chirp, which leads to focusing. In the same way, parts of the *positive dispersion* can also be transferred into the laser pulse (by positive chirping), and the molecular wavepacket starts with a dispersion corresponding to some later time in the regular propagation.

In the experiment, the width (full width at half maximum, FWHM) of the almost-Gaussian experimental pump pulse profile was 635 cm^{-1}, centered at 580 nm (17,240 cm^{-1}). The vibrational levels $v = 7, 8, 9$, and 10 of the B state were covered by the pulses' FWHM. The pulse duration was confirmed to be about $\Delta\tau = 300$ fs in a cross-correlation experiment with the probe pulse at 600 nm. The Fourier transform-limited duration of the 580 nm pump pulse was calculated to be $\Delta\tau_0 = 23$ fs. A positive chirp of $\beta' = 2.9$ fs cm was derived by its elongation and by the FROG technique. The positive chirp was accomplished by introducing plates of fused silica into the pump beam. The amount of chirp and the resulting pulse stretching given by the amount of material were calculated and compared with the experiment. A perfect agreement was found.

The result of an experiment performed with a positively chirped pulse is shown in Figure 5.10a. The spectrum shows modulations with $T/3$ of the fundamental period T. This tripled oscillation frequency is the clear signature of the 1/6 revival. The frequency is superimposed on the fundamental oscillation frequency. The T and $T/3$ contribution depend sensitively on the probe position, according to

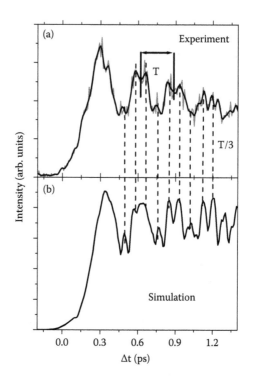

FIGURE 5.10 (a) Experimental pump-probe spectrum of Br_2:Ar excited by a pump pulse centered at 580 nm of width 635 cm^{-1} (thin line), smoothed by a bandpass filter (thick line). The excitation pulse had a chirp of 2.9 fs cm. The spectrum was probed at 600 nm via a two-photon transition. (b) Simulation of the spectrum using the free Br_2 B state potential (R_{win} = 0.322 nm, ΔR_{win} = 0.011nm). (Reproduced from M. Gühr et al., *Phys. Chem. Chem. Phys.*, 6:5353–5361, 2004, by permission of the PCCP owner societies.)

the simulations. For the simulation of the advanced 1/6 revival in Figure 5.10a, the probe window R_{win} was chosen to optimize reproduction of the experimental transient.

A precondition to observe a 1/6 revival structure with a threefold oscillation frequency is a coherent coupling of four vibrational levels. The $T/3$ modulation is visible to about 1.2 ps, indicating a four-level dephasing (decoherence) time of $T_{deph}^4 \approx 1.2$ ps.

5.3.6 ELECTRONIC DECOHERENCE

The positively chirped pulse is prolonged by a factor of ten to 300 fs and exceeds the molecular vibrational period of 265 fs. During the long pulse–molecule interaction, no fundamental molecular vibrational period is visible in Figure 5.10. The time evolution generates the interference structure around 1.2 ps observable as the periods T and $T/3$. This occurs, however, only if the imprinted phase of the vibrational levels follows the chirp. The phase of each vibrational level is given by the chirp parameter of the pulse. The exact timing of the vibrational levels predetermines the evolution of the spatial and temporal interference structures of the wavepacket. An electronic dephasing process shifts either the relative energy between electronic ground state and excited state (ΔE) or their relative internuclear position (ΔR) and in general both [54]. If such an event would occur during the excitation of the levels, then the phase sequence of the vibrational levels would be distorted. A change of ΔE and ΔR in a statistical way during the 300 fs pulse duration over the molecular ensemble probed (as usually meant by dephasing) would smear out the evolution of the interference structures.

Since we achieve the interference pattern of the full chirp in the experiment according to Figure 5.10, it is clear that in our example the electronic dephasing (or decoherence) time $T_{\text{deph}}^{\text{el}}$ is longer than the pulse duration of 300 fs. Thus, monitoring the evolution and persistence of the vibrational interference pattern versus increased chirp parameter β' (and thus pulse length) also provides a means to derive $T_{\text{deph}}^{\text{el}}$. Typically, $T_{\text{deph}}^{\text{el}}$ is expected to be the shortest of the dephasing times with values shorter than 100 fs in multidimensional systems.

5.3.7 COMPARISON WITH OTHER DEPHASING TIMES OF HALOGENS IN MATRICES

5.3.7.1 Vibrational Dephasing Times of Halogens in Matrices

The vibrational dephasing time of a free Br_2 molecule's B state in a cooled gas jet extends into the range of several hundred picoseconds and is mainly induced by vibration-rotation coupling [4, 134]. Ultrafast data for vibrational dephasing times of Br_2 in condensed phase other than those presented here are not available. A variety of experimental and theoretical data exist for the I_2 molecule in different hosts.

The vibrational dephasing in the ground state of I_2 in solid argon [18, 20, 69, 70] and krypton [68] was studied with CARS by the Apkarian group. For solid argon, a vibrational dephasing time around 10 ps was reported in Ref. [70]. The dephasing rate increases linearly with v for I_2:Kr. The corresponding dephasing time decreases from about 10 ps at $v = 4$ to 2.5 ps at $v = 14$ (see Figure 5 in Ref. [69]). The variation of the vibrational dephasing for I_2:Kr in the ground state with vibrational quantum number v and temperature was tested in Ref. [68]. The rate increases with the second and fourth power of v, and the times are 100 ps for $v = 1$ and 33 ps for $v = 6$ (see Figure 8 in [68]). The authors discuss the different I_2 behavior in solid Ar and Kr and state that it reflects the interaction potential of the ground state with the cage. The cage is tight in the case of Ar, and thus the molecule "feels" a very repulsive potential, whereas the cage is looser in Kr, and the molecule samples a flat potential "locally quadratic and quartic" [68]. The tighter Ar cage enhances the dephasing. At comparable temperatures and vibrational numbers, the dephasing rates for Ar are approximately one order of magnitude larger than in the Kr solid. The ground-state wavepacket studies of I_3^- in different liquids, performed by Ruhman's group yield much shorter dephasing times compared with I_2 in RGS. They lie in the 1-ps range [45, 136].

Pump-probe spectroscopy conducted in RGSs delivered dephasing times of several picoseconds for wavepackets on the B state of I_2 [5–8, 20, 50, 52, 124, 140, 141, 144, 145]. However, the dispersion was not separated from vibrational dephasing (as applied in Section 5.3.3), and therefore an exact determination of $T_{\mathrm{deph}}^{\mathrm{vib}}$ itself is not possible. Changing from the ground state X to the electronically excited B state increases the molecule–host interaction: the molecular bond lengths grow roughly about 10% in the B $\longleftarrow$ X transition. Furthermore, the halogen rare gas equilibrium distance depends on the molecular state [53, 99], as will be demonstrated later in the context of coherent phonons. Thus, the iodine atoms interact more strongly with the nearest cage atoms. Keeping that in mind, the vibrational dephasing time in the B state of Br_2:Ar of 3 ps determined here fits into the trend of the I_2:RGS ground-state results.

Apart from a change in the molecular state of I_2, the host local density was varied in a series of experiments by the Zewail group [11, 12, 30, 85, 86, 90, 135], motivating theoretical studies by Engel, Meier, and coworkers [36–38, 71, 91]. The authors determined vibrational energy relaxation times T_1 and vibrational dephasing times T_2 as a function of rare gas pressure. The behavior of T_2 (corresponding to $T_{\mathrm{deph}}^{\mathrm{vib}}$ in the nomenclature used here) with rising pressure is nontrivial. Two processes induce a dephasing of the vibrational levels: collisions and vibration-rotation coupling. The free I_2 molecular rotation is not blocked, in contrast to the model system described here. In the binary collision model, T_2 scales linearly with the collision time [102, 103], whereas the trend is reversed for the vibration-rotation dephasing [120]. The vibrational dephasing time scales from typically infinite values at 0 bar rare gas pressure to 1 ps at 2-kbar pressure (examples for He as buffer gas).

The results for Ar buffer gas [86] are now compared with those in the solid phase. The solid Ar environment used in our experiments has a number density of $27\ \mathrm{nm}^{-3}$ according to the mass density at 4 K of $1.771\ \mathrm{g/cm}^3$ and the mass number of 40. Extrapolating Zewail's experiment linearly to the solid Ar density yields a time constant shorter than 250 fs [86]. This value is one order of magnitude shorter than $T_{\mathrm{deph}}^{\mathrm{vib}} = 3$ ps for Br_2:Ar. The wavepacket excitation in the I_2:(gas Ar) case was accomplished near the potential minimum. In our experiment, the wavepacket was excited halfway between the minimum and the dissociation limit. Thus, the dephasing times of Br_2 in solid Ar near the minimum of the B state are expected to be even longer than 3 ps, and the difference to the high pressure buffer gas is more dramatic.

The effect of a highly symmetric environment on molecular processes has been proved before for the molecular predissociation. The predissociation rate of I_2 in buffer gases increases linearly with density [85, 90]. Even in liquid environments, this law is obeyed [115, 117, 145]. However, in the highly symmetric cages of the even denser rare gas matrices, the predissociation rate is reduced [50, 52].

The vibrational motion in the double-substitutional site is well guided by the environment, and the molecules move like a piston in a cylinder. At same number densities in high-pressure gas phase or liquids, the environment shows less order, and stronger collisions with nearest neighbors will occur more frequently, leading to higher vibrational dephasing times.

5.3.7.2 Electronic Dephasing Times of Halogens in Matrices

The electronic dephasing time $T_{deph}^{el} > 300$ fs (Section 5.3.6) is now compared with the linewidths in an excitation spectrum of Br_2:Ar from Ref. [22]. The width of the ZPL is 15 cm^{-1} at $v' = 10$ (excitation wavelength of 580 nm). Taking this as the FWHM of a Lorentz profile, a time constant of 350 fs is derived. The linewidth for the $v' = 1$ level of 6 cm^{-1} corresponds to a time constant of 900 fs. Besides electronic dephasing, vibrational dephasing and inhomogeneous broadening can contribute to the observed linewidth and its broadening with increasing v'. Thus, 350 fs are again a lower limit for the electronic dephasing time T_{deph}^{el}. This is consistent with our result for the electronic coherence.

Concerning ultrafast spectroscopic techniques, no experiments on the electronic dephasing of Br_2 in solvents are available. Recently, the method of phase-locked-pulse-pair (PLPP) spectroscopy [114,116] has been applied to Cl_2 in solid Ar. The experiments show that the electronic dephasing time in the B $\longleftarrow$ X transition is longer than 660 fs [42]. The widths of ZPLs in the excitation spectrum of Cl_2:Ar [21] are similar to those of Br_2:Ar.

The CARS spectra on I_2 in solid Ar and Kr do not show any coherences involving a propagation on the excited B state. Therefore, it was stated that the electronic coherence between the B and X states is shorter than one molecular period of 300 fs [18,68–70,121]. An excitation spectrum for I_2 in rare gas solids has never been published according to our knowledge.

In addition to the experiments mentioned, the electronic coherence of I_2 was determined in photon echo experiments [48,92]. They were carried out in gaseous rare gas environments by Dantus and coworkers [34]. Typical number densities in those experiments ranged up to 10^{-3}/nm^3 (i.e., four orders of magnitude lower than for the solid Ar density of 27/nm^3). The electronic dephasing time at such densities in Ar was determined to be about 40 ps, and the corresponding rate increases linearly with higher density. A linear extrapolation to the number density of solid Ar yields an electronic dephasing time of less than 2 fs. This is more than two orders of magnitude shorter than our result for bromine in solid Ar. The authors used a collision-based model to explain the trend in the low-density experiments. The electronic phase in the transition B $\longleftarrow$ X is disturbed when an Ar atom passes at a certain distance due to the Br_2:Ar pair potential difference in the B and X states. The interaction time in the collision is estimated from the Ar velocity (at room temperature) and an interaction length of only 1.2 nm (much larger than the Van der Waals radii). This ballistic model seems inappropriate for the solid Ar case. Many atoms are included in the 1.2-nm shell around the chromophore (the lattice constant is 0.35 nm in an Ar crystal); nevertheless, dephasing is slow. The qualitative difference to gases and liquids is that the atoms in the low-temperature solid are *ordered* and spatially *fixed*. Only a change of their position induced, for example, by the phonons, results in a variation of the molecular transition energy, which leads to electronic dephasing.

5.4 COHERENT PHONON DYNAMICS

Besides the intramolecular vibrational dynamics, another modulation is observed in pump-probe spectra of Br_2 in solid argon and for I_2 in solid krypton. Those oscillations are attributed to *coherent host dynamics* and have been omitted until now in the presentation of the ultrafast dynamics results.

The first peak in a typical experimental transient of Br_2:Ar (solid line in Figure 5.11a) shows a vibrational splitting. Afterward, the splitting vanishes in the second round-trip, and one can follow the vibrational wavepacket dynamics on the B state with a typical oscillation period of 340 fs. However, after about 3 ps, a new oscillatory pattern with a period of 500 fs can be observed. The modulation is very weak and hardly visible on the transient. To increase the modulation contrast, the spectrum is normalized on the 70-data point average, shown as the dashed line in Figure 5.11a. The oscillatory pattern is smeared out in this average. Dividing the spectrum by this average increases the oscillatory contrast at late delay times (Figure 5.11b). Indeed, the pattern after 3 ps becomes more pronounced. The normalization preserves the phase of the oscillation, as indicated by the solid vertical lines. The modulation of 500 fs is stable, however, the noise on the 500-fs

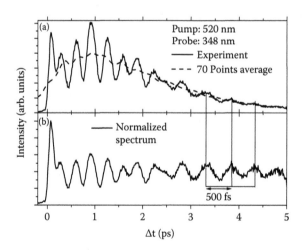

FIGURE 5.11 Pump-probe spectrum of Br_2:Ar with $\lambda_{pump} = 520$ nm and $\lambda_{probe} = 348$ nm. The solid line in (a) gives the experimental transient, while the dashed line was calculated by a 70-point average of the experiment. (b) The experimental spectrum normalized by the average from (a). A 500-fs oscillation sets in at about 2.5 ps.

modulation at delay times close to 5 ps increases due to the signal decay in the original data. The 500-fs modulation can also be observed with different pump and probe wavelength combinations in the B state [55] and in the A state [51].

In the system I_2:Kr, very similar coherent oscillations appear. Figure 5.12a shows a pump-probe spectrum of the I_2 :Kr B state wavepacket when pumping at $\lambda_{pump} = 530$ nm and probing at $\lambda_{probe} = 508$ nm (one-photon probe transition to the first charge-transfer manifold of I_2:Kr). As in the case of Br_2 in an argon crystal, two different oscillatory patterns occur: The dynamics from 0 to 4 ps shows an oscillation period around 420 fs. It is damped and hence vanishes at 4 ps. This period can be consistently interpreted as the trace of the B state intramolecular vibrational wavepacket in I_2:Kr. At 4 ps, a new pattern with a 650-fs period starts. To improve the modulation contrast, the spectrum is once more normalized to an average. The normalized spectrum is shown in Figure 5.12b. The 650-fs period behaves in a similar way as the 500-fs period in the Br_2:Ar case. It does not decay, and its phase is stable. This is illustrated by fitting the 650-fs modulation by

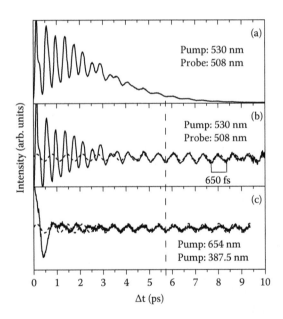

FIGURE 5.12 Pump-probe spectra of I_2:Kr. (a) The original experimental spectrum when exciting ($\lambda_{pump} = 530$ nm) and probing ($\lambda_{probe} = 508$ nm) the B state are shown. (b) The spectrum from (a) normalized to a 70-data point average. Besides the vibrational dynamics of the B state with periods around 420 fs, a stable modulation with 650-fs period becomes visible after 4 ps. The solid line in (c) shows the normalized A state pump-probe spectrum with $\lambda_{pump} = 654$ nm and $\lambda_{probe} = 387.5$ nm. After the initial A state wavepacket dynamics, again a 650-fs modulation sets in at 3 ps. The phase of the late dynamics is the same in (b) and (c) (dashed line). A cosine curve (dashed) is plotted in (b) and (c). (Reprinted with permission M. Gühr et al., *Phys. Rev. Lett.*, 91:085504, 2003. Copyright 2003 by American Physical Society.)

a single sinusoidal function (dashed curve in Figure 5.12b). As in the case of Br_2:Ar, the apparent noise level at long delays rises due to the normalization procedure.

The A state of I_2 in solid Kr can be excited at $\lambda_{pump} = 654$ nm without any B excitation. Figure 5.12c shows a normalized A state transient spectrum probed at $\lambda_{probe} = 387.5$ nm. Because the A state is excited above its gas-phase dissociation limit, it shows a typical long excursion of the vibrational wavepacket in the first picosecond. After 3 ps, this vibrational dynamics has decayed, and again a stable 650-fs modulation dominates the dynamics. As in the case of Br_2:Ar, the modulations shown in (b) and (c) have the same phase. The time zero for the B and A state transients in the case of I_2:Kr can be seen directly in different pump-probe spectra because the signals appear directly on positive and negative time delays. The negative time delays correspond to an exchange in the role of pump and probe pulses. The minimum between the first peaks of the respective spectra to positive and negative time delays corresponds to the time zero. The time zero of spectrum (c) in Figure 5.12 was determined by simultaneously measuring B and A states with $\lambda_{probe} = 387.5$ nm and polarization-selective techniques.

The 650-fs oscillation in I_2:Kr spectra and the 500-fs oscillation in Br_2:Ar spectra are attributed to host dynamics for the following reasons:

• The oscillation does not show any sign of dispersion expected from anharmonicity (see Section 5.3.2). The intramolecular states for I_2:Kr and Br_2:Ar are anharmonic oscillators with well-known anharmonicity; therefore, the dynamics cannot be assigned to any molecular state.

• The two frequencies for Br_2:Ar ($f_P = 2$ THz) and I_2:Kr ($f_P = 1.5$ THz) are quasi monochromatic, and the linewidth is only limited by the observation time window, as a further analysis shows [53, 55]. For all intramolecular wavepackets, energy relaxation occurs. Taking the anharmonicity of molecular electronic states into account, the vibrational relaxation will result in shortening of the wavepacket period in the course of the propagation. This once more excludes an intramolecular origin of the oscillations.

• The 650-fs (I_2:Kr) or 500-fs (Br_2:Ar) oscillation is observed when exciting at many different energies in the B state of the respective molecule (see Figs. 2 and 7 in Ref. [55]) and when exciting the electronic A state (see Figure 5.12). Thus, it is independent of the initial molecular electronic or vibrational state.

The 650- or 500-fs oscillation cannot be assigned to intramolecular vibrational wavepacket dynamics when considering the evidence given above. If the dynamics are not molecular vibrations, then the only origin can be a coherent vibration of the host that imprints its dynamics on the molecular pump-probe spectrum.

Now, three questions arise:

• What kind of host motion is visible in the molecular pump-probe spectrum?

- How can this host motion be excited?
- How does the host motion influence the molecular pump-probe spectrum?

5.4.1 HOST MOTION ASSIGNMENT

Figure 5.13 shows the dispersion relation for solid ^{84}Kr [122] (solid circles) and solid ^{36}Ar [41] (solid squares) on the left side. The data were obtained using the technique of neutron scattering in single crystals. The Ar isotope used in those experiments was ^{36}Ar because of its high neutron scattering cross-section. However, in our experiments Ar in natural abundance, consisting mostly of ^{40}Ar, was used. The phonon frequencies have to be shifted by $\sqrt{36/40}$ to account for the mass difference. The frequency of the ZBP for ^{40}Ar is marked by an arrow on the left side of Figure 5.13. Only the longitudinal parts of the < 100 > branch are shown. Comparison of the phonon dispersion relations with the Fourier transformations of the pump-probe spectra reveals that the coherent host peaks match the frequency of the phonons with reduced wave vector $k = 1$, which are called Zone Boundary

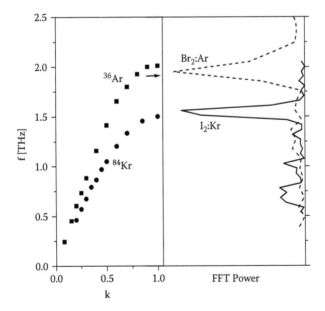

FIGURE 5.13 Left side: < 100 > branch of the phonon dispersion relation of argon (solid squares) and of krypton (solid circles). Only the frequencies of the longitudinal phonons are given. The arrow indicates the frequency of the zone boundary phonon of a ^{40}Ar host. Right side: The coherent host dynamics maxima of Br_2 :^{40}Ar (dashed line) and I_2:Kr (solid line). The lines match exactly the frequency of the zone boundary phonon in the dispersion relation given on the right side. (Reproduced from M. Gühr and N. Schwentner, *Phys. Chem. Chem. Phys.* 6:760–767, 2005. With permission of the PCCP owner societies.)

Phonons (ZBP). These have the shortest possible wavelength of all phonons. The ZBP are located at a frequency of 1.5 THz ($50\,cm^{-1}$) in the case of solid krypton and at about 2 THz ($67\,cm^{-1}$) for solid argon. Thus, the coherent dynamics observed in the pump-probe spectra can be attributed to coherent ZBPs of the host crystal, as documented for I_2:Kr in Ref. [53] and in Ref. [55] for Br_2:Ar. The same type of coherent ZBP trace manifesting in a 2-THz or 500-fs oscillation for solid argon was observed on Cl_2:Ar pump-probe spectra obtained by Fushitani et al. [43]. In addition, coherent ZBP oscillations were found on some pump-probe spectra for I_2 in solid Xe with a frequency of 1.25 THz, corresponding to a period of 800 fs [20].

How are the coherent phonons excited, and why do they show up in the pump-probe spectrum of the molecule? The ZBPs have a vanishing group velocity $v_g = d\omega/dk$ since the slope in the phonon dispersion relation is zero near the edge of the first Brillouin zone of the crystal. This fact is crucial in clarifying these questions.

5.4.2 Excitation Scheme for Coherent ZBP

To elaborate on an excitation scheme of the coherent ZBP, it is again useful to collect some of the experimental evidence. First, one observes for the cases of iodine in solid krypton and bromine in solid argon only one host-induced frequency. The experimental observations also exclude phonon dynamics that are forced by the vibrational motion of the molecule. In such a *forced-oscillation* scenario, the environment should show a mode resonant with the exciting motion. The intramolecular vibrational oscillation period T was changed over a large range from $T = 1$ ps for excitations near the dissociation limits of I_2 and Br_2 down to 250 fs for Br_2 and about 350 fs for I_2. Under all excitation conditions, a coherent ZBP motion with its typical frequency could be observed, and no beating phenomena occurred. Thus, a forced oscillation cannot account for a phonon excitation.

The phase of the phonon oscillations also deserves attention. The relative phase is stable when changing the excitation energy of the B state and when exciting the A state and comparing it with a B state excitation (see Figure 5.12). This overall phase stability implies a well-defined phase with respect to $t = 0$; otherwise, the oscillation would be averaged out. The defined phase at $t = 0$ calls for an *impulsive* excitation of the environment correlated to the optical molecular excitation.

In principle, one can think about two different mechanisms of impulsive excitation. Considering the large internuclear elongations in the first wavepacket, oscillation near the gas-phase dissociation limit might lead to the following assumption. The molecule hits the matrix environment very hard during the first vibrational excursion and is afterward essentially decoupled from the crystalline environment. This interpretation is rejected because the experiment shows coherent host dynamics also for excitation energies significantly below the gas-phase dissociation limit, and it is essentially independent of the vibrational energy of the electronic state. For example, in the case of I_2 in solid krypton, the coherent phonon oscillation can be observed at $\lambda_{pump} = 540$ nm, which is $1480\,cm^{-1}$ below the B state dissociation limit. At $\lambda_{pump} = 540$ nm, the energy relaxation is slow [6], and the molecule reaches its maximal elongation several times, which would lead to a sequence of cage collisions. This would again result in a forced oscillation scheme, which was previously excluded.

An alternative impulsive excitation scheme is proposed here based on the Displacive Excitation of Coherent Phonons (DECP). An excellent review of coherent phonons and, in particular DECP, was given by Dekorsy et al. [35]. The scheme was originally introduced for the excitation of zone center phonons in the case of semimetals and semiconductors [31, 32, 81, 146]. If those materials are irradiated by an ultrafast laser pulse, then an interband transition from bonding to antibonding orbitals occurs. The electronically excited system reaches an equilibrium on timescales much faster than the nuclear response times. Due to the change of forces between the nuclei, the system begins to oscillate around the new equilibrium geometry. Only A_1 modes maintaining the crystal symmetry were excited.

Such an impulsive excitation drives the system to a new equilibrium position, while initially at $t = 0$, the environment is still arranged in the electronic ground-state equilibrium position. According to this, the oscillation starts with the extreme amplitude and the atoms will oscillate around their new equilibrium position finally. Therefore, the DECP excitation results in a $\pm \cos(2\pi f_P t)$ characteristic of the coherently excited phonons. When extrapolating the phonon oscillation to the excitation time $t = 0$, the phonon amplitude has a maximum or minimum.

Now, the DECP scheme is applied to doped solid rare gases. Instead of changing the electronic orbital configuration in the solid host, we change the electronic orbital of a chromophore embedded in the host similar to the "bubble" mechanism proposed by Chergui for NO in different matrices [33, 64–66, 110, 131–133]. The molecules I_2 or Br_2 are initially in the $v = 0$ level of the electronic ground-state X $^1\Sigma_g$. The femtosecond pump pulse excites the molecule to a $^3\Pi_{\Omega u}$ state, such as A ($\Omega = 1$) or B ($\Omega = 1$). The internuclear distance of the molecule is not changed during the transition. Nevertheless, the electronic orbitals have changed since the molecule ends up in another electronic potential. The environment, however, is still in its equilibrium around the electronic ground state. Considering typical phonon periods of several hundred femtoseconds the excitation of a molecule by a 20- to 50-fs light pulse is indeed *impulsive* because it is much shorter than any phonon period. How does the solid-state environment react to the change of the molecular orbital in an A $\longleftarrow$ X or B $\longleftarrow$ X transition?

To answer this question, the potential of a single rare gas atom in the vicinity of the halogen molecule was calculated. The potential minima indicate the equilibrium distance of the rare gas atom from the molecule. The framework of the calculation is the so-called diatomics in molecules (DIM) formalism. It was originally designed to calculate the potential energy surfaces of triatomics [128] and is now extensively applied in doped rare gas environments with dopants such as I_2 [13, 14, 138], F_2 [95], HCl [46, 80, 96, 97], Cl_2 [46], B [72, 73], Hg_2 [47], and NO [93]. The DIM formalism is based on the pair potentials between all atoms.

The interaction potential between halogen and rare gas atoms is known from scattering experiments. For the calculations performed here, the scattering data from Ref. [27] for the I_2:Kr system and from Ref. [26] for the Br_2:Ar system were used. The interaction between the halogen atom and the rare gas atom is not isotropic because the halogen has a hole in a p orbital (equivalent to five electrons in the p orbital). The detailed description of the calculation can be found in Refs. [56] and [51].

Figure 5.14c shows the lines of vanishing force for the Kr atom in the vicinity of an I_2 molecule. The line of vanishing gradient for the B state (long dashed line)

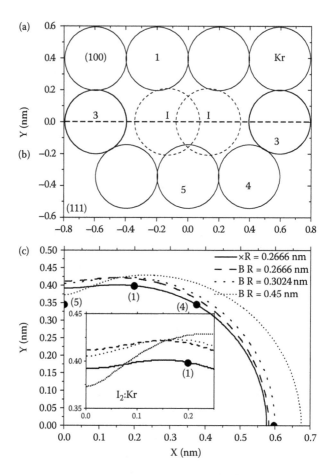

FIGURE 5.14 (a) (100) plane of the I_2:Kr system. (b) (111) plane of the I_2:Kr system. Rare gas atoms are separated into groups: 1, coherent phonon atoms; 3, head-on atoms; 4, window atoms; 5, belt atoms. (c) Equilibrium lines (local potential minima) of a Kr atom around the I_2 molecule for different electronic states and internuclear distances. Only one quadrant is shown for symmetry reasons. Solid line: electronic ground state X with $R_{e,X} = 0.2666$ nm; dashed line: excited state B with $R_{e,X} = 0.2666$ nm; dotted line: excited state B with $R_{e,B} = 0.3024$ nm; small dotted line: B state with large internuclear distance $R = 0.45$ nm. (d) Same as (c) but for Br_2:Ar system. (Panels (a) through (c) reprinted with permission from M. Gühr et al., *Phys. Rev. Lett.*, 91:085504, 2003. Copyright 2003 by American Physical Society. Panel (d) reproduced from M. Gühr and N. Schwentner, *Phys. Chem. Chem. Phys.* 6:760–767, 2005, by permission of the PCCP owner societies.

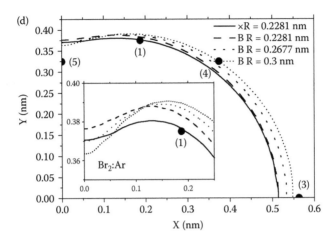

FIGURE 5.14 (Continued.)

lies outside that for the ground state (solid line). Thus, when exciting the molecule from its electronic ground state to the B state, the molecule expands in the "eyes" of a rare gas atom. It can be seen for I_2:Kr (Figure 5.14c) as well as for Br_2:Ar (Figure 5.14d).

In addition, one has to search for those atoms that are decoupled from the molecular vibrational motion since the coherent ZBP is not influenced by the molecular vibration. Figure 5.14a and 5.14b show the (100) and (111) planes of the fcc Kr crystal, respectively. The I_2 molecule on a double-substitutional site replaces two nearest-neighbor atoms. The atoms are separated into four groups.

For example, the head-on atoms denoted as 3 are sitting on the elongation of the molecular axis. They are visible in the (100) and (111) plane. The head-on atoms are supposed to strongly couple to the molecular vibration, as one can see in Figure 5.14c. As the molecule vibrates in the B state, internuclear elongations of 0.45 nm are easily reached. However, at this elongation the Kr atom centered at approximately 0.59 nm will be repelled by the interaction with the molecule, as the dotted line shows. This "tightness" in the $< 110 >$ direction gives rise to a strong vibrational energy relaxation. Furthermore, "shock waves" can be generated in this direction, traveling at ultrasonic speeds [23]. Thus, one can conclude that the head-on atoms are strongly coupled to the intramolecular dynamics and thus not relevant for the DECP process.

Another group denoted by 5 in Figure 5.14b is called *belt atoms* because four of them are sitting in the plane that cuts the molecular axis perpendicularly, forming a "belt" around the molecular "waist." As can be seen in the equilibrium lines in Figure 5.14c and 5.14d, those atoms are repelled from the molecule in the electronic transition B $\longleftarrow$ X. However, when the molecule is vibrating in the B state, the belt atoms are strongly coupled to that motion. As the molecular internuclear distance elongates, the belt atoms are attracted toward the molecular axis. As the internuclear distance shrinks in the next half cycle of a vibration, the belt atoms are repelled from the axis. In terms of volume conservation, the head-on atoms (3) increase their distance from the chromophore center, and the belt atoms in turn are pushed inward closer to the molecule. The mechanism has been observed in molecular dynamic simulations [23, 101, 140]. The long outward excursion followed by a slow inward motion of the chromophore atoms was attributed to the belt motion of the molecule (see Ref. [56]). Thus, we conclude that the belt atoms are strongly coupled to molecular vibration and therefore are not dominant in the coherent phonon we observe.

A third group of atoms is denoted by number 4. These are often called *window atoms* because four of them are forming a rectangular window on each side of the molecule. During the elongation of the molecular axis the halogen atoms pass the window through its center. The coupling to the molecular vibration is strong and comparable to the head-on atoms (3).

Finally, there is one group of atoms that is essentially decoupled from the molecular vibrational motion. This group is sitting in the (100) plane of the fcc lattice symmetrical to the molecular center and is denoted by 1. As the molecule undergoes the B $\longleftarrow$ X transition, this group of atoms experiences a force in the Y direction. The inset in Figures 5.14c for I_2 in Kr and 5.14d for Br_2 in solid Ar magnifies the region of interest. The Ar atoms (1) are repelled to a position further away from the molecular axis in the Y direction (Figure 5.14). This new position is very close to a point where all B state equilibrium lines for different internuclear distances of the chromophore intersect. The Kr or Ar atom of group 1 is essentially free of any force induced by the I_2 or Br_2 oscillation. Thus, the coherent phonon oscillation has to be attributed essentially to the group of atoms (1). The Ar atom (1) and its equivalents in the (100) plane seem to be further away from the zero-force line crossings than in the case of I_2:Kr. Therefore, one expects a stronger influence of the intermolecular Br_2 vibration on the coherent Ar crystal phonon than in the case of I_2:Kr. This might result in a faster dephasing of the Ar crystal coherent phonons compared with the Kr phonons.

The problem of the cage reaction on the molecular electronic transition is now reduced to one type of cage atom. This is an approximation since the influence of all other Ar atoms in the vicinity is neglected. Nevertheless, the calculation gives the right trend concerning the force imposed on a rare gas atom by molecular electronic excitation or vibration.

A detailed calculation of how the environment atoms move in the course of time is not yet available for I_2 and Br_2 in RGS. However, for Cl_2:Ar a calculation has been made [119], and the results have been published together with experimental data for this system [43]. They confirm the picture presented here.

In summary, the impulsive excitation of rare gas phonons was attributed to the expansion of the molecules' electronic cloud during the transition from the electronic ground state to an excited covalent state. Furthermore, we can find a group of rare gas atoms in the (100) plane decoupled from the intramolecular vibration. Since the electronic excitation is accomplished by a 20- to 50-fs light pulse, the phonons are excited impulsively. Most probably, many different phonon modes are excited in the vicinity of the molecule. In this context, it is important to realize the symmetry of the phonon excitation. The molecule might be simplified as a cylinder. The electronic expansion changes the radius of the cylinder, as can be seen in Figure 5.14. Thus, phonons are excited symmetrically in one direction with a wavevector **k** and in the other direction with a wavevector of −**k**. The total wavevector of the phonons cancels. Therefore, photons of a very small wavevector $k = 2\pi/\lambda$ of typical values around $1 \cdot 10^7$ m^{-1} in the visible light range can excite ZBPs of k around 10^{10} m^{-1}. The momentum conservation is not violated due to the counterpropagation of the phonons. The known examples for DECP processes usually lead to excitation of zone center phonons with a small wavevector (see, e.g., [44, 57, 58, 88, 123]). The ability to generate coherent ZBPs makes the molecular excitation method unique.

After phonons have been excited in the impulsive process, given by the electronic expansion of the molecule, the phonons travel away from the excitation source (molecule) according to their group velocity. The phonons form a kind of wavepacket that propagates and disperses because its constituents propagate with different velocities. The part of the acoustic wavepacket belonging to zone boundary-type phonons stays at the excitation source since, as stated above, $v_g = 0$ at the boundary of the Brillouin zone. Thus, the dispersion relation of the crystal provides a sort of a filter mechanism. Zone center phonons propagate with the velocity of sound, which is 1.64 nm/ps for longitudinal phonons in Ar and 1.375 nm/ps for longitudinal phonons in Kr (see, e.g., Ref. [74] or [40]). Thus, after about 500 fs the zone center phonons have reached the second solvation shell and leave the vicinity of the chromophore. If the phonons are probed later on, only the ZBPs remain and contribute to the coherent signal.

5.4.3 DETECTION SCHEME FOR COHERENT ZBPs

So far, the excitation and decoupling of the coherent ZBPs have been explained. The mechanism of the phonon–molecule interaction that allows the phonon to show up in the pump-probe spectrum of the chromophore now has to be answered. The coherent vibration of the lattice excited is usually detected by small reflectivity changes of an ultrafast (probe) pulse [44, 57, 58, 88], an absorption change [24, 28, 147], or diffraction of an ultrafast X-ray pulse [10, 123]. The rare gas crystals are optically inactive in the range of visible and ultraviolet light used in this study. The first absorption band of Kr lies at 10 eV and in Ar at 12 eV. Therefore, the phonon has to influence the internuclear transitions to show up in the pump-probe spectrum. This aspect is explained in this chapter via the solvation shift of the charge transfer (CT) states of the dopant. The observed pump-probe spectra consist of two components. One is attributed to the intramolecular vibration, the other one to the coherent phonon contribution. Figure 5.11 gives a good example for the case of Br$_2$:Ar; the pump-probe spectrum is dominated by the intramolecular vibrational motion up to 2 ps and afterward is modulated by the coherent ZBP, which shows up in the 500-fs period.

The intensity of the intramolecular contribution decays monotonically, and simultaneously its modulation contrast is increasingly washed out. Then, the decaying intramolecular signal is modulated by the phonon contribution. If this statement holds, the measured signal intensity I_S at late times should be the product of the time-averaged intramolecular contributions $< I_V >$ and the phonon modulation amplitude I_P:

$$I_S = < I_V > I_P \qquad (5.15)$$

To test this hypothesis, we calculate the structureless mean value $< I_V >$ of Figure 5.11a and divide the measured signal I_S in Figure 5.11a by $< I_V >$ to obtain Figure 5.11b. Indeed, a clear sinusoidal modulation is observed after 2 ps with well-defined frequency and essentially constant or weakly monotonically decaying amplitude. This result for I_P is a confirmation that the decomposition according to Eq. (5.15) catches the essence of the spectra at late times.

Furthermore it is now possible to work out this decomposition with respect to the region between 0 and 2 ps. In that domain, the structures in Figure 5.11b are still dominated by the intramolecular dynamics. The key to distinguish the contributions is a variation of the probe photon energy.

The wavelength $\lambda_{probe} = 348$ nm in Figure 5.11 was chosen to emphasize the intramolecular as well as the phonon contribution, a variation of λ_{probe} now allows for enhancement of either the intramolecular or the phonon contribution. This effect is used to identify the detection scheme for the phonon contribution. Probing deeper in the B state with a shorter probe wavelength ($\lambda_{probe} = 345$ nm) for the same $\lambda_{pump} = 520$ nm yields the normalized spectrum $I_S^{345} / < I_S^{345} >$, which contains essentially only the intramolecular period of 310 fs, and the modulation amplitude decays smoothly around 3 ps to the mean value (Figure 5.15a). Probing higher with a probe wavelength $\lambda_{probe} = 354$ nm leads to $I_S^{354} / < I_S^{354} >$ with the phonon period of 500 fs between 2 and 4 ps, and at shorter times, the oscillations are rather complex (see Figure 5.15b). The reasons for the complex behavior are beatings between the intramolecular and the phonon periods. Thus the more general decomposition scheme for all times is given by

$$I_S = I_V(1 + \alpha I_P) \qquad (5.16)$$

The weight factor α for the strength of the phonon modulation varies with λ_{probe} and can be exploited to separate out I_V as well as I_P. Figure 5.15a shows that $\alpha = 0$ for $\lambda_{probe} = 345$ nm, and this spectrum is close to the intramolecular dynamics I_V. Obviously, α is large for $\lambda_{probe} = 354$ nm (Figure 5.15b). Thus, dividing the spectrum of Figure 5.15b ($I_S^{354} / < I_S^{354} >= I_S$) by the spectrum of Figure 5.15a ($I_S^{345} / < I_S^{345} >= I_V$) should yield

$$\frac{I_S^{354}}{I_S^{345}} \frac{< I_S^{345} >}{< I_S^{354} >} = C(1 + \alpha_{354} I_P) \qquad (5.17)$$

with a structureless $C = < I_S^{345} > < I_S^{354} >$. The ratios are given in Figure 5.15c, and its modulation is expected to originate from the phonon contribution I_P. Indeed, we obtain now a single sinusoidal modulation with the phonon period of

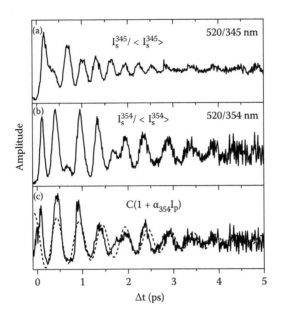

FIGURE 5.15 (a) and (b) Normalized pump-probe spectra of Br_2 in solid Ar with $\lambda_{pump} = 520$ nm and $\lambda_{probe} = 345$ and 354 nm, respectively. (c) Solid line: spectrum calculated by dividing spectrum (b) by (a) (see text). Dashed line: 2-THz oscillation with exponential decay. (Reproduced from M. Gühr and N. Schwentner, *Phys. Chem. Chem. Phys.* 6:760–767, 2005 by permission of the PCCP owner societies.)

500 fs and smoothly decaying amplitude (as soon as both intensities are above noise, i.e., from 100 fs on). The possibility to exclusively extract the phonon part, especially in the regions dominated by the intramolecular dynamics, clearly demonstrates the validity of the decomposition scheme. The scheme clarifies that optimization of the phonon contribution has to compromise between an I_V signal significantly above noise and a large α. The $\lambda_{probe} = 354$ nm favors a large α; however, I_V decays rather quickly. This is due to a probe window lying high with respect to the relaxing B population. Therefore, the signal in Figure 5.15b is lost in the noise beyond 3.5 ps. $\lambda_{probe} = 348$ nm provides a moderate α with a pronounced intramolecular region; however, the population in the phonon-modulated region remains longer close to the probe window, and the phonon contribution is visible up to 5 ps in Figure 5.15b.

The phonon contribution in Figure 5.11a has been shown to appear as a modulation of the mean B state signal intensity according to Eq. (5.15), and it becomes visible at long delay times. At those times, the intramolecular modulation vanishes due to dispersion and dephasing in combination with energy relaxation. A smoothly decaying spectrum develops, and in the wavepacket's center of gravity relaxes below the energy at which it is probed in the B state (E_{win}). Thus, only the high-energy wing remains visible. The probe distance R_{win} was defined as the

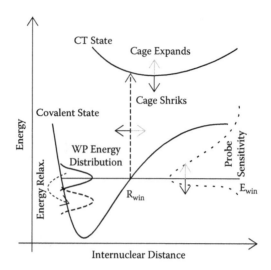

FIGURE 5.16 Phonon detection mechanism. The phonon modulates the charge transfer (CT) state solvation energy. Thus, the probe resonance (R_{win}, E_{win}) and the schematic sensitivity curve (dashed, right-hand side) shift accordingly, leading to a changing overlap of the probe sensitivity with the relaxing vibrational wavepacket (WP) in the covalent state. (Reproduced from M. Gühr and N. Schwentner, *Phys. Chem. Chem. Phys.* 6:760–767, 2005 by permission of the PCCP owner societies.)

internuclear distance, at which the wavepacket on the B state is resonantly probed to the charge transfer state. Suppose, some process periodically shifts the energy of the charge transfer states up and down, as indicated by the arrows from the charge transfer state in Figure 5.16. As a consequence, R_{win} would shift right- and leftward. Thus, the probe window energy E_{win} would shift up and down. What are the consequences of the probe window shift on the pump-probe spectrum? To answer this question, one must take a closer look at the probe process. The wavepacket is probed not only at energies corresponding to E_{win}, but also above and below as indicated by the dotted sensitivity curve in Figure 5.16. This is due to the broad Franck–Condon overlap of B and the charge transfer state and the finite energetic width of the probe pulse. Above E_{win}, the sensitivity can be expressed classically and is proportional to $1/\sqrt{\Delta E}$, where ΔE is the energy difference to E_{win} [143]. Below E_{win}, the sensitivity decays rather quickly. Energy conservation would give a sharp cutoff in the classical picture, which is reflected in the quantum mechanical picture by a steeply decaying Franck–Condon factor.

The pump-probe signal is proportional to the overlap of the probe sensitivity curve with the vibrational wavepacket. If the charge transfer state shifts a bit, E_{win} and the whole sensitivity curve shift by a small amount. The overlap does not change much for a wavepacket located on the maximum of the sensitivity curve because the probe sensitivity versus energy is flat. In contrast, when the wavepacket is located below E_{win} near the edge of the sensitivity curve, the overlap changes quite dramatically with a variation of energy due to the steep decay in this range. The wavepacket can be prepared in this range (compare Figure 5.15b), or its energy can relax to this region with time (Figure 5.15a). If the wavepacket is prepared significantly above E_{win}, then a periodic shift of the sensitivity curve causes a weak modulation of the pump-probe spectrum since the $1/\sqrt{\Delta E}$ decay is smooth. Only after energy relaxation has transferred it below E_{win} can the modulation be observed with high sensitivity on the pump-probe spectrum. The effect was demonstrated for the B state of Br_2. The explanation for the A state is analogous because the experimental features are similar to those in the B state.

Now, the source for the shift of the probe window is discussed. The solvation energy of the charge transfer state with 3300 cm^{-1} amounts to 10% of the transition energy and is huge compared with that of the covalent B state. A local compression of the lattice around Br_2 is going to increase this solvation energy of the charge transfer state by the shrinking of cavity diameter d in the Onsager model. The model allows estimatation of the energy shift ΔE of the dipole μ sitting in a spherical cavity with diameter d in a polarizable host with the dielectric constant ϵ:

$$\Delta E = \frac{1}{4\pi\epsilon_0}\frac{\mu^2}{d^3}\frac{8(\epsilon - 1)}{2\epsilon + 1} \qquad (5.18)$$

This equation was proposed in Ref. [98] and extensively applied for charge transfer states in RGSs in [59]. Stronger solvation pushes the charge transfer states downward in energy, shifts the probe position R_{win} inward and E_{win} down in energy to accommodate for the fixed probe photon energy. Coherent oscillation of matrix atoms in the Br_2 vicinity generates such density modulations that can be decoded from the B state pump-probe signal.

With this scenario in mind, one can also interpret the behavior at early times in Figure 5.15 when the B state dynamics causes strong fluctuations in the signal. The signal intensity I_S scales with the fraction of population of the wavepacket that just passes the probe window [i.e., I_V in Eq. (5.16)]. In addition, it can be enhanced by a fraction αI_P due to a local compression if this increases the sensitivity. This explains why α in Eq. (5.16) is probe wavelength dependent and in general even time dependent. In Figure 5.15a, the probe energy is located deep in the potential well due to a short probe wavelength. Therefore, a major part of the B state wavepacket is accessible all the time, and an increased sensitivity by compression has a marginal effect on the sensitivity $\alpha = 0$. The wavepacket is barely reached at its high-energy wing in Figure 5.15b with the steep part of the sensitivity curve from the very beginning due to the long λ_{probe}. Thus, α is very high; a shift of the probe window by compression strongly enhances the signal and the phonon modulation becomes visible. However, soon the wavepacket will be lost for detection due to the additional energy relaxation, and the B state signal disappears in the noise. For the mean λ_{probe} of Figure 5.11a, a major part of the wavepacket is caught at early times, leading to a small α, while after significant energy relaxation, α starts to rise, and a long-standing coherent phonon signal can be deduced from the surviving B state signal.

So far, the argumentation was carried out for the case of Br_2 molecules in solid argon. The explanation for I_2 in solid krypton is similar. Figure 5.12 presents pump-probe spectra of the A and B state. As in the case of Br_2, the intramolecular vibrational motion shows up first, followed by a phonon-modulated background. The phonon modulation is stable in frequency and amplitude, thus favoring a decomposition given in Eq. (5.16). In the case of I_2:Kr, the model of the oscillating cage and the periodically shifting charge transfer states can be even quantified [53]. The energy relaxation of the iodine B state in solid Kr can be determined to a high accuracy, as was demonstrated in Refs. [6] and [8]. When looking at Figure 5.12a, a history of the wavepacket energy as a function of time can be generated by simulating the energy loss from the initial energy at $t = 0$ and the energy relaxation rates dE/dt. The oscillation around the decaying integrated signal can be reproduced in the simulation when setting the phonon-induced shift of the charge transfer states to $40\ cm^{-1}$. This is indeed very small compared with the overall matrix shift of about $3100\ cm^{-1}$.

When attributing the effect exclusively to the change of the cage diameter in the Onsager model [Eq. (5.18)], a diameter change of 0.002 nm will occur. This is a very small amplitude of the ZBP located at the chromophore. It is only 0.5% of the whole cage diameter d of the two nearest-neighbor distances. Thus, one can conclude that the coherent ZBP amplitude is very small. Perhaps this is one of the preconditions for it to stay coherent on a timescale of several picoseconds.

Concerning the coherence lifetime of the ZBP, no other experimental data are available. The spectral resolution in neutron scattering did not allow for a prediction of the lifetime [41, 122].

5.4.4 COMPARISON WITH ALTERNATIVE MODELS

5.4.4.1 Comparison with Previous I_2:Kr and Cl_2:Ar Experiments and Simulations

The observation of coherent host motions in molecular pump-probe spectroscopy is an interesting feature. Furthermore, it is very important for the problem of electronic state coupling. A prominent example is the predissociation of the I_2 electronic B state. Predissociation is enforced by the collision with rare gas atoms by orders of magnitudes in comparison with the free molecule. It is, however, very much reduced in rare gas matrices [52, 144] compared with high-pressure gas environments of similar density [12, 85, 90] and liquids [30, 115, 117, 145]. The reason lies in the symmetry dependence of the coupling matrix elements [13, 14, 145]. Due to the high symmetry of the rare gas crystal, many contributions of the coupling cancel. When exciting host motions, some cancellation effects may be reduced, and the coupling can be strengthened.

The group of Apkarian invested a lot of effort to clarify the motion of the cage atoms in the vicinity of the I_2 chromophore in different environments. The first experimental evidence for what we identify now as a coherent ZBP [53, 55] was published for I_2:Kr in Ref. [140]. The experimental data shown there were produced by one-color pump-probe spectroscopy of I_2 molecules in solid Kr. It was stated that the observed 650-fs oscillation must be attributed to a modulation of the B state population in the probe transition to the charge transfer states in accordance with the interpretation given above. The modulation of the charge transfer states was attributed to a local mode of the host, and the host motion should be resonantly driven by the vibrational motion of the molecule. At the specific pump wavelengths used in Ref. [140], a 2:1 resonance of the I_2 intramolecular vibration period and the 650-fs coherent host vibration period was found. The host motion was assigned to the belt atoms, denoted as number 5 in our nomenclature (see Figure 5.14b). Those belt atoms are supposed to be strongly coupled to the intramolecular vibration. The simulation of the local belt mode resonantly driven by the molecule, however, did not show any coherence (see Figure 8 in [140]). The contradiction of coherence in the experiment and its lack in the simulation could not be resolved. However, an impulsive creation (either electronic or vibronic in the first excursion of the intramolecular wavepacket) was excluded [140]: "It would be difficult to imagine that such a collective coherence could last for ca. 8 ps if it were impulsively created" (p. 265)

In a following simulation on Cl_2:Ar [101], the role of the host motions in pump-probe spectra was further examined. Normal modes of the cage–molecule system were calculated. The change of the Cl_2–Ar interaction potential when undergoing the B ⟵ X transition was completely neglected. Thus, the possibility of impulsive excitation of phonons proposed in Section 5.4.2 was not contained in the simulation.

With the first results of CARS on the electronic ground state of I_2, a new tool to monitor host motion was at hand [18, 69, 70, 142]. Long-lasting coherent vibrations of the host–molecule system could be found. As an example, a coherent mode at 41.5 cm^{-1} was observed for 100 ps. For this mode, an impulsive excitation mechanism similar to the one presented in [53] was proposed. In CARS, it has to occur between the pump and Stokes pulse. In a normal-mode analysis of the I_2(X):Ar system, such a mode with energy 41.5 cm^{-1} did not appear.

Clear evidence for an impulsive excitation of the 650-fs host motion was presented [53]. The excitation quantum energy (and thus the vibrational period) was varied over a large range. For instance, the I_2 molecule was excited not only to its electronic B state but also to the A state close to the free molecule dissociation limit (see Figure 5.12). In that case, the vibrational frequency was about 500 fs; thus, a 2:1 resonance of the pure molecular vibration and the host vibration of 650 fs was clearly ruled out.

The impulsive excitation mechanism was nevertheless not accepted. An alternative to the impulsive excitation mechanism was suggested [20] and is briefly explained here. The molecule is initially excited to the B state that is assumed to predissociate. This predissociation has to occur a well-defined 700 fs after the wavepacket excitation. The dissociative states in which some population ends up after the predissociation are bound by the cage for steric reasons. Thus, they are sensitive to motions of the cage atoms. Their equilibrium distance is much larger than the covalent equilibrium distances. The normal mode of the molecule–cage system with the molecule sitting in cage-bound states modulates the probe transition by changing the shape and absolute energy of the cage-bound state. The cage-bound states have to be probed by a two-photon transition to the charge-transfer states, used in the laser induced fluore scence detection. A pump-probe spectrum consists of the cage-bound state spectrum *added* on the B state pump-probe spectrum in this picture.

In Ref. [55], clear evidence is given to reject the mechanism explained above. The arguments are next summarized briefly.

First, the coherent host motion contribution (coherent ZBP in our interpretation) is proportional to the B or A state decaying signal (see Figure 5.12 for I_2:Kr and Figure 5.11 for Br_2:Ar). It would be a nonplausible coincidence if the amplitude of cage-bound state dynamics would scale with the amount of B or A population that is left in the probe window.

Second, we have shown that our signals, especially at short delays, have to be explained by a time-correlated connection of phonon and intramolecular dynamics [Eq. (5.16)], which clearly contradicts the additive mechanism proposed in Ref. [20]. Returning to Figure 5.15, one can clearly observe the increase or suppression of single vibrational peaks in Fig. 5.15a due to the phonon influence presented in Fig. 5.15b. The shoulder of the first vibrational peak in Fig. 5.15a is strongly enhanced due to the higher density of the cage induced by the ZBP, whereas the second vibrational peak from Fig. 5.15a is suppressed due to less cage density induced by the ZBP. This behavior cannot be explained by adding two signals but is clearly a sign of the ZBP modulation on a vibrational spectrum. The argument was also given for the case of I_2:Kr [55].

Third, the coupled molecule–cage normal mode depends on the properties of the rare gas *and* the molecule. In contrast, it was observed for I_2:Kr [53], Br_2:Ar [55], and recently also Cl_2:Ar [43] that the period of the coherent host dynamics exclusively depends on the rare gas crystal. This is supported by pump-probe spectra of I_2 in solid Xe (by the Apkarian group), where a coherent host oscillation period of 800 fs (the ZBP period) was found (see Figure 14 in [20]). Thus, the assignment to cage-bound state signals has to be rejected, and indeed all coherent host oscillations show the ZBP period of the rare gas crystal used.

The coherent ZBP frequency seems to be exclusively connected to the detection of the charge transfer state LIF in pump-probe experiments. If one monitors the LIF of I*I* states (in the infared range [16, 67, 145]), another type of coherent host dynamics can be observed [20]. Probably, the I*I* states are more sensitive to other types of cage motion than the charge transfer states.

Finally, the spatial extension of the coherent ZBP has to be addressed. The excitation mechanism proposed in Section 5.4.2 is highly local. The phonon is most likely confined to the vicinity of the chromophore. It forms a wavepacket with certain width Δk near the zone boundary. The flat dispersion curve allows for its high spectral purity and its very slow dispersion. The ZBP is the most local of all phonons. It has the shortest possible wavelength. Thus, a few atoms are sufficient to carry the zone boundary character of the phonon.

5.4.4.2 Problems in Calculations of Coherent ZBP

Longitudinal zone center phonons propagate with a group velocity v_g of 1.64 nm/ps in Ar and 1.38 nm/ps in Kr. In addition, molecular excitation in the matrix can result in supersonic sound waves or shock waves traveling with up to 3 nm/ps in solid Ar [23,65]. This velocity corresponds to 5.5 lattice constants a ($a = 0.53$ nm) propagation in 1 ps. Furthermore, the shock waves travel along favored directions in the rare gas crystal [23].

In numerical simulations, an fcc cell of side length $M \cdot a$ contains $4 * M^3$ atoms. For $M = 5$, the cell includes 500 atoms. In trajectory simulations, the chromophore is embedded in the middle of a rare gas cell. Now, the choice of boundary conditions is crucial. Shock waves excited by the chromophore at $t = 0$ (e.g., by the expansion of the electron cloud) are reflected at the cell walls and travel back to the excitation source. Taking the velocity given above into account, the fastest supersonic waves reach the molecule again after passing $5 \cdot a$, which corresponds to about 1 ps in solid Ar. Thus, such artificially reflected contributions can destroy the small coherent amplitudes of cage atoms in the vicinity of the chromophore in the simulation.

The trajectories of I_2 in rare gas matrices have been calculated so far in cells of 108 atoms ($M = 3$) [13, 14] and 500 atoms ($M = 5$) [139] with the DIM approach and periodic boundary conditions. Increasing the cage size is difficult with up-to-date computing power. Even for 500 atoms, only 16 trajectories could be calculated in a reasonable time in Ref. [139]. No signs of coherent ZBP were found in theory so far, which is comprehensible with the underlying cell sizes.

The only realistic solution for simulations would be the application of absorbing boundary conditions to effectively remove the high-velocity matrix excitations from the simulation cell.

ACKNOWLEDGMENT

First, I would like to thank Prof. N. Schwentner for continuous support. He contributed crucially with his ideas and critical questions to all results presented, and the discussions with him were a pleasure and in every respect fruitful. Furthermore, I acknowledge with pleasure Dr. M. Bargheer as a great colleague and discussion partner during and after his doctoral work at the Freie Universität Berlin. H. Ibrahim participated in the wavepacket-focusing experiments, and Dr. M. Fushitani did beautiful experiments on Cl_2 in RGS. I would like to thank both for their great help. I would like to thank Prof. J. Manz (Freie Universität Berlin) for many insightful discussions about time-dependent quantum mechanics. Professor D. Tannor (Weizmann Institute) is acknowledged for permission to use and modify his wavepacket propagation code and the great opportunity to discuss results with him. Finally, I would like to thank all members of Prof. Schwentner's group and the technical staff at the Institut für Experimentalphysik, Freie Universität Berlin.

The project was funded by the Deutsche Forschungsgemeinschaft in the framework of the Sfb 450 "Analysis and Control of Ultrafast Photoinduced Reactions."

REFERENCES

1. I. Sh. Averbukh and N. F. Perelman. Fractional regenerations of wavepackets in the course of long term evolution of highly excited quantum systems. *Sov. Phys. JETP*, 69:464–469, 1989.
2. I. Sh. Averbukh and N. F. Perelman. Fractional revivials: iniversality in the long-term evolution of quantum wavepacket s beyond the correspondence principle dynamics. *Phys. Lett. A*, 139:449–453, 1989.
3. I. Sh. Averbukh and N. F. Perelman. The dynamics of wavepacket s of highly-excited states of atoms and molecules. *Sov. Phys. Usp.*, 34:572–591, 1991.
4. I. Sh. Averbukh, M. J. J. Vrakking, D. M. Villeneuve, and A. Stolow. Wavepacket isotope separation. *Phys. Rev. Lett.*, 77:3518–3521, 1996.
5. M. Bargheer, P. Dietrich, K. Donovang, and N. Schwentner. Extraction of potentials and dynamics from condensed phase pump-probe spectra: application to I_2 in Kr matrices. *J. Chem. Phys.*, 111:8556–8564, 1999.
6. M. Bargheer, M. Gühr, P. Dietrich, and N. Schwentner. Femtosecond spectroscopy of fragment-cage dynamics: I_2 in Kr. *Phys. Chem. Chem. Phys.*, 4:78–81, 2002.
7. M. Bargheer, M. Gühr, and N. Schwentner. Depolarization as a probe for ultrafast reorientation of diatomics in condensed phase: ClF vs I_2 in rare gas solids. *J. Chem. Phys.*, 117:5–8, 2002.
8. M. Bargheer, M. Gühr, and N. Schwentner. Collisions transfer coherence. *Israel J. Chem.*, 44:9–17, 2004.
9. M. Bargheer, J. Pietzner, P. Dietrich, and N. Schwentner. Ultrafast laser control of iononic-bond formation: ClF in Ar solids. *J. Chem. Phys.*, 115:9827–9833, 2001.
10. M. Bargheer, N. Zhavoronkov, Y. Gritsai, J. C. Woo, D. S. Kim, M. Woerner, and T. Elsaesser. Coherent atomic motions in a nanostructure studied by femtosecond X-ray diffraction. *Science*, 306:1771–1773, 2004.
11. J. S. Baskin, M. Chachisvilis, M. Gupta, and A. H. Zewail. Femtosecond dynamics of solvation: microscopic friction and coherent motion in dense fluids. *J. Phys. Chem. A*, 102:4158–4171, 1998.
12. J. S. Baskin, M. Gupta, M. Chachisvilis, and A. H. Zewail. Femtosecond dynamics of microscopic friction: nature of coherent versus diffusive motion from gas to liquid density. *Chem. Phys. Lett.*, 275:437–444, 1997.
13. V. S. Batista and D. F. Coker. Nonadiabatic molecular dynamics simulation of photodissociation and geminate recombination of I_2 in liquid xenon. *J. Chem. Phys.*, 105:4033–4054, 1996.
14. V. S. Batista and D. F. Coker. Nonadiabatic molecular dynamics simulation of ultrafast pump-probe experiments on I_2 in solid rare gases. *J. Chem. Phys.*, 106:6923–6941, 1997.
15. T. Baumert, V. Engel, C. Röttgermann, W. T. Strunz, and G. Gerber. Femtosecond pump-probe study of the spreading and recurrence of a vibrational wavepacket in Na_2. *Chem. Phys. Lett.*, 191:639–644, 1992.
16. A. V. Benderskii, R. Zadoyan, and V. A. Apkarian. Caged spin-orbit excited $I^{*2}P_{1/2}+I^{*2}P_{1/2}$ atom pairs in liquids and in cryogenic matrices: spectroscopy and dipolar quenching. *J. Chem. Phys.*, 107:8437–8445, 1997.
17. Z. Bihary, R. B. Gerber, and V. A. Aparian. Vibrational self-consistent field approach to anharmonic spectroscopy of molecules in solids: application to iodine in argon matrix. *J. Chem. Phys.*, 115:2695–2701, 2001.

18. Z. Bihary, M. Karavitis, and V. A. Apkarian. Onset of decoherence: six-wave mixing measurements of vibrational decoherence on the excited state of I_2 in solid argon. *J. Chem. Phys.*, 120:8144–8156, 2004.

19. Z. Bihary, M. Karavitis, R. B. Gerber, and V. A. Apkarian. Spectral inhomogeneity induced by vacancies and thermal phonons and associated pbservables in time- and in frequency-domain nonlinear spectroscopy: I_2 isolated in matrix argon. *J. Chem. Phys.*, 115:8006–8013, 2001.

20. Z. Bihary, R. Zadoyan, M. Karavitis, and V. A. Apkarian. Dynamics and the breaking of a driven cage: I_2 in solid Ar. *J. Chem. Phys.*, 120:7576–7589, 2004.

21. V. E. Bondybey. Photophysics of low-lying electronic states of Cl_2 in rare-gas solids. *J. Chem. Phys.*, 64:3615–3620, 1976.

22. V. E. Bondybey, S. S. Bearder, and C. Fletcher. Br_2 $B^3\Pi(0_u^+)$ excitation spectra and radiative lifetimes in rare gas solids. *J. Chem. Phys.*, 64:5243–5246, 1976.

23. A. Borrmann and C. C. Martens. Nanoscale shock wave generation by photodissociation of impurities in solids: a molecular dynamics study. *J. Chem. Phys.*, 102:1905–1916, 1995.

24. C. J. Brabec, G. Zerza, G. Cerullo, S. De Silvestri, S. Luzzati, J. C. Hummelen, and S. Sariciftci. Tracing photoinduced electron transfer process in conjugated polymer/fullerene bulk heterojunctions in real time. *Chem. Phys. Lett.*, 340:232–236, 2001.

25. J. Cao and K. R. Wilson. A simple physical picture for quantum control of wavepacket localization. *J. Chem. Phys.*, 107:1441–1450, 1997.

26. P. Casavecchia, G. He, R. K. Sparks, and Y. T. Lee. Interaction potentials for $Br(^2P)+Ar$, kr, and $Xe(^1S)$ by the crossed molecular beams method. *J. Chem. Phys.*, 75:710–721, 1981.

27. P. Casavecchia, G. He, R. K. Sparks, and Y. T. Lee. Rare gas-halogen atom interaction potentials from crossed molecular beams experiments: $I(^2p_{3/2})+$kr, $Xe(^1S_0)$. *J. Chem. Phys.*, 77:1878–1885, 1982.

28. G. Cerullo, G. Lanzani, M. Muccini, C. Taliani, and S. De Silvestri. Real-time vibronic coupling dynamics in a prototypical conjugated oligomer. *Phys. Rev. Lett.*, 83:231–234, 1999.

29. G. Cerullo and S. De Silvestri. Ultrafast optical parametric amplifiers. *Rev. Sci. Instr.*, 74:1–18, 2003.

30. M. Chachisvilis, I. Garcia-Ochoa, A. Douhal, and A. H. Zewail. Femtochemistry in nanocavities: dissociation, recombination and vibrational cooling of iodine in cyclodextrin. *Chem. Phys. Lett.*, 293:153–159, 1998.

31. T. K. Cheng, S. D. Brorson, A. S. Kazretoonian, J. S. Moodera, G. Dresselhaus, M. S. Dresselhaus, and E. P. Ippen. Impulsive excitation of coherent phonons observed in reflection in bismuth and antimony. *Appl. Phys. Lett.*, 57:1004–1006, 1990.

32. T. K. Cheng, J. Vidal, H. J. Zeiger, G. Dresselhaus, M. S. Dresselhaus, and E. P. Ippen. Mechanism for displacive excitation of coherent phonons in Sb, Bi, Te, and Ti_2O_3. *Appl. Phys. Lett.*, 59:1923–1925, 1991.

33. M. Chergui. Trends in femtosecond lasers and spectroscopy. In: *Structural Dynamics in Quantum Solids*, Academie des sciences, Paris, 2001, pp. 1453–1467.

34. M. Comstock, V. V. Lozovoy, and M. Dantus. Femtosecond photon echo measurements of electronic coherence relaxation between the $X^1\Sigma_{g+}$ and $B^3\Pi_{0u+}$ states of I_2 in the presence of He, Ar, O_2, C_3H_8. *J. Chem. Phys.*, 119:6546–6553, 2003.

35. T. Dekorsy, G. C. Cho, and H. Kurz. Coherent phonons in condensed media. In *Light Scattering in Solids VIII*, Springer, Berlin, 2000.

36. V. A. Ermoshin, V. Engel, and A. K. Kazanky. Phase and energy relaxation of vibrational motion and its manifestation in femtosecond pump-probe experiments on I_2 in rare gas enironment. *J. Phys. Chem. A*, 105:7501–7507, 2001.

37. V. A. Ermoshin, A. K. Kazansky, and V. Engel. Quantum-classical molecular dynamics simulation of femtosecond spectroscopy on I_2 in inert gases: mechanisms for the decay of pump-probe signals. *J. Chem. Phys.*, 111:7807–7817, 1999.

38. V. A. Ermoshin, V. Engel, and C. Meier. Collision-induced bound state motion in I_2. A classical molecular dynamics study. *J. Chem. Phys.*, 113:6585–6591, 2000.

39. J. Franck and E. Rabinowitch. Some remarks about free radicals and the photochemistry of solutions. *Trans. Faraday Soc.*, 30:120–131, 1934.

40. I. Ya. Fugol. Excitons in rare-gas crystals. *Adv. Phys.*, 27:1–87, 1978.

41. Y. Fujii, N. A. Lurie, R. Pynn, and G. Shirane. Inelastic neutron scattering from solid ^{36}Ar. *Phys. Rev. B*, 10:3647–3659, 1974.

42. M. Fushitani, M. Bargheer, M. Gühr, and N. Schwentner. Pump-probe spectroscopy with phase locked pulses in the condensed phase: decoherence and control of vibrational wavepackets. *Phys. Chem. Chem. Phys.*, 7:760–767, 2005.

43. M. Fushitani, N. Schwentner, M. Schroeder, and O. Kühn. Cage motions induced by electronic and vibrational excitations: Cl_2 in Ar. *J. Chem. Phys.*, in print, 2005.

44. G. A. Garrett, T. F. Albrecht, J. F. Whitaker, and R. Merlin. Coherent THz phonons driven by light pulses and the sb problem: what is the mechanism? *Phys. Rev. Lett.*, 77:3661–3664, 1996.

45. E. Gershgoren, J. Vala, R. Kosloff, and S. Ruhman. Impulsive control of ground surface dynamics of I_3^- in solution. *J. Phys. Chem. A*, 105:5081–5095, 2001.

46. I. H. Gersonde and H. Gabriel. Molecular-dynamics of photodissociation in matrices including nonadiabatic processes. *J. Chem. Phys.*, 98:2094–2106, 1993.

47. C. R. Gonzales, S. Fernandez-Alberti, J. Echave, and M. Chergui. Simulations of the absorption band of the d state of Hg_2 in rare gas matrices. *Chem. Phys. Lett.*, 367:651–656, 2003.

48. B. I. Grimberg, V. V. Lozovoy, M. Dantus, and S. Mukamel. Ultrafast nonlinear spectroscopic techniques in the gas phase and their density matrix representation. *J. Chem. Phys. A*, 106:697–718, 2002.

49. M. Gruebele, G. Roberts, M. Dantus, R. M. Bowman, and A. H. Zewail. Femtosecond temporal spectroscopy and direct inversion to the potential: application to iodine. *Chem. Phys. Lett.*, 166:459–469, 1990.

50. M. Gühr. Schwingungsrelaxation und Prädissoziation von Jodmolekülen in Edelgasmatrizen. Diplomarbeit, Fachbereich Physik der Freien Universität Berlin, 2001.

51. M. Gühr. *Coherent Dynamics of Small Molecules in Rare Gas Crystals*, Cuvillier Verlag, Göttingen, 2005.

52. M. Gühr, M. Bargheer, P. Dietrich, and N. Schwentner. Predissociation and vibrational relaxation in the B state of I_2 in a Kr matrix. *J. Phys. Chem. A*, 106:12002–12011, 2002.

53. M. Gühr, M. Bargheer, and N. Schwentner. Generation of coherent zone boundary phonons by impulsive excitation of molecules. *Phys. Rev. Lett.*, 91:085504, 2003.

54. M. Gühr, H. Ibrahim, and N. Schwentner. Controlling vibrational wavepacket revivals in condensed phase: dispersion and coherence for Br_2:Ar. *Phys. Chem. Chem. Phys.*, 6:5353–5361, 2004.

55. M. Gühr and N. Schwentner. Coherent phonon dynamics: Br_2 in solid Ar. *Phys. Chem. Chem. Phys.*, 6:760–767, 2005.

56. M. Gühr and N. Schwentner. Effective chromophore potential, dissipative trajectories and vibrational energy relaxation: Br_2 in Ar matrix. *J. Chem. Phys.*, 123, 244506, 2005.

57. M. Hase, I. Ishioka, M. Kitajima, S. Hishita, and K. Ushida. Dephasing of coherent THz phonons in bismuth studied by femtosecond pump-probe technique. *Appl. Surf. Sci.*, 197-198:710–714, 2002.

58. M. Hase, M. Kitajima, S. Nakashima, and K. Mizoguchi. Dynamics of coherent anharmonic phonons in bismuth using high density photoexcitation. *Phys. Rev. Lett.*, 88:067401, 2002.

59. J. Helbing and M. Chergui. Solvation of ion-pair states in nonpolar media: I_2 in solid neon, argon and krypton. *J. Chem. Phys.*, 115:6158–6172, 2001.

60. E. J. Heller. The semiclassical way to molecular spectroscopy. *Acc. Chem. Res.*, 14:368, 1981.

61. G. Herzberg. *Molecular Spectra and Molecular Structure I. Spectra of Diatomic Molecules*, Van Nostrand Reinhold, New York, 1950.

62. J. Heufelder, H. Ruppe, S. Rutz, E. Schreiber, and L. Wöste. Fractional revivals of vibrational wavepackets in the NaK $A^1\Sigma_+$ state. *Chem. Phys. Lett.*, 269:1–8, 1997.

63. K. P. Huber and G. Herzberg. *Molecular Spectra and Molecular Structure IV. Constants of Diatomic Molecules*, Van Nostrand Reinhold, New York, 1979.

64. C. Jeannin, M. T. Porella-Oberli, S. Jimenez, F. Vigliotti, B. Lang, and M. Chergui. Femtosecond dynamics of electronic "bubbles" in solid argon: viewing the inertial response and the bath coherences. *Chem. Phys. Lett.*, 316:51–59, 2000.

65. S. Jimenez, M. Chergui, G. Rojas-Lorenzo, and J. Rubayo-Soneira. The medium response to an impulsive redistribution of charge in solid argon: molecular dynamics simulations and normal mode analysis. *J. Chem. Phys.*, 114:5264–5272, 2001.

66. S. Jimenez, A. Paquarello, R. Car, and M. Chergui. Dynamics of structural relaxation upon Rydberg excitation of an impurity in an Ar crystal. *Chem. Phys.*, 233:343–352, 1997.

67. M. Karavitis and V. A. Apkarian. The $I*^2P_{1/2}$-$I*^2P_{1/2}$ contact pair emission in condensed media: a molecular spring-gauge for cavity sizing. *J. Phys. Chem. B*, 106:8466–8470, 2002.

68. M. Karavitis and V. A. Apkarian. vibrational coherence of I_2 in solid Kr. *J. Chem. Phys.*, 120:292–299, 2004.

69. M. Karavitis, D. Segale, Z. Bihary, M. Pettersson, and V. A. Apkarian. Time-resolved CARS measurements of the vibrational decoherence of I_2 isolated in an Ar matrix. *Low Temp. Phys.*, 29:814–821, 2003.

70. M. Karavitis, R. Zadoyan, and V. A. Apkarian. Time resolved coherent anti-Stokes Raman scattering of I_2 isolated in matrix argon: vibrational dynamics on the ground electronic state. *J. Chem. Phys.*, 114:4131–4140, 2001.

71. A. K. Kazansky, V. A. Ermoshin, and V. Engel. Phase-energy approach to collision-induced vibrational relaxation. *J. Chem. Phys.*, 113:8865–8868, 2000.

72. T. Kiljunen, J. Eloranta, J. Ahokas, and H. Kunttu. Magnetic properties of atomic boron in rare gas matrices: an electron paramagnetic resonance study with ab initio and diatomics-in-molecules molecular dynamics analysis. *J. Chem. Phys.*, 114:7144–7156, 2001.

73. T. Kiljunen, J. Eloranta, J. Ahokas, and H. Kunttu. Optical properties of atomic boron in rare gas matrices: an ultraviolet-absorption/laser induced fluorescence study with ab initio diatomics-in-molecules molecular dynamics analysis. *J. Chem. Phys.*, 114:7157–7165, 2001.

74. M. L. Klein and J. A. Venables. *Rare Gas Solids, Vols. 1 and 2*, Academic Press, London, 1976.

75. O. Klein. Zur Berechnung von Potentialkurven für zweiatomige Moleküle mit Hilfe von Spektraltermen. *Z. Physik*, 76:226–235, 1932.

76. D. Kohen, C. C. Marston, and D. J. Tannor. Phase space approach to theories of quantum dissipation. *J. Chem. Phys.*, 107:5236–5253, 1997.

77. D. Kohen and D. J. Tannor. Classical-quantum correspondence in the Redfield equation and its solutions. *J. Chem. Phys.*, 107:5141–5153, 1997.

78. B. Kohler, V. V. Yakovlev, J. Che, J. L. Krause, M. Messina, K. Wilson, N. Schwentner, R. M. Whitnell, and Y. Yan. Quantum control of wavepacket evolution with tailored femtosecond pulses. *Phys. Rev. Lett.*, 74:3360–3363, 1995.

79. J. Krause, R. M. Whitnell, K. R. Wilson, Y. Yan, and S. Mukamel. Optical control of molecular dynamics: molecular cannons, reflectrons, and wave-packet focusers. *J. Chem. Phys.*, 99:6562–6579, 1993.

80. A. I. Krylov and R. B. Gerber. Photodissociation dynamics of HCl in solid Ar: cage exit, nonadiabatic transitions, and recombination. *J. Chem. Phys.*, 106:6574–6587, 1997.

81. A. V. Kutsnetsov and C. J. Stanton. Theory of coherent phonon oscillations in semiconductors. *Phys. Rev. Lett.*, 73:3243–3246, 1994.

82. L. D. Landau and E. M. Lifschitz. *Lehrbuch der theoretischen Physik*, Vol. 1: *Mechanik*. Akademie-Verlag, Berlin, 1979.

83. M. Leibscher, I. Sh. Averbukh, P. Rozmej, and R. Arvieu. Semiclassical catastrophes and cumulative angular squeezing of a kicked quantum rotor. *Phys. Rev. A*, 69:032102, 2004.

84. C. Leichtle, I. Sh. Averbukh, and W. P. Schleich. Multilevel quantum beats: an analytical approach. *Phys. Rev. A*, 54:5299–5312, 1996.

85. Ch. Lienau and A. H. Zewail. Solvation ultrafast dynamics of reactions. 11. Dissociation and caging dynamics in the gas-to-liquid transition region. *J. Phys. Chem.*, 100:1829–1849, 1996.

86. Q. Liu, C. Wan, and A. H. Zewail. Solvation ultrafast dynamics of reactions. 13. Theoretical and experimental studies of wavepacket reaction coherence and its density dependence. *J. Phys. Chem.*, 100:18666–18682, 1996.

87. T. Lohmuller, V. Engel, J. A. Beswick, and C. Meier. Fractional revivals in the rovibrational motion of I_2. *J. Chem. Phys.*, 120:10442–10449, 2004.

88. K. Mizoguchi M. Hase, H. Harima, S. Nakashima, M. Tanbi, K. Sakai, and M. Hangyo. Optical control of coherent optical phonons in bismuth films. *Appl. Phys. Lett.*, 69:2474–2476, 1996.

89. N. Mankoc-Borstnik, L. Fonda, and B. Borstnik. Coherent rotational states and their creation and time evolution in molecular and nuclear systems. *Phys. Rev. A*, 35:4132–4146, 1987.

90. A. Materny, Ch. Lienau, and A. H. Zewail. Solvation ultrafast dynamics of reactions. 12. Probing along the reaction coordinate and dynamics in supercritical argon. *J. Phys. Chem.*, 100:18650–18665, 1996.

91. C. Meier and J. A. Beswick. Femtosecond pump-probe spectroscopy of I_2 in a dense rare gas environment: a mixed quantum/classical study of vibrational decoherence. *J. Chem. Phys.*, 121:4550–4558, 2004.

92. S. Mukamel. *Principles of Nonlinear Optical Spectroscopy*, Oxford University Press, New York, 1995.

93. F. Y. Naumkin and D. J. Wales. Diatomics-in-molecules potentials incorporating ab initio data: application to ionic, Rydberg-excited, and molecule-doped rare gas clusters. *Comp. Phys. Comm.*, 145:141–155, 2002.

94. A. Nitzan, S. Mukamel, and J. Jortner. Energy-gap law for vibrational-relaxation of a molecule in a dense medium. *J. Chem. Phys.*, 63:200–207, 1975.

95. M. Y. Niv, M. Bargheer, and R. B. Gerber. Photodissociation and recombination of F_2 molecule in Ar-54 cluster: nonadiabatic molecular dynamics simulations. *J. Chem. Phys.*, 113:6660–6672, 2000.

96. M. Y. Niv, A. I. Krylov, and R. B. Gerber. Photodissociation, electronic relaxation and recombination of HCl in Ar-n(HCl) clusters — non-adiabatic molecular dynamics simulations. *Farad. Diss.*, 108:243–254, 1997.

97. M. Y. Niv, A. I. Krylov, R. B. Gerber, and U. Buck. Photodissociation of HCl adsorbed on the surface of an Ar-12 cluster: Nonadiabatic molecular dynamics simulations. *J. Chem. Phys.*, 110:11047–11053, 1999.

98. L. Onsager. Electric moments of molecules in liquids. *J. Am. Chem. Soc.*, 58:1486–1493, 1936.

99. M. Ovchinnikov and V. A. Apkarian. Condensed phase spectroscopy from mixed-order semiclassical molecular dynamics: Absorption, emission, and resonant Raman spectra of I_2 isolated in solid Kr. *J. Chem. Phys.*, 105:10312–10331, 1996.

100. M. Ovchinnikov and V. A. Apkarian. Quantum interference inresonant Raman spectra of I_2 in condensed media. *J. Chem. Phys.*, 106:5775–5778, 1997.

101. M. Ovchinnikov and V. A. Apkarian. Mixed-order smiclassical dynamics in coherent state representation: the connection between phonon sidebands and guest–host dynamics. *J. Chem. Phys.*, 108:2277–2284, 1998.

102. D. W. Oxtoby. Hydrodynamic theory for vibrational dephasing in liquids. *J. Chem. Phys.*, 70:2605–2610, 1979.

103. D. W. Oxtoby, D. Levesque, and J. J. Weis. Molecular-dynamics simulation of dephasing in liquid-nitrogen. *J. Chem. Phys.*, 68:5528–5533, 1978.

104. J. Parker and C. R. Stroud. Coherence and decay of Rydberg wavepackets. *Phys. Rev. Lett.*, 56:716–719, 1986.

105. M. D. Poulsen, E. Peronne, H. Stapelfeldt, C. Z. Bisgaard, S. S. Viftrup, E. Hamilton, and T. Seideman. Nonadiabatic alignment of asymmetric top molecules: rotational revivals. *J. Chem. Phys.*, 121:783–791, 2004.

106. E. Rabinowitch and W. C. Wood. Collision mechanism and the primary photochemical process in solutions. *Trans. Faraday Soc.*, 32:1381–1387, 1936.

107. E. Rabinowitch and W. C. Wood. Properties of illuminated iodine solutions. I. Photochemical dissociation of iodine molecules in solution. *Trans. Faraday Soc.*, 32:547–556, 1936.

108. A. Rees. The calculation of potential-energy curves from band-spectroscopic data. *Proc. Phys. Soc. London A*, 59:998–1008, 1947.

109. R. W. Robinett. Quantum wavepacket revivals. *Phys. Lett.*, 392:1–119, 2004.

110. G. Rojas-Lorenzo, J. Rubayo-Soneira, F. Vigliotti, and M. Chergui. Ultrafast structural dynamics in electronically excited solid neon. II. Molecular-dynamics simulations of the electronic bubble formation. *Phys. Rev. B*, 67:115119, 2003.

111. F. Rosca-Pruna and M. J. J. Vrakking. Revival structures in picosecond laser-induced alignment of I_2 molecules. I. Experimental results. *J. Chem. Phys.*, 116:6567–6578, 2002.

112. S. Rutz, S. Greschik, E. Schreiber, and L. Wöste. Femtosecond wavepacket propagation in spin-orbit coupled electronic states of the Na_2 molecule. *Chem. Phys. Lett.*, 257:365–373, 1996.

113. R. Rydberg. Graphische Darstellung einiger bandenspektroskopischer Ergebnisse. *Z. Phys.*, 73:376–385, 1931.

114. N. F. Scherer, R. J. Carlson, A. Matro, M. Du, A. J. Ruggiero, V. Romero-Rochin, J. A. Cina, and G. R. Flemming. Fluorescence-detected wavepacket interferometry: time resolved molecular spectroscopy with sequences of femtosecond phase locked pulses. *J. Chem. Phys.*, 95:1487–1511, 1991.

115. N. F. Scherer, D. M. Jonas, and G. R. Flemming. Femtosecond wavepacket and chemical reaction dynamics in iodine in solution: tunable probe study of motion along the reaction coordinate. *J. Chem. Phys.*, 99:153–168, 1993.

116. N. F. Scherer, A. Matro, L. D. Ziegler, M. Du, R. J. Carlson, J. A. Cina, and G. R. Flemming. Fluorescence-detected wavepacket interferometry. II. Role of rotations and determination of the susceptibility. *J. Chem. Phys.*, 96:4180–4194, 1992.

117. N. F. Scherer, L. D. Ziegler, and G. R. Flemming. Heterodyne-detected time-domain measurement of I_2 predissociation and vibrational dynamics in solution. *J. Chem. Phys.*, 96:5544–5547, 1992.

118. W. P. Schleich. *Quantum Optics in Phase Space*, Wiley VCH, Berlin, 2001.

119. M. Schröder. PhD thesis, Fachbereich Chemie, Freie Universität Berlin, 2004.

120. K. S. Schweizer and D. Chandler. Vibrational dephasing and frequency-shifts of polyatomic-molecules in solution. *J. Chem. Phys.*, 76:2296–2314, 1982.

121. D. Segale, M. Karavitis, E. Fredj, and V. A. Apkarian. Quantum coherent dissipation: a glimpse of the "cat." *J. Chem. Phys.*, 122:111104, 2005.

122. J. Skalyo, Y. Edoh, and G. Shirane. Inelastic neutron scattering from solid krypton at 10 K. *Phys. Rev. B*, 9:1797–1803, 1974.

123. K. Sokolowski-Tinten, C. Blome, J. Blums, A. Cavalleri, C. Dietrich, A. Tarasevitch, I. Uschmann, E. Förster, M. Kammler, M. Horn-von-Hoegen, and D. von der Linde. Femtosecond X-ray measurement of coherent lattice vibrations near the Lindemann stability limit. *Nature*, 422:287–289, 2003.

124. M. Sterling, R. Zadoyan, and V. A. Apkarian. Interrogation and control of condensed phase chemical dynamics with lineary chirped pulses: I_2 in solid Kr. *J. Chem. Phys.*, 104:6497–6506, 1996.

125. D. J. Tannor. *Introduction to Quantum Mechanics, a Time-Dependent Perspective*, University Science Books, Sausalito, CA, 2007.

126. D. J. Tannor, R. Kosloff, and S. A. Rice. Coherent pulse sequence induced control selectivity of reactions: exact quantum mechanical calculations. *J. Chem. Phys.*, 85:5805, 1986.

127. D. J. Tannor and S. A. Rice. Control of selectivity of chemical reaction via control of wavepacket evolution. *J. Chem. Phys.*, 83:5013, 1985.

128. J. C. Tully. Diatomics-in-molecules potential-energy surfaces. 1. First-row triatomic hydrides. *J. Chem. Phys.*, 58:1396–1410, 1973.

129. S. I. Vetchinkin and V. V. Eryomin. The structure of wavepacket fractional revivals in a morse-like anharmonic system. *Chem. Phys. Lett.*, 222:394–398, 1994.

130. S. I. Vetchinkin, A. S. Vetchinkin, and V. V. Eryomin. Gaussian wavepacket dynamics in an anharmonic system. *Chem. Phys. Lett.*, 215:11–16, 1993.

131. F. Vigliotti, L. Bonacina, and M. Chergui. Structural dynamics in quatum solids. II. Real-time probing of the electronic "bubble" formation in solid hydrogens. *J. Chem. Phys.*, 116:4553–4562, 2002.

132. F. Vigliotti, L. Bonacina, and M. Chergui. Ultrafast structural dynamics in electronically excited solid neon. I. Real-time probing of the electronic bubble formation. *Phys. Rev. B*, 67:115118, 2003.
133. F. Vigliotti, L. Bonacina, M. Chergui, G. Rojas-Lorenzo, and J. Rubajo-Soneira. Ultrafast expansion and vibrational coherences of electronic "bubbles" in solid neon. *Chem. Phys. Lett.*, 362:31–38, 2002.
134. M. J. J. Vrakking, D. M. Villeneuve, and A. Stolow. Observation of fractional revivals of a molecular wavepacket. *Phys. Rev. A*, 54:R37–R40, 1996.
135. C. Wan, M. Gupta, J. S. Baskin, Z. H. Kim, and A. H. Zewail. Caging phenomena in reactions: femtosecond observation of coherent, collisional confinement. *J. Chem. Phys.*, 106:4353–4356, 1997.
136. Z. H. Wang, T. Wasserman, F. Gershgoren, and S. Ruhman. Vibrational dephasing of I_3^- in cooled ethanol solutions — where is the inhomogeneity? *J. Mol. Liq.*, 86:229–236, 2000.
137. T. Wilhelm, J. Piel, and E. Riedle. Sub-20-fs pulses tunable across the visible from a blue-pumped single-pass noncollinear parametric converter. *Opt. Lett.*, 22:1494–1496, 1997.
138. N. Yu and D. F. Coker. Ion pair state emission from I_2 in rare gas matrices: effects of solvent induced symmetry breaking. *Mol. Phys.*, 102:1031–1044, 2004.
139. N. Yu, C. J. Margulis, and D. F. Coker. Influence of solvation environment on excited state avoided crossings and photodissociation dynamics. *J. Phys. Chem. B*, 105:6728–6737, 2001.
140. R. Zadoyan, J. Almy, and V. A. Apkarian. Lattice dynamics from the "eyes" of the chromophore, real-time studies of I_2 isolated in rare gas matrices. *Faraday Discuss.*, 108:255–269, 1997.
141. R. Zadoyan, P. Ashjian, C. C. Martens, and V. A. Apkarian. Femtosecond dynamics of coherent photodissociation-recombination of I_2 isolated in matrix Ar. *Chem. Phys. Lett.*, 218:504–514, 1994.
142. R. Zadoyan, D. Kohen, D. A. Lidar, and V. A. Apkarian. The manipulation of massive ro-vibronic superpositions using time-frequency-resolved coherent anti-Stokes Raman scattering (TFRCARS): from quantum control to quantum computing. *Chem. Phys.*, 266:323–351, 2001.
143. R. Zadoyan, N. Schwentner, and V. A. Apkarian. Wavepacket diagnosis with chirped probe pulses. *Chem. Phys.*, 233:353–363, 1998.
144. R. Zadoyan, M. Sterling, and V. A. Apkarian. Dynamical spectroscopy of many body interactions, coherent vibrations and predissociation of $I_2(B)$ in solid Kr. *J. Chem. Soc., Faraday. Trans.*, 92:1821–1829, 1996.
145. R. Zadoyan, M. Sterling, M. Ovchinnikov, and V. A. Apkarian. Predissociation dynamics of I_2 in liquid CCl_4 observed through femtosecond pump-probe measurements: electronic caging through solvent symmetry. *J. Chem. Phys.*, 107:8446–8460, 1997.
146. H. J. Zeiger, J. Vidal, T. K. Cheng, E. P. Ippen, G. Dresselhaus, and M. S. Dresselhaus. Theory for displacive excitation of coherent phonons. *Phys. Rev. B*, 45:768–778, 1992.
147. G. Zerza, C. J. Brabec, G. Cerullo, S. De Silvestri, and N. S. Sariciftci. Ultrafast charge transfer in conjugated polymer–fullerene composites. *Synthetic Metals*, 119:637–638, 2001.

6 Coherent Vibrational Dynamics of Exciton Self-Trapping in Quasi-One-Dimensional Systems

Susan L. Dexheimer

CONTENTS

6.1 INTRODUCTION

The localization of electronic excitations via electron–lattice interactions is an important process in a wide range of condensed matter systems and has a dramatic impact on the optical and transport properties of materials. Examples include polaron formation in semiconductors and molecular crystals, formation of photoinduced defect states such as metastable defects in amorphous silicon, DX

223

centers in III–V semiconductors, color centers in alkali halide crystals, and non-linear electronic excitations in low-dimensional materials [1, 2]. These localized states not only have clear technological importance but also reflect fundamental interactions in the physics of condensed matter systems through the interplay of electron–electron and electron–phonon interactions.

Quasi-one-dimensional materials are ideal systems for studying the localiza-tion process because their reduced dimensionality can lead to strong electron–phonon interactions, and the linear structure of the materials simplifies the dynam-ical configuration space in that the dominant motion is expected to occur along the linear axis. A key example is the self-trapping process for excitons: in the case of an ideal one-dimensional lattice, the transition from the extended free exciton state to the localized self-trapped exciton state (also known as an exciton–polaron or neutral bipolaron) is theoretically predicted to be a barrierless process, giving rise to extremely rapid dynamics for the photoinduced structural rearrangement. The application of femtosecond coherent phonon techniques has allowed the lattice mo-tions associated with the exciton self-trapping process in quasi-one-dimensional systems to be directly time resolved, providing new insight into the physics of localization.

This chapter focuses specifically on transient absorption studies of photoexci-tation dynamics in a class of quasi-one-dimensional materials, the halide-bridged mixed-valence transition metal linear chain (or MX) complexes, which have proven to be excellent model systems for investigating the physics of electron–lattice inter-actions in low-dimensional systems. As discussed in the chapter, the MX materials can be systematically tuned in a way that allows direct control over the relevant electronic and lattice properties by varying their chemical composition. In addi-tion, the vibrational periods for the metal–halide vibrational modes along the chain axis are relatively long, allowing the motions to be resolved in detail with fem-tosecond impulsive excitation techniques. These studies of exciton self-trapping in the MX materials provide an unusually clear observation of excited-state vi-brational wavepacket dynamics associated with a photoinduced structural change. The dynamics of MX materials have also attracted the attention of theorists and have provided fertile ground for interaction of experimental and theoretical work.

The properties of the MX materials are discussed in Section 6.2, focusing first on the structural and electronic properties and their tunability in Section 6.2.1. Optical and vibrational spectroscopy of the PtX materials are reviewed in Section 6.2.2, and the properties of the photoinduced localized excitations are discussed in Section 6.2.3. A series of experiments involving coherent phonon techniques is discussed in Section 6.3. Detailed studies of the PtBr(en) complex, including analysis and discussion of the ground-state resonantly enhanced stimulated impul-sive Raman response and the excited-state vibrational dynamics associated with self-trapping, are presented in Sections 6.3.1 to 6.3.3. Coherent phonon studies of the mechanism of exciton self-trapping that exploit the structural tunability of the MX materials are discussed in Section 6.3.4. Further studies at low tempera-ture that address vibrational dephasing and provide striking evidence of acoustic mode dynamics are reviewed in Section 6.3.5 and are followed by a discussion in

Section 6.3.6 of multiple pulse excitation experiments that use resonantly enhanced stimulated impulsive Raman excitation of the excited state to carry out vibrational spectroscopy on the self-trapped exciton state.

6.2 PROPERTIES OF THE MX MATERIALS

6.2.1 STRUCTURE

A wide range of halide-bridged, mixed-valence transition metal linear chain complexes have been characterized in terms of their atomic structure and their optical and vibrational spectra [3]. The basic structural motif of the MX complexes is a covalently bonded linear chain of alternating transition metal ions (M) and halide ions (X) that defines the quasi-one-dimensional geometry. The transition metal ions are also coordinated by equatorial ligands in a nearly square planar geometry, and counterions may also be present between parallel discrete chains in the crystal structure. Among the most extensively studied MX materials are the platinum ethylenediamine (en) complexes, for example, $\{[Pt(en)_2][Pt(en)_2X_2](ClO_4)_4\}$, with X = Cl, Br, or I; sometimes this complex is abbreviated as PtX(en). These complexes form crystals that consist of extended chains of the $[Pt(en)_2][Pt(en)_2X_2]$ repeat unit. In the $\{[Pt(en)_2][Pt(en)_2X_2]\cdot(ClO_4)_4\}$ complexes, ethylenediamine $(C_2H_8N_2)$ ligands fill the transverse bonding sites on the metal ions, and perchlorate counterions serve to balance the overall charge in the crystal structure, as well as to spatially separate the chains so that they experience minimal interchain interaction. The structure of the $\{[Pt(en)_2][Pt(en)_2Cl_2]\cdot(ClO_4)_4\}$ complex as determined by X-ray diffraction is presented in Figure 6.1, which shows three repeat units of the linear chain structure together with the packing arrangement of the chains and counterions in the crystal structure. Other counterions, such as PF_6^-, can also be used in PtX(en) crystals; this substitution typically results in only minor variation of the chain-axis properties.

In general, a one-dimensional lattice is unstable with respect to a Peierls distortion that reduces the translational symmetry [4]. In the PtX complexes, the Peierls distortion is manifested in the chain structure as alternating metal–halide bond lengths and is accompanied by a charge density wave electronic ground state with mixed-valence Pt ions. The structure of the ground state of the chain complexes can be represented schematically as shown in Figure 6.2a, which shows both the periodic bond length (Peierls) distortion and the periodic charge disproportionation (or mixed-valence character) characteristic of the charge density wave ground state. The structure of the self-trapped exciton state that is formed following optical excitation, as discussed in more detail below, is shown schematically in the highly localized two-site limit in Figure 6.2b. Both the degree of lattice distortion and charge disproportionation in the ground state of the complex and the spatial extent of the relaxed self-trapped exciton state depend on the relative strengths of the electron–electron and electron–phonon interactions. In the MX complexes, the electron–electron interaction is dominated by the on-site coulomb repulsion if two electrons occupy the d_{z^2} chain-axis orbital of a metal ion, and the electron–phonon interaction is largely due to coupling of the electronic density on a metal ion to

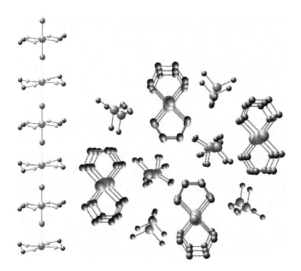

FIGURE 6.1 Structure of the {[Pt(en)$_2$][Pt(en)$_2$Cl$_2$]·(ClO$_4$)$_4$} complex determined by X-ray diffraction (en = ethylenediamine; H atoms are omitted from the structure). Three repeat units of the [Pt(en)$_2$][Pt(en)$_2$Cl$_2$] chain are shown on the left. Each Pt ion is coordinated by two ethylenediamine units in a nearly square planar geometry, and Cl ions bridge the Pt ions along the chain. The crystal packing arrangement of the quasi-one-dimensional chains and the ClO$_4^-$ counterions is shown on the right. For clarity, only a short segment of the extended-chain structure is shown. From Ref. [18], copyright 1999 by the American Physical Society.

the symmetric stretching motion of the negatively charged halide ions. For the PtX complexes, the electron–phonon coupling is the dominant effect, resulting in a charge density wave ground state in which Pt ions with a nominal valence of +3 undergo charge disproportion to give alternating $+(3 + \delta)$ and $+(3 - \delta)$ charge states, with a corresponding periodic distortion of the metal–halide bond lengths. The degree of charge disproportionation is characterized by the parameter

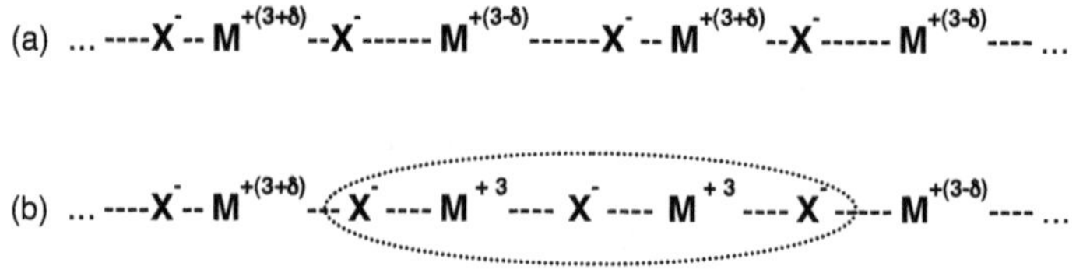

(a) ... ----X⁻-- M$^{+(3+\delta)}$--X⁻------ M$^{+(3-\delta)}$------X⁻-- M$^{+(3+\delta)}$--X⁻------ M$^{+(3-\delta)}$---- ...

(b) ... ----X⁻- M$^{+(3+\delta)}$--X⁻ ---- M^{+3} ---- X⁻ ---- M^{+3} ---- X⁻ --- M$^{+(3-\delta)}$---- ...

FIGURE 6.2 (a) Schematic structure of the charge density wave ground state of the halide-bridged, mixed-valence platinum linear chain complexes, neglecting the transverse ligands and counterions. The chain structure has a periodic variation in Pt-halide bond length and a periodic variation in Pt-ion valence, with the effective Pt ion charge denoted by $3 \pm \delta$. (b) Schematic structure of the self-trapped exciton state resulting from structural relaxation of the chain following excitation of the optical intervalence charge transfer transition, and shown in the highly localized limit.

δ, which varies from $\delta = 0$ for the completely valence-delocalized case with a uniform valence state of Pt^{+3} for all sites to $\delta = 1$ for the limit of maximum charge transfer, giving the completely valence-localized case corresponding to alternating Pt^{+2} and Pt^{+4} sites [5].

The PtX complexes demonstrate the compositional tunability of this class of materials: as the bridging halide is varied through the series Cl, Br, and I, the relative strength of the electron–phonon interaction ranges from very strong coupling in the highly distorted, valence-localized PtCl system to weak coupling in the valence-delocalized PtI system. The effect of the relative strength of the electron–phonon coupling can be seen clearly in the ratio ρ of the short Pt–X bond length to the long Pt–X bond length: structures determined by X-ray diffraction give values of $\rho = 0.743, 0.828$, and 0.890 for the $\{[Pt(en)_2][Pt(en)_2Cl_2]\cdot(ClO_4)_4\}$, $\{[Pt(en)_2][Pt(en)_2Br_2]\cdot(ClO_4)_4\}$, and $\{[Pt(en)_2][Pt(en)_2I_2]\cdot(ClO_4)_4\}$ complexes, respectively [3]. Varying the metal through the series Pt, Pd, and Ni also dramatically reduces the lattice distortion as a result of the relative increase in the electron–electron interaction. In fact, in some nickel–halide complexes, the electron–phonon interaction is dominated by the electron–electron interaction, leading to the absence of a Peierls distortion and the formation of a spin-density wave state [6]. The degree of charge disproportionation δ in the PtX(en) complexes has been addressed by modeling of their X-ray photoelectron spectra in comparison with those of the corresponding monovalent monomers [7, 8]. This analysis yields values of $\delta = 0.91$ for PtCl(en), $\delta = 0.64$ for PtBr(en), and $\delta = 0.36$ for PtI(en), reflecting the increase in valence delocalization across the X = Cl, Br, I series.

Although the electronic and structural properties are largely determined by the metal and halide ions, varying the transverse ligands and counterions provides an additional, though more subtle, means of controlling the properties of the complexes, and a variety of MX chain complexes has been synthesized and characterized [3, 9]. Because the PtX complexes give the widest range of the lattice distortion and charge-density-wave behavior relevant to self-trapped exciton formation in these systems, they are the focus of the studies reviewed here. These materials have been studied for some time, although recent work has provided important refinements of synthetic procedures. In particular, earlier preparations of these materials were found to be subject to extrinsic defects resulting from heterogeneity of halide bridging ions [10], as well as polaronic defects resulting from nonstoichiometric densities of metal ion charge states [11]. Recent preparations that avoid these defects have resulted in significantly improved material properties [11, 12].

6.2.2 OPTICAL AND VIBRATIONAL PROPERTIES

The electronic and the optical properties of the PtX complexes are determined largely by the relative strengths of the electron–electron and electron–phonon interactions. The conductivity of the complexes varies from that characteristic of insulators for highly distorted complexes to that of narrow-gap semiconductors in the more valence-delocalized systems. The dominant feature in the optical absorption spectrum of the metal–halide complexes is a broad,

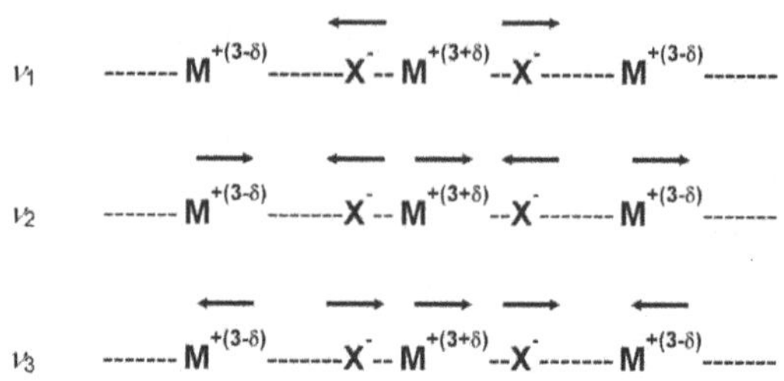

ν_1 $\text{-------}\ M^{+(3-\delta)}\ \text{--------}\ \overset{\cdot}{X}\ \text{--}\ M^{+(3+\delta)}\ \text{--}\overset{\cdot}{X}\ \text{-------}\ M^{+(3-\delta)}\ \text{-------}$

ν_2 $\text{-------}\ M^{+(3-\delta)}\ \text{--------}\ \overset{\cdot}{X}\ \text{--}\ M^{+(3+\delta)}\ \text{--}\overset{\cdot}{X}\ \text{-------}\ M^{+(3-\delta)}\ \text{-------}$

ν_3 $\text{-------}\ M^{+(3-\delta)}\ \text{--------}\ \overset{\cdot}{X}\ \text{--}\ M^{+(3+\delta)}\ \text{--}\overset{\cdot}{X}\ \text{-------}\ M^{+(3-\delta)}\ \text{-------}$

FIGURE 6.3 Optically active chain-axis vibrational modes for the metal–halide linear chain complex. From Ref. [38], copyright 2000 by the American Chemical Society.

intense optical intervalence charge transfer transition in which electron density is transferred between inequivalent metal sites. As expected from the structure of the complexes, the optical intervalence charge transfer transition is strongly polarized along the chain axis. The spectral shape of this absorption band has been modeled in terms of a one-dimensional charge transfer band with strong excitonic character [13]. The energy of the onset of the intervalence charge transfer transition correlates directly with the degree of the Peierls distortion, with higher transition energies for the more distorted complexes: ~ 2.5 eV for $\{[Pt(en)_2][Pt(en)_2Cl_2] \cdot (ClO_4)_4\}$, ~ 1.5 eV for $\{[Pt(en)_2][Pt(en)_2Br_2] \cdot (ClO_4)_4\}$, and ~ 1.2 eV for $\{[Pt(en)_2][Pt(en)_2I_2] \cdot (ClO_4)_4\}$ [8].

The vibrational properties of the metal–halide chain complexes have been extensively characterized by Raman and far-infrared spectroscopies [9, 12, 14–18]. The vibrational spectra of the complexes have been interpreted in detail and reveal strong vibrational modes along the chain axis. The unit cell contains four atoms along the chain axis (two metal and two halide ions), giving rise to four vibrational modes corresponding to motions along the chain axis: one Raman-active mode (ν_1) and two infrared (IR)-active modes (ν_2 and ν_3), which are shown schematically in Figure 6.3, as well as an acoustic mode. Typical frequencies for the ν_1 mode range from 315 cm^{-1} (or a vibrational period of ~ 100 fs) for PtCl(en) complexes to 125 cm^{-1} (or a vibrational period of ~ 270 fs) for PtI complexes, with the variation in frequency resulting both from the change in mass of the halide bridging ligand, and from the extent of bonding interactions along the chain axis [5]. The higher-frequency ν_2 (asymmetric stretch) IR-active mode ranges from ~ 350 cm^{-1} for PtCl complexes to ~ 150 cm^{-1} for PtI complexes. The low-frequency ν_3 IR-active mode may be expected to be substantially weaker, with expected frequencies between 80 and 150 cm^{-1}.

Since the equilibrium metal–halide bond lengths depend on the charge distribution about the metal ions, the coupling of the Raman-active ν_1 (symmetric stretch) mode to the optical intervalence charge transfer transition is very strong. Resonance Raman spectra of the MX complexes show an intense response from the symmetric stretch mode, including a strong overtone progression [9]. Because of the symmetries of the chain-axis modes, only the symmetric stretch chain-axis mode is coupled to the optical intervalence charge transfer transition. Vibrational modes associated with transverse motions of the chain elements as well as modes associated with the transverse ligands are also present in the MX complexes; however, because the intervalence charge transfer transition is strongly polarized along the chain axis, these modes have substantially weaker coupling to the optical transition.

6.2.3 LOCALIZED EXCITATIONS

In general, the extent of localization versus delocalization of an electronic excitation in a deformable medium depends on the strength of the electron–lattice interaction relative to the electronic bandwidth and the lattice deformation energy [19]. A particularly interesting limiting case occurs in one dimension in that there is no adiabatic barrier to self-trapping [19, 20]. In this case, an initially delocalized electronic excitation is expected to undergo extremely rapid localization dynamics in which the lattice deformation that stabilizes the localized state forms on a time scale on the order of a single period of the relevant vibrational motions.

Important aspects of photoexcitations in MX complexes have been addressed in earlier theoretical and experimental work. The dynamics are dominated by the strong electron–phonon coupling, which leads to the formation of localized excitations. The nonlinear excitations in MX complexes have been extensively modeled with numerical calculations based on tight-binding Peierls–Hubbard Hamiltonians [3, 5, 21–26]. These calculations predict the formation of a self-trapped exciton following photoexcitation. Unlike the case for another prototypical one-dimensional system, polyacetylene, the self-trapped exciton state in the MX complexes is predicted to be metastable. Additional lattice relaxation processes, as well as transfer of charge along the metal ions in the chain, can lead to the formation of longer-lived nonlinear excitations, including solitons, polarons, and bipolarons. The localized excitations can have a finite mobility along the chain axis and may recombine either radiatively or nonradiatively to the ground state.

Peierls–Hubbard model calculations have provided significant insight into the nature of the localized excitations in the MX materials; however, it is not, in general, possible to determine unique, independent values for all of the Hamiltonian parameters, and as a result, it is not possible to definitively predict absolute energies and detailed characteristics of the various nonlinear excitations. The modeling has proved especially useful for identifying trends associated with the structural tunability of the complexes. An important example is the predicted variation of the spatial extent of the excitations with the degree of valence delocalization and the strength of the electron–phonon coupling [5]. In the strongly distorted, highly valence-localized PtCl(en) complex, the excitations are predicted to be strongly localized within a few unit cells of the chain structure, with the size of the self-trapped exciton predicted to approach the highly localized limit schematically shown in Figure 6.2b. In contrast, in the more weakly distorted, valence-delocalized PtI(en) complex, the excitations are considerably delocalized, with the self-trapped exciton predicted to extend over more than 20 unit cells. The spatial extent of the localized excitations in PtBr(en) are expected to be intermediate between these two extremes.

Early time-resolved optical measurements on MX materials provided initial indications of rapid dynamics following photoexcitation: picosecond time-resolved photoluminescence measurements showed large Stokes shifts as well as a short luminescence lifetime (< 100 ps) attributed to the self-trapped exciton state [27–29]. Transient absorption measurements showed multiple decay components on picosecond timescales attributed to various nonlinear excitations [30–32]. Evidence for localized electronic excitations associated with defect states was provided by electron paramagnetic resonance studies, in which photoexcitations trapped at low temperature exhibit hyperfine couplings indicative of unpaired spin density extending over a limited number of lattice sites [33]. The femtosecond time-resolved optical measurements in the impulsive excitation regime discussed in the following section have allowed the formation of the self-trapped exciton and the associated lattice dynamics to be resolved, providing a more detailed picture of the localization dynamics.

6.3 COHERENT VIBRATION EXPERIMENTS

Since exciton self-trapping is driven by direct interaction of the electronic excitation with the vibrational degrees of freedom of the lattice, femtosecond vibrationally impulsive excitation techniques are ideal for observing the physics of the self-trapping process. In general, resonant optical excitation with a pulse short compared with the characteristic vibrational periods of a material will result in generation of vibrational wavepackets in both the excited electronic state and the ground electronic state [34–36].

In the MX materials, excitation of the optical intervalence charge transfer band, which transfers electron density between inequivalent metal sites, creates a highly nonequilibrium lattice configuration since the equilibrium metal–halide bond lengths are strongly influenced by the metal ion valence state. We expect that the evolution of the excited-state vibrational wavepacket following excitation of the optical intervalence charge transfer transition will reflect the atomic motions associated with the creation of the localized lattice deformation that stabilizes the self-trapped exciton state. Impulsive excitation of the intervalence charge transfer transition is also expected to generate a ground-state vibrational wavepacket response via the resonantly enhanced stimulated impulsive Raman mechanism, and the strong Raman activity of the symmetric stretch mode of the MX complexes leads to a dramatic ground-state stimulated impulsive Raman response. Impulsive excitation pump-probe measurements that reveal both the excited-state and ground-state vibrational dynamics in the PtX(en) complexes are discussed below, focusing first on detailed studies of the PtBr(en) complex, including analysis of the ground-state and excited-state vibrational coherences, followed by analogous studies on the PtCl(en) and PtI(en) complexes to investigate the dependence of the self-trapping dynamics on relative strength of the electron–phonon coupling. Further discussion then focuses on low-temperature studies of acoustic mode dynamics in exciton self-trapping and impulsive excitation spectroscopy of the equilibrated self-trapped exciton state.

6.3.1 PtBr(en) Impulsive Excitation Pump-Probe Response

The coherent vibrational response of a single-crystal sample of the {[Pt(en)$_2$] [Pt(en)$_2$ Br$_2$]·(PF$_6$)$_4$} complex following impulsive excitation of the optical intervalence charge transfer transition together with a fit to components that include both excited-state and ground-state vibrational dynamics are shown in Figure 6.4 [37, 38]. The excitation pulse, generated by a 1-kHz Ti:sapphire regenerative amplifier, is centered at 800 nm, on the low-energy side of the intervalence charge transfer band. The pulse duration is 35 fs, short compared with the 190-fs period of the 175-cm^{-1} Raman-active symmetric stretch vibrational mode of the ground state of the complex. In the measurement presented in Figure 6.4, the time-resolved differential transmittance is measured in a degenerate pump-probe configuration in which identical pulses are used for the pump and probe, and the probe pulse is

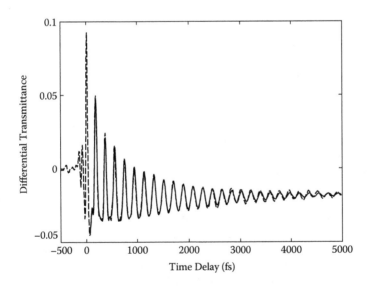

FIGURE 6.4 Measurement of the time-resolved differential transmittance of [Pt(en)$_2$] [Pt(en)$_2$Br$_2$]·(PF$_6$)$_4$ following impulsive excitation near the onset of the intervalence charge transfer band using 35-fs pulses centered at 800 nm. The measurement was taken in the degenerate pump-probe configuration with a detection wavelength of 830 nm. Data points are displayed as solid circles connected by dashed lines. The solid line represents a fit to the data using the LPSVD method discussed in the text, with parameters shown in Table 6.1. From Ref. [38], copyright 2000 by the American Chemical Society.

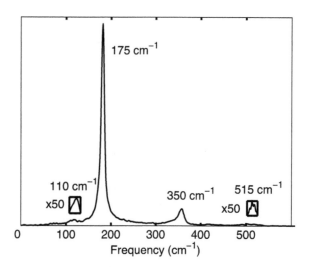

FIGURE 6.5 Fourier power spectrum of the oscillatory part of the time-resolved differential transmittance of $[Pt(en)_2][Pt(en)_2Br_2]\cdot(PF_6)_4$ shown in Figure 6.4, showing frequency components at 110, 175, 350, and 515 cm^{-1}. From Ref. [38], copyright 2000 by the American Chemical Society.

spectrally filtered after it is transmitted through the sample. Because the optical response is strongly polarized, both the pump and probe pulses are polarized along the chain axis. The figure presents the response at a detection wavelength of 830 nm, which lies on the red side of the pulse spectrum and overlaps the red-shifted induced absorption spectrum of the self-trapped exciton state.

A number of components can be identified in the response. In addition to the strong oscillatory modulation due to vibrational wavepackets, there is an underlying overall decrease in transmittance that corresponds to the induced absorbance associated with the formation of the self-trapped exciton state. The frequency content of the coherent vibrational response can be determined by Fourier analysis, and the Fourier power spectrum of the oscillatory part of the response, after subtraction of an exponentially varying baseline, is shown in Figure 6.5. A strong peak at ~ 175 cm^{-1} is the dominant feature in the Fourier power spectrum. This component corresponds to the fundamental frequency of the symmetric stretch chain-axis mode as determined by resonance Raman spectroscopy. Additional Fourier peaks are present at frequencies of 350 and ~ 515 cm^{-1}, which are also observed in the Raman spectrum and correspond to the second and third harmonics of the symmetric stretch mode. In contrast, the peak at ~ 110 cm^{-1} in the Fourier power spectrum of the time-resolved oscillatory response corresponds to a new feature that is not present in the ground-state Raman spectrum. As discussed in detail below, the ~ 110-cm^{-1} component, which is much more prominent at detection wavelengths lying farther into the red-shifted absorption band of the self-trapped exciton, is assigned to the excited-state vibrational response through its spectral and temporal dependence and can be identified as the lattice motion that carries the system toward the self-trapped structure.

The components of the time-dependent response were also characterized using linear prediction/singular value decomposition (LPSVD) analysis. LPSVD analysis is a signal-processing technique that models a time series response as a sum of exponentially damped cosine waves and includes noise filtering [39]. The signal is represented by the sum:

$$S(t) = \sum_i a_i e^{-t/\tau_i} \cos(2\pi \nu_i t + \phi_i) \qquad (6.1)$$

Exponential signal decays appear in the linear prediction model as zero-frequency components. Linear prediction analysis provides an approach complementary to Fourier analysis and directly gives oscillation damping times as well as the frequency, amplitude, and phase. Linear prediction analysis avoids some of the artifacts associated with Fourier transformation, most notably distortion of Fourier power spectra by common phase artifacts as well as by spurious periodic frequency shifts that can result from time windowing of oscillatory time series containing closely spaced frequency components. It is worth noting that the various signal processing techniques for time-frequency analysis have their own particular advantages as well as potential artifacts [40]; in the work presented here, the response was analyzed using a number of independent approaches, including Fourier transformation, LPSVD, and direct nonlinear least-squares fitting. While functional forms other than the simple exponential decays inherent to the linear prediction model may be considered in modeling the vibrational dynamics, we find that this model generally provides an excellent characterization of the time-resolved response.

The result of the linear prediction fit is shown superimposed on the measured time-resolved differential transmittance in Figure 6.4, demonstrating the excellent fit to the response. In this case, the data were fit starting at a delay time of 65 fs (corresponding to approximately two pulse widths) to avoid contributions from coherent electronic interactions and spectral distortions due to cross-phase modulation that may occur during the temporal overlap of the pump and probe pulses. The resulting fit parameters are presented in Table 6.1, and the individual components reconstructed from the linear prediction fit parameters are shown in Figure 6.6. Four oscillatory components are evident at frequencies that correspond, within experimental uncertainty, to the frequencies identified in the Fourier power spectrum. In addition, two zero-frequency components, which correspond to simple exponentials, appear in the fit results. The faster component follows the formation of the induced absorbance signal assigned to the self-trapped exciton state. The slower component, although not well determined in this 5-ps data trace, likely includes contributions both from the decay of the self-trapped exciton population and from ground-state population changes since the 830-nm detection wavelength overlaps both the red-shifted self-trapped exciton absorption and the low-energy edge of the intervalence charge-transfer transition.

234

TABLE 6.1
Linear prediction/singular value decomposition parameters for the fit to the $[Pt(en)_2][Pt(en)_2Br_2]\cdot(PF_6)_4$ differential transmittance data presented in Figure 6.4. From Ref. [38], copyright 2000 by the American Chemical Society

Frequency (cm^{-1})	τ (fs)	ϕ (degrees)	Amplitude
0	370	—	0.014
0	26000	—	0.022
103	240	149	0.0085
176	1300	9	0.018
351	510	17	0.012
510	210	80	0.0043

The excited-state response can be more clearly observed in time-resolved measurements at detection wavelengths extending farther into the red-shifted self-trapped exciton absorption band [38, 41]. Near-infrared probe pulses were generated using broadband continuum generation in a 2-mm sapphire plate

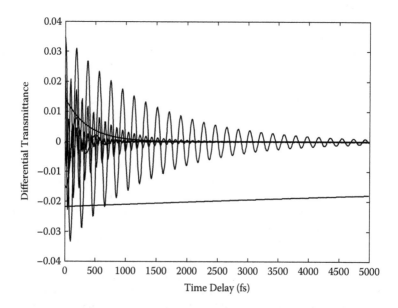

FIGURE 6.6 Individual damped oscillatory and exponential components reconstructed from the LPSVD analysis of the $[Pt(en)_2][Pt(en)_2Br_2]\cdot(PF_6)_4$ data trace displayed in Figure 6.4. The components correspond to the parameters given in Table 6.1. From Ref. [38], copyright 2000 by the American Chemical Society.

followed by compression with a prism pair to improve the time resolution, and wavelength-resolved measurements were carried out by spectrally filtering the probe pulse after transmission through the sample. The time-resolved differential transmittance measured with continuum probe pulses at detection wavelengths of 830 and 880 nm is presented in Figure 6.7. The response at 830 nm shows the effect of the somewhat lower time resolution of the continuum probe measurements relative to the degenerate pump-probe measurements. Although the same signal contributions are present in both measurements, the lower time resolution in the continuum probe measurement results in a reduced oscillation amplitude for the fundamental symmetric stretch frequency and significantly damps the harmonic response. The time-resolved differential transmittance at a detection wavelength of 880 nm clearly shows a large induced absorbance signal associated with the formation of the self-trapped exciton state. Again neglecting signal contributions from pulse overlap effects around zero delay time, the formation of the underlying induced absorbance signal fits to an exponential time constant of ~ 250 fs, and the decay of the induced absorbance, reflecting the decay of the self-trapped exciton population, fits to a time constant of ~ 7 ps. Even though the detection wavelength of 880 nm is well to the red of the peak of the optical intervalence charge transfer transition and well outside the spectral range of the pump pulse, the signal is still modulated by oscillations at the chain-axis symmetric stretching mode at 175 cm^{-1}, although the amplitude of the oscillations is significantly lower than that detected at 830 nm. The signal shows marked structure at short times that results from beating between the 175 cm^{-1} mode and a large-amplitude, strongly damped component at ~ 110 cm^{-1}, the same frequency detected at lower amplitude in the degenerate pump-probe response at a detection wavelength of 830 nm.

The dynamics associated with the formation of the self-trapped exciton are evident in Figure 6.8, which presents continuum probe measurements of the early time response at a series of detection wavelengths throughout the absorption band of the self-trapped exciton, together with the excited state components determined by linear prediction analysis [41], as discussed in detail in Section 6.3.3. The Fourier power spectra of the oscillatory components of these data traces are presented in Figure 6.9, showing a significant contribution from the ~ 110-cm^{-1} component in the early time response. A strong Fourier component is still present from the 175-cm^{-1} symmetric stretch mode, and its second harmonic is also detected at 350 cm^{-1}, although reduced in amplitude as a result of the lower time resolution. The parameters for the excited-state components determined by linear prediction analysis of the data in Figure 6.8a are presented in Table 6.2. The excited-state linear prediction components include an oscillatory component that appears at a frequency, within the experimental uncertainties, of ~ 110 cm^{-1}, consistent with the Fourier power spectra, together with a zero-frequency component that reflects the formation time for the induced absorbance signal.

To interpret the oscillatory components in the pump-probe measurements presented above in terms of detailed vibrational dynamics, the origin of each of the individual components must be identified. Since the electronic dephasing is expected to be rapid owing to the finite temperature (~ 293 K) and the condensed

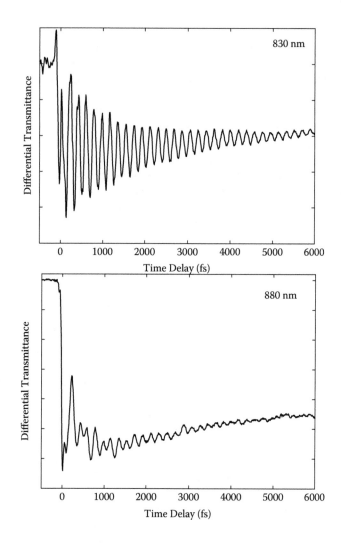

FIGURE 6.7 Time-resolved differential transmittance measurements of $[Pt(en)_2]$ $[Pt(en)_2Br_2]\cdot(PF_6)_4$ following impulsive excitation near the onset of the intervalence charge transfer band using 35-fs pulses centered at 800 nm, probed with a compressed continuum at detection wavelengths of (a) 830 nm and (b) 880 nm. From Ref. [38], copyright 2000 by the American Chemical Society.

phase environment, the observed oscillatory modulations can be characterized in terms of distinct contributions from the ground electronic state and the excited electronic state. In the systems discussed here, it is possible to distinguish between ground-state and excited-state vibrational processes by comparison of the observed frequencies with those in the resonance Raman spectrum and by careful investigation of the spectral dependence of the oscillatory response.

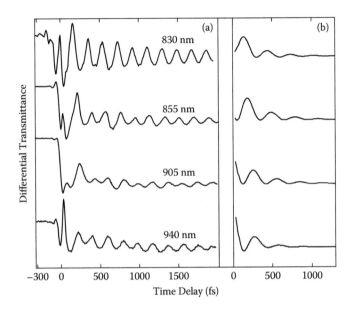

FIGURE 6.8 (a) Time-resolved differential transmittance of $[Pt(en)_2][Pt(en)_2Br_2].(PF_6)_4$ following impulsive excitation with 35-fs pulses centered at 800 nm at a series of probe wavelengths selected from a compressed broadband femtosecond continuum. Structure at delay times during which the pump and probe pulses are overlapped includes contributions from nonlinear processes not associated with the self-trapping dynamics. (b) The sum of the excited-state components, which include the exponential formation and 110-cm^{-1} oscillation, extracted by LPSVD analysis of each trace in part (a), showing the systematic phase shift with detection wavelength. From Ref. [38], copyright 2000 by the American Chemical Society.

6.3.2 PtBr(en): Ground-State Coherent Vibrational Dynamics

The strong oscillatory component at 175 cm^{-1} is clearly consistent with a vibrational wavepacket in the ground electronic state; its frequency matches the chain-axis symmetric stretching mode identified by resonance Raman spectroscopy, and the observed damping time is consistent with the measured Raman linewidth. Since the CW Raman response of the symmetric stretch chain-axis mode has a strong resonant enhancement within the intervalence charge transfer absorption band [9], significant excitation of this mode by the resonantly enhanced impulsive stimulated Raman mechanism is expected on optical excitation with a finite-duration pump pulse. As discussed in detail in Ref. [42], ground-state wavepacket oscillations can be generated via impulsive Raman excitation for a wide range of excitation pulse durations. A simple model calculation [42] predicts oscillatory modulation of significant amplitude for excitation pulse durations ranging from a few percent of the vibrational period to about 100% of the pulse duration, with maximal ground-state wavepacket oscillation amplitude for an excitation pulse duration $\sim$ 30% of

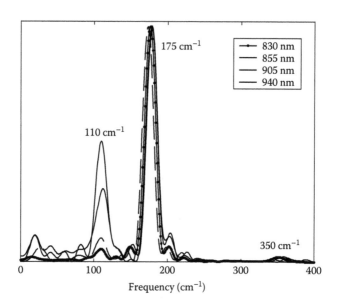

FIGURE 6.9 Fourier power spectra of the oscillatory part (i.e., after subtraction of exponentially varying components) of the PtBr(en) differential transmittance traces in Figure 6.8a. Fourier spectra are normalized to the amplitude of the peak at 175 cm^{-1}. From Ref. [41], copyright 2000 by the American Chemical Society.

TABLE 6.2
Linear prediction/singular value decomposition parameters for the excited-state components extracted from the wavelength-resolved [Pt(en)$_2$][Pt(en)$_2$Br$_2$]·(PF$_6$)$_4$ differential transmittance measurements in Figure 6.8a. Zero-frequency components correspond to exponential delays. Adapted from Ref. [41], copyright 2000 by the American Chemical Society

Detection Wavelength (nm)	Frequency (cm^{-1})	τ (fs)	ϕ (degrees)	Amplitude
830	0	190	—	0.022
	115	280	155	0.016
855	0	290	—	0.016
	107	290	112	0.011
905	0	290	—	0.028
	114	290	21	0.010
940	0	190	—	0.032
	109	180	14	0.023

the vibrational period. For the PtBr(en) measurements presented in Section 6.3.1, the excitation pulse duration of 35 fs is $\sim 20\%$ of the period of the 175-cm^{-1} symmetric stretch chain-axis mode, and a large amplitude resonantly enhanced impulsive stimulated Raman response is expected.

A striking feature of the coherent vibrational response in PtBr(en) is the observation of ground-state wavepacket oscillations at detection wavelengths far to the red of the excited optical transition. The detection of ground-state vibrational oscillations in spectral regions outside the ground-state absorption spectrum is expected to occur for resonant impulsive stimulated Raman excitation and can be understood in terms of two established mechanisms. Impulsive excitation of coherent vibrational motion can result in modulation of both the real and the imaginary components of the optical response, and both of these effects can contribute to an observable oscillatory modulation of the temporal response in a wavelength-resolved differential transmittance measurement. As discussed in detail in Ref. [36], coherent vibrations create a time-dependent modulation of the real part of the index of refraction, resulting in a periodic frequency shift of the transmitted probe beam. This periodic frequency shift translates into a periodic modulation of the detected probe energy when the probe pulse is passed through a spectral filter following transmission through the sample in a spectrally resolved pump-probe measurement. Since this effect results from the real part of the index of refraction, it can result in oscillatory modulation at probe wavelengths away from the ground-state absorption spectrum. The oscillatory response at the ground-state vibrational frequencies can also include contributions from modulation of the imaginary part of the optical response, i.e., the optical absorption. In particular, for resonant optical excitation, the impulsive stimulated Raman mechanism generates a ground-state vibrational wavepacket that includes higher-lying vibrational levels that are not populated in thermal equilibrium. Absorption of the probe pulse from these levels gives rise to a red-shifted component to the impulsively excited optical response that lies outside the equilibrium ground-state absorption spectrum, a process that has been modeled in detail [42, 43].

Another striking feature of the oscillatory response in PtBr(en) is the significant harmonic content that results from the strong electron–phonon coupling. In addition to the fundamental chain-axis symmetric stretching mode at 175 cm^{-1}, the second and third harmonic frequencies at 350 and 515 cm^{-1} appear in both the oscillatory response and the resonance Raman spectrum. Strong harmonic content is consistent with a large excited-state displacement of the vibrational mode [44], in this case, the symmetric stretch chain-axis motion. The unusual structure of the degenerate pump-probe signal shown in Figure 6.4 results from excitation of the fundamental and second harmonic nearly in phase: alternate peaks of the second harmonic interfere constructively with the peaks of the fundamental and destructively with the troughs of the fundamental. This interference leads to an effective sharpening of the wavepacket response. As expected [45], a faster dephasing rate is observed for progressively higher harmonic components, so that detection of higher harmonics in these measurements is limited by their rapid dephasing times as well as their shorter periods.

6.3.3 PTBR(EN): EXCITON SELF-TRAPPING DYNAMICS

The dynamics associated with the formation of the self-trapped exciton occur in the excited electronic state following photoexcitation of the intervalence charge transfer transition. The initially excited, delocalized free exciton state is expected to undergo a rapid transition to the localized self-trapped exciton state, driven by the strong electron–lattice interaction. The formation of the self-trapped exciton is evident from the appearance of its characteristic optical absorption, which corresponds to a transition from the self-trapped exciton potential surface to a higher-lying electronic state and is red shifted relative to the ground-state intervalence charge transfer absorption. The formation of this optical absorption appears as an approximately 300-fs exponential component in the PtBr(en) differential transmittance response. The lattice motion associated with the self-trapping is evident in the large-amplitude, strongly damped, 110-cm^{-1} oscillation that accompanies the formation of the induced absorbance. As noted, the 110-cm^{-1} frequency does not appear in the ground-state Raman spectrum; it is observed only in femtosecond time-resolved measurements. The excited-state response is shown in Figure 6.8b, which presents the sum of the linear prediction components associated with the excited state (i.e., the strongly damped 110-cm^{-1} oscillatory component and the zero-frequency exponential formation of the self-trapped exciton-induced absorbance), as reconstructed from the linear prediction fit parameters in Table 6.2.

The evolution of the wavepacket in the excited state is evident in the dependence of the phase of the 110-cm^{-1} oscillation on detection wavelength, as seen in the measurements in Figure 6.8 taken at probe wavelengths throughout the absorption band of the self-trapped exciton. The shortest detection wavelength, 830 nm, falls within the spectrum of the pump pulse and therefore includes contributions from the wavepacket as it is initially prepared in the Franck–Condon region. The initial phase of the oscillation is close to 180°, consistent with the observation of an induced absorbance (i.e., negative differential transmittance) signal originating from the excited-state potential surface and terminating on a higher-lying potential surface. As the wavepacket propagates in the excited state, the peak of the induced absorbance, which reflects the position of the wavepacket, appears at progressively longer delay times for successively longer detection wavelengths. At a detection wavelength of 940 nm, the oscillation has shifted nearly 180° out of phase relative to its value in the Franck–Condon region, indicating that the wavepacket is approaching the outer turning point in the excited state.

The behavior of the 110-cm^{-1} oscillation is consistent with a lattice motion that carries the system from the initial nuclear configuration in the excited state, which, following the Franck–Condon principle, corresponds to the Peierls-distorted structure of the ground electronic state, toward the structure that stabilizes the localized exciton. As can be seen by comparing the ground-state structure and the schematic self-trapped exciton structure shown in Figure 6.2, this transition is expected to involve motion of the halide bridging ions from their Peierls-distorted positions toward the midpoint between the neighboring Pt ions. This motion includes a significant component analogous to the symmetric stretch normal mode of the ground state shown in Figure 6.3. However, the observed excited-state oscillation is strongly shifted to lower frequency relative to the 175-cm^{-1} ground-state symmetric stretch mode. The formation of the characteristic induced absorbance of the self-trapped exciton occurs on a timescale of a single vibrational period of this oscillation, consistent with the theoretical prediction of a barrierless transition to the self-trapped state in a one-dimensional system. Interestingly, the 110-cm^{-1} wavepacket oscillation appears as an approximately 100% modulation of this induced absorbance, consistent with the role of this mode as a significant contributor to the transformation.

Two striking features of the excited-state oscillatory response in PtBr(en) are its large frequency shift relative to the ground-state Raman-active vibrational mode and its rapid dephasing rate. In principle, the presence of a frequency shift and vibrational dephasing accompanying self-trapping could be expected from a simple idealized picture involving coupling to a single vibrational mode: the structural transformation involves a range of phonon wavevectors, so that dispersion in the phonon mode would result in a shift in vibrational frequency, and the superposition of a range of phonon frequencies would result in dephasing of the initially generated vibrational coherence. However, in the MX materials, the estimated dispersion of the symmetric stretch optical phonon mode is not sufficient to account for the magnitude of the observed frequency shift; moreover, the change in charge distribution associated with excitation is expected to significantly modify the vibrational properties [16, 46]. As seen in the further experiments discussed in Sections 6.3.5 and 6.3.6, it is clear that additional factors, including coupling to acoustic modes, frequency shifts on equilibration, and thermally induced dephasing processes, contribute to the observed dynamics.

Time-resolved photoluminescence provides a complementary approach for detecting the excited-state dynamics. Tomimoto et al. [47, 48] reported fluorescence upconversion measurements on $\{[Pt(en)_2][Pt(en)_2Br_2]\cdot(ClO_4)_4\}$ that resolve the ~ 300-fs oscillatory modulation characteristic of the PtBr(en) excited state. These authors initially attributed the oscillations to vibrational motion subsequent to self-trapping of the exciton [47]; however, their observed response (aside from the model-dependent shifts introduced in the time axes to set the zero time delay for the luminescence) appears consistent with the 110-cm^{-1} excited-state component of the impulsive excitation transient absorption studies discussed above. More recently [48], these authors modeled the time-resolved photoluminescence response using a model developed for time-dependent luminescence of a strongly coupled localized center with linear electron–phonon coupling [49], although one limitation of this approach is that, because it assumes an initially localized electronic state, the model cannot reflect dynamics that occur during the exciton localization process.

6.3.4 STRUCTURAL TUNABILITY: HALIDE DEPENDENCE OF THE SELF-TRAPPING DYNAMICS

The structural tunability of the MX complexes provides an opportunity for gaining further insight into the physics of the self-trapping process. The characteristics of the charge density wave ground state of the MX complexes vary dramatically with chemical composition, with the identity of the halide bridging ion having the most significant effect. As described in Section 6.2.1, these properties can be seen in the series of PtX(en) complexes, from the highly distorted, valence-localized PtCl(en) complex, the intermediate PtBr(en) complex, and the less-distorted, more valence-delocalized PtI(en) complex. In each of these complexes, impulsive excitation of the intervalence charge transfer band results in coherent vibrational dynamics, including components associated with the exciton self-trapping dynamics.

A representative measurement on a single crystal of $\{[Pt(en)_2][Pt(en)_2Cl_2] \cdot (ClO_4)_4\}$ and the results of a linear prediction fit are presented in Figure 6.10 [50]. The complex is excited near the onset of its intervalence charge transfer band using pulses 25 fs in duration centered at 600 nm generated by spectral filtering and compression of a hollow-fiber continuum produced using an amplified Ti:sapphire femtosecond laser system operating at a repetition rate of 1 kHz. The resulting

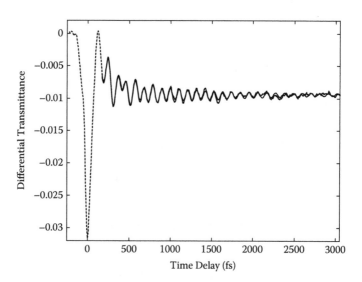

FIGURE 6.10 Time-resolved differential transmittance of $[Pt(en)_2][Pt(en)_2Cl_2] \cdot (ClO_4)_4$ following impulsive excitation of the intervalence charge transfer band using 25-fs pulses centered at 600 nm, probed with 35-fs pulses centered at 800 nm. The feature near t = 0 includes contributions from nonlinear processes that occur during the temporal overlap of the pump and probe pulses. The solid line corresponds to the linear prediction fit discussed in the text. From Ref. [50], copyright 2001, Springer-Verlag.

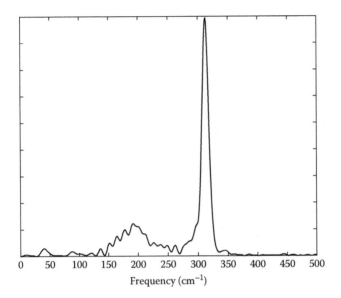

FIGURE 6.11 Fourier power spectrum of the oscillatory part of the time-resolved differential transmittance of $[Pt(en)_2][Pt(en)_2Cl_2] \cdot (ClO_4)_4$ differential transmittance presented in Figure 6.10, showing frequency components at 180 and 314 cm^{-1}. From Ref. [50], copyright 2001, Springer-Verlag.

induced absorbance is probed with a 35-fs pulse centered at 800 nm, within the red-shifted absorption band of the self-trapped exciton in PtCl(en), and wavelength resolution is achieved by spectrally filtering the probe beam after transmission through the sample.

The time-resolved differential transmittance shows the formation of an induced absorbance, corresponding to the characteristic absorption of the self-trapped exciton state, on a timescale of $\sim$ 200 fs. The induced absorbance is strongly modulated by wavepacket oscillations. The Fourier power spectrum of the response, following subtraction of an exponentially varying background, reveals two frequency components as shown in Figure 6.11. A strong Fourier component appears at 314 cm^{-1}, which is the frequency of the symmetric stretch chain-axis mode of PtCl(en) observed by resonance Raman spectroscopy [9], consistent with resonantly enhanced impulsive stimulated Raman excitation of this fully symmetric ground-state mode. A new component, not present in the ground-state Raman spectrum, is detected at a lower frequency of $\sim$ 180 cm^{-1} in the Fourier spectrum.

The presence of these frequency components is confirmed by the linear prediction fit presented along with the measured response in Figure 6.10. The linear prediction fit also reveals an approximately 200-fs exponential formation time for the underlying induced absorbance signal, providing the formation time of the self-trapped exciton state. The oscillatory component at 180 cm^{-1} (185-fs period) is found to be heavily damped, with a time constant comparable to the formation time of the self-trapped exciton-induced absorbance signal, and has an amplitude that gives rise to $\sim 100\%$ modulation of the induced absorbance. In contrast, the 314-cm^{-1} oscillation damps on a picosecond timescale, consistent with its assignment to coherent vibrational excitation in the ground electronic state. As was also seen in the PtBr(en) measurements, the formation of the self-trapped exciton absorbance occurs on the timescale of a single vibrational period of the excited-state oscillation, although this corresponds to a faster timescale in the PtCl(en) complex, and the excited-state vibrational oscillation is significantly shifted in frequency from the symmetric stretch ground-state vibration.

Impulsive excitation transient absorption measurements carried out on the more weakly coupled PtI(en) complex are presented in Figure 6.12. In these experiments [51], the optical intervalence charge transfer transition is pumped using 35-fs pulses at 800 nm, and the resulting induced absorbance is probed at wavelengths extending to 1 μm using a compressed continuum. The components of the response were determined using linear prediction and were confirmed by nonlinear least-squares fitting. In PtI(en), the red-shifted self-trapped exciton absorbance appears on a timescale of ~ 400 fs and is accompanied by a strongly damped wavepacket oscillation at a frequency of 106 cm^{-1}. Oscillatory components are also evident at frequencies of 127 and 256 cm^{-1}, consistent with the ground-state symmetric stretch frequency and its second harmonic as seen in the CW Raman spectrum. The slightly slower timescale for self-trapping is consistent with the lower characteristic vibrational frequency in PtI(en). Two components at frequencies close to the 106- and 127-cm^{-1} frequencies discussed here were observed in transient absorption following impulsive excitation of PtI(en) at 600 nm by Sugita et al. [52]; however,

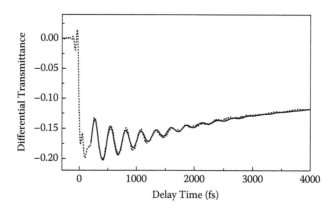

FIGURE 6.12 Time-resolved differential transmittance of $[\text{Pt(en)}_2][\text{Pt(en)}_2\text{I}_2]\cdot(\text{ClO}_4)_4$ following impulsive excitation of the intervalence charge transfer band using 35-fs pulses centered at 800 nm, probed with a compressed broadband continuum at a detection wavelength of 940 nm. The solid line corresponds to a fit to exponential and exponentially damped oscillatory components discussed in the text.

TABLE 6.3
Structural parameters and observed coherent vibrational
frequencies for the series of PtX(en) complexes with bridging
halide group X = Cl, Br, and I

	Cl	Br	I
Lattice distortion, ρ	0.743	0.828	0.890
Charge disproportionation, δ	0.91	0.64	0.36
Excited-state frequency	$180\ cm^{-1}$	$110\ cm^{-1}$	$106\ cm^{-1}$
Ground-state symmetric stretch frequency	$314\ cm^{-1}$	$175\ cm^{-1}$	$127\ cm^{-1}$
Frequency ratio	0.57	0.63	0.83

we assign the higher-frequency 127-cm^{-1} component to the ground state and the lower-frequency 106-cm^{-1} component to the excited state in accordance with the observed CW Raman spectrum and our previous modeling of vibrational dynamics involved in self-trapping in the MX materials [41].

These studies of the series of PtX(en) complexes indicate a consistent physical mechanism for the self-trapping dynamics: in all cases, the self-trapped exciton absorbance appears on the timescale of a single vibrational period, in concert with the dephasing of an excited-state vibrational oscillation. An intriguing result from this work is the systematic variation through the PtX series of the extent of the frequency shift between the excited-state vibrational mode involved in self-trapping and the ground-state symmetric stretch mode, as summarized in Table 6.3. (Note that the motions of the halide ions in the symmetric stretch mode have a close correspondence to the motions needed to form the self-trapped structure, so strong involvement of the symmetric stretch mode is expected in the self-trapping dynamics.) In the strongly distorted, highly valence-localized PtCl(en) complex, the observed excited-state frequency is substantially shifted in frequency to a value that is only 60% of that of the ground-state symmetric stretch mode. This effect becomes progressively smaller through the halide series, and we find that in the PtI(en) complex, which has the least distortion and the highest degree of valence delocalization, the excited-state frequency is nearly 90% of the ground-state value. Interestingly, smaller frequency shifts correlate with a greater degree of valence delocalization and a lesser degree of initial lattice distortion and, correspondingly, with a larger expected spatial extent of the final self-trapped exciton state [5]. In the PtX(en) complexes, the expected spatial extent of the self-trapped exciton increases as the halide bridging group is varied through the series X = Cl, Br, and I, so that the range of phonon wavevectors involved in the structural change as well as the extent of the change in the charge distribution of the excited electronic state is expected to vary accordingly. The trend in the variation of the excited-state frequency relative to the ground-state frequency seen in Table 6.3 reflects the changes associated with the self-trapping transition: in the PtI(en) complex, the charge is already significantly delocalized, and lattice is less distorted in the ground state, so that the changes in charge distribution and halide positions associated with the self-trapped exciton correspond to a relatively smaller difference from the ground-state properties, while the reverse is true for the PtCl(en) complex. These results provide motivation for further studies, discussed below, that address the spatial extent of the photoexcitations and the involvement of acoustic phonon modes as well as frequency shifts associated with equilibration of the self-trapped exciton state.

6.3.5 LOW-TEMPERATURE STUDIES: ACOUSTIC PHONON DYNAMICS AND VIBRATIONAL DEPHASING

Although clear excited-state processes are observed in the room temperature experiments described above, thermally induced effects may be expected to contribute both to the degree of initial lattice disorder and to thermally induced dephasing processes, potentially obscuring additional components of the dynamics. Impulsive excitation studies on PtBr(en) at low temperature have provided additional insight into the nature of the vibrational dephasing in the excited state and have provided striking evidence for the involvement of acoustic phonon modes in the self-trapping process [53].

A comparison of the response of a single-crystal sample of PtBr(en) complex following impulsive excitation of the intervalence charge transfer transition using pulses 35 fs in duration centered at 800 nm at room temperature (293 K) and at 77 K is presented in Figure 6.13a. The room temperature response consists of the components described above, while the low-temperature response reveals an additional large-amplitude, strongly damped, low-frequency oscillatory component. In Figure 6.13b, the low-temperature response is characterized by fitting to exponential formation and decay components together with exponentially damped oscillations, revealing a new component at a frequency of ~ 11 cm^{-1}, within the range characteristic of acoustic phonons.

In principle, both optical and acoustic phonons can contribute to the electron–phonon interactions that drive the self-trapping process [54]. Acoustic phonon modes can contribute to the formation of the lattice deformation that stabilizes the self-trapped state, and acoustic waves that carry away the excess lattice energy associated with the deformation have been predicted to propagate as the exciton wavefunction collapses from the delocalized to the localized state [19, 55, 56], with dominant contributions expected from acoustic phonons of wavelength on the order of the size of the localized state. Using an acoustic velocity estimated from previous modeling of MX materials [9, 17], we find that our observed frequency corresponds to a wavelength on the order of ten unit cells. Remarkably, this value is consistent with the spatial extent of the self-trapped exciton wavefunction in PtBr(en) estimated by Peierls–Hubbard calculations [5], consistent with the observed low-frequency modulation resulting from the acoustic wave associated with the self-trapping transition. Measurements as a function of excitation density further support this assignment.

A second interesting observation in the low-temperature measurements is the change in the dephasing time for the excited-state vibrational wavepacket response

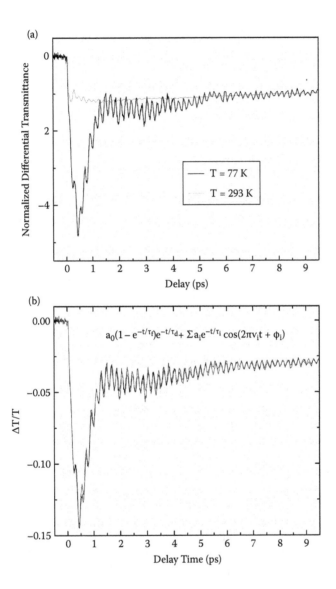

FIGURE 6.13 (a) Normalized time-resolved differential transmittance of $[Pt(en)_2]$ $[Pt(en)_2Br_2]\cdot(ClO_4)_4$ following excitation of the optical intervalence charge transfer transition with a 35-fs pulse centered at 800 nm at room temperature (293 K) and at 77 K. The response is probed within the red-shifted absorption band of the self-trapped exciton state at a detection wavelength of 940 nm using a compressed continuum. (b) Time-resolved differential transmittance of PtBr(en) at 77 K, together with a fit to a sum of exponentially damped oscillations and exponential formation and decay components.

corresonding to the optical phonon motion relative to that observed in the room temperature measurements. The analysis shown in Figure 6.13 also reveals an increased dephasing time for the 110-cm^{-1} excited-state vibrational coherence at low temperature, indicating that the rapid vibrational dephasing dynamics observed in the room temperature measurements are, in part, thermally induced rather than entirely intrinsic to the excited-state structural change.

6.3.6 MULTIPLE-PULSE EXCITATION: VIBRATIONAL SPECTROSCOPY OF THE SELF-TRAPPED EXCITON

In the impulsive excitation measurements described above, the wavepacket motion associated with the initially generated excited state damps during the formation of the self-trapped exciton, so that information on the vibrational properties of the exciton is lost as the self-trapped exciton forms. An important issue that remains for understanding the physics of the localization process is the nature of the internuclear potential that stabilizes the electronic excitation. We have investigated the vibrational properties of the self-trapped exciton in PtBr(en) in its equilibrated structure using a pump-pump-probe sequence to detect vibrational wavepacket oscillations generated by impulsive excitation of the excited-state species [57]. This time domain excited-state resonant impulsive Raman measurement allows us to detect vibrational modes in a time and frequency range that is effectively inaccessible by established picosecond Raman and femtosecond stimulated Raman methods [58].

A representative measurement on $[Pt(en)_2][Pt(en)_2Br_2] \cdot (ClO_4)_4$ is shown in Figure 6.14a [57]. An initial pump pulse (800 nm, 120 fs, $T = -1$ ps) excites the intervalence charge transfer transition to generate a population of excitons. The duration of the first pump pulse is chosen to be relatively long, ~ 120 fs, to generate only a relatively small-amplitude wavepacket response. A delay of 1 ps between the first and second pump pulses ensures that the population of excitons has relaxed to the equilibrated self-trapped structure and that the small wavepacket response generated by the initial pump pulse has decayed away. The equilibrated self-trapped exciton population is then impulsively excited by a second pump pulse 45 fs in duration. The second pump pulse, which is generated from the compressed output of an optical parametric amplifier, is centered at 1.31 μm, within the red-shifted absorption band of the self-trapped exciton. The resulting wavepacket motion is detected as a function of time delay between the second pump and the probe pulse, which is wavelength resolved following transmission through the sample. The differential transmittance response originating from the excited state is detected using simultaneous modulation of both pump beams.

The response can be characterized by a sum of exponentials and exponentially damped cosine waves using linear prediction analysis and nonlinear least-squares fitting, and the resulting components are presented in Figure 6.14b. In addition to excitation of the Raman-active ground-state symmetric stretch mode at 170 cm^{-1} and its second harmonic at 340 cm^{-1}, we detect a new oscillatory component at 125 cm^{-1}. We assign this new frequency component at 125 cm^{-1} to the *equilibrated*

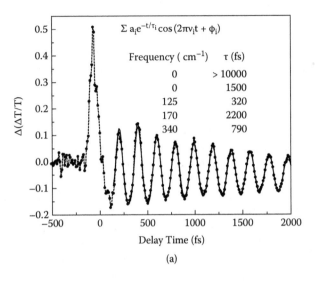

(a)

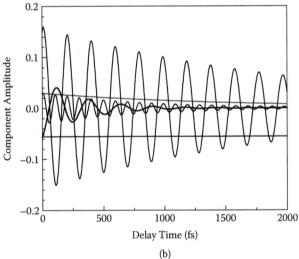

(b)

FIGURE 6.14 (a) Time-resolved double-modulated differential transmittance of [Pt(en)$_2$] [Pt(en)$_2$Br$_2$]·(ClO$_4$)$_4$ following double-pulse excitation to impulsively excite the equilibrated self-trapped exciton state, together with a fit to the expression shown. (b) Individual damped oscillatory and exponential components reconstructed from the fit parameters. From Ref. [57], copyright 2007, Springer-Verlag.

self-trapped exciton state. We note that this component is detected only in the double-modulated measurement; it is not present in pump-probe measurements carried out with only a single pump pulse at either 800 nm or 1.31 μm, and it is not present in the ground-state Raman spectrum. The wavepacket modulation from the ground-state symmetric stretch mode in the double-modulated measurement is consistent with ground-state excitation via stimulated emission pumping as the 1.31-μm wavelength of the second pump lies within the luminescence spectrum of the self-trapped exciton.

The observation of a shift in vibrational frequency associated with equilibration of the self-trapped exciton is an intriguing result that provides further insight into the self-trapping process. In the single pump pulse experiments discussed above, in which the oscillatory response following impulsive excitation of the intervalence charge transfer transition is detected, the formation of the self-trapped exciton in PtBr(en) is accompanied by a strongly damped wavepacket oscillation of frequency 110 cm^{-1} that decays as the characteristic red-shifted absorbance of the self-trapped exciton appears on a timescale of 300 fs, and we identified this oscillation as the lattice motion that carries the system toward the self-trapped structure. In the new work presented here, we find that in the distorted structure of the equilibrated self-trapped state, the vibration shifts to higher frequency, 125 cm^{-1}. Both the 110-cm^{-1} frequency associated with the initially excited state and the 125-cm^{-1} frequency associated with the equilibrated self-trapped exciton structure are significantly lower than the 170-cm^{-1} frequency of the ground-state symmetric stretch vibration. Shifts in the excited-state frequencies are expected to reflect the charge distribution and degree of delocalization of the excited-state electronic wavefunction. In the initially excited state (i.e., the final state of the intervalence charge transfer transition), the electronic wavefunction is expected to be relatively delocalized over the metal–halide chain, while the ions are initially in the positions associated with the Peierls-distorted ground state. As the system evolves, the lattice distorts and equilibrates to form the stabilized self-trapped structure, and the charge distribution associated with the electronic wavefunction becomes more localized, consistent with an upward shift of the characteristic vibrational frequency. This shift in vibrational frequency from the initial state to the equilibrated state would also be expected to contribute to the dephasing of the 110-cm^{-1} oscillation that damps during the localization process.

6.4 CONCLUSIONS

The halide-bridged, mixed-valence, transition metal quasi-one-dimensional MX complexes have proven to be an exceptionally useful system for studying the physics of localized electronic states. A number of studies reviewed here involving coherent vibrational techniques have provided detailed information on the dynamics of formation of localized excitations, including the nature of the phonon dynamics and the interactions that stabilize the self-trapped state.

ACKNOWLEDGMENTS

It is a pleasure to thank my former and current students Aaron Van Pelt and Fran Morrissey for their contributions to the experiments discussed here, and Jim Brozik, Wayne Buschmann, and Basil Swanson for synthesis and characterization of the samples of the MX materials used in the studies. Various aspects of this work were supported by the National Science Foundation under grants DMR-0706407, DMR-0305403, MRI-0079774, and CAREER award DMR-9875765 and by the donors to the Petroleum Research Fund, administered by the American Chemical Society.

REFERENCES

1. K. S. Song and R. T. Williams. *Self-Trapped Excitons*, 2nd ed., *Springer Series in Solid-State Sciences*, Vol. 105, Springer, New York, 1996.
2. D. Redfield and R. H. Bube. *Photoinduced Defects in Semiconductors*, Cambridge University Press, New York, 1996.
3. B. Scott, S. P. Love, G. S. Kanner, S. R. Johnson, M. P. Wilkerson, M. Berkey, B. I. Swanson, A. Saxena, X. Z. Huang, and A. R. Bishop. Control of selected physical properties of MX solids: an experimental and theoretical investigation. *J. Mol. Struct.*, 356:207–229, 1995.
4. R. E. Peierls. *Quantum Theory of Solids*, Oxford University Press, Oxford, 1955.
5. J. T. Gammel, A. Saxena, I. Batistic, A. R. Bishop, and S. R. Phillpot. Two-band model for halogen-bridged mixed-valence transition-metal complexes. I. Ground state and excitation spectrum. *Phys. Rev. B*, 45:6408–6434, 1992.
6. K. Toriumi, Y. Wada, T. Mitani, S. Bandow, M. Yamashita, and Y. Fujii. Synthesis and crystal structure of a novel one-dimensional halogen-bridged nickel(III)-X-nickel(III) compound, {[Ni(R,R-chxn)$_2$Br]Br$_2$}. *J. Am. Chem. Soc.*, 111:2341–2342, 1989.
7. M. Yamashita, N. Matsumoto, and H. Kida. Studies of mixed-valence complexes of platinum and palladium. IV. X-ray photoelectron spectra of some of the platinum complexes. *Inorg. Chim. Acta*, 31:L381–L382, 1978.
8. Y. Wada, T. Mitani, M. Yamashita, and T. Koda. Charge transfer exciton in halogen-bridged mixed-valent Pt and Pd complexes: analysis based on the Peierls-Hubbard model. *J. Phys. Soc. Jpn.*, 54:3143–3153, 1985.
9. R. J. H. Clark. Raman and resonance Raman spectroscopy of linear chain complexes. In: (*Advances in Infrared and Raman Spectroscopy*) R. J. H. Clark and R. E. Hester, Eds., Wiley, 1984, pp. 95–132.
10. S. C. Huckett, R. J. Donohoe, L. A. Worl, A. D. F. Bulou, C. J. Burns, J. R. Laia, D. Carroll, and B. I. Swanson. Mixed-halide MX chain complexes. *Chem. Mater*, 3:123, 1991.
11. J. A. Brozik, B. L. Scott, and B. I. Swanson. Synthetic control of intrinsic ground-state defects in a mixed valence quasi-one-dimensional Pt halide chain. *Inorg. Chem. Acta*, 294:275–280, 1999.
12. W. E. Buschmann, S. D. McGrane, and A. P. Shreve. Chemical tuning of nonlinearity leading to intrinsically localized modes in halide-bridged mixed-valence platinum materials. *J. Phys. Chem. A*, 107:8198–8207, 2003.
13. K. Iwano and K. Nasu. Theory for spectral shape of optical-absorption in halogen-bridged mixed-valent metal-complexes. *J. Phys. Soc. Jpn.*, 61:1380–1389, 1992.

14. L. Degiorgi, P. Wachter, M. Haruki, and S. Kurita. Far-Infrared optical investigations on quasi-one-dimensional halogen-bridged mixed-valence compounds. *Phys. Rev. B*, 40:3285–3293, 1989.

15. L. Degiorgi, P. Wachter, M. Haruki, and S. Kurita. Phonons in one-dimensional Peierls-Hubbard systems. *Phys. Rev. B*, 42:4341–4350, 1990.

16. S. P. Love, L. A. Worl, R. J. Donohoe, S. C. Huckett, and B. I. Swanson. Origin of the fine-structure in the vibrational-spectrum of $[Pt(C_2H_8N_2)_2]$ $[Pt(C_2H_8N_2)_2Cl_2](ClO_4)_4$—vibrational localization in a quasi-one-dimensional system. *Phys. Rev. B*, 46:813–816, 1992.

17. S. P. Love, S. C. Huckett, L. A. Worl, T. M. Frankcom, S. A. Ekberg, and B. I. Swanson. Far-infrared spectroscopy of halogen-bridged mixed-valence platinum-chain solids: isotope-substitution studies. *Phys. Rev. B*, 47:11107–11123, 1993.

18. B. I. Swanson, J. A. Brozik, S. P. Love, G. F. Strouse, A. P. Shreve, A. R. Bishop, W.-Z. Wang, and M. I. Salkola. Observation of intrinsically localized modes in a discrete low-dimensional material. *Phys. Rev. Lett.*, 82:3288–3291, 1999.

19. Y. Toyozawa. Electrons, holes, and excitons in deformable lattice. In: Relaxation of Elementary Excitations, (R. Kubo and E. Hanamura, Eds.), Springer-Verlag, New York, pp. 3–18, 1980.

20. D. Emin and T. Holstein. Adiabatic theory of an electron in a deformable continuum. *Phys. Rev. Lett.*, 36:323–326, 1976.

21. D. Baeriswyl and A. R. Bishop. Localised polaronic states in mixed-valence linear chain complexes. *J. Phys. C: Solid State Phys.*, 21:339–356, 1988.

22. A. Mishima and K. Nasu. Nonlinear lattice relaxation of photogenerated charge-transfer excitation in halogen-bridged mixed-valence metal complexes. I. Soliton and self-trapped exciton. *Phys. Rev. B*, 39:5758–5762, 1989.

23. A. Mishima and K. Nasu. Nonlinear lattice relaxation of photogenerated charge-transfer excitation in halogen-bridged mixed-valence metal complexes. II. Polaron channel. *Phys. Rev. B*, 39:5763–5766, 1989.

24. S. M. Weber-Milbrodt, J. T. Gammel, A. R. Bishop, and J. E. Y. Loh. Two-band model for halogen-bridged mixed-valence transition-metal complexes. II. Electron-electron correlations and quantum phonons. *Phys. Rev. B*, 45:6435, 1992.

25. M. Suzuki and K. Nasu. Nonlinear lattice-relaxation process of excitons in quasi-one-dimensional halogen-bridged mixed-valence metal complexes: self-trapping, solitons, and polarons. *Phys. Rev. B*, 45:1605–1610, 1992.

26. K. Iwano. Relaxation processes yielding nonlinear excitations in the extended Peierls-Hubbard model for photoexcited MX chains. *J. Phys. Soc. Jpn.*, 66:1088–1096, 1997.

27. H. Tanino, W. W. Ruhle, and K. Takahashi. Time-resolved photoluminescence study of excitonic relaxation in one-dimensional systems. *Phys. Rev. B*, 38:12716–12718, 1988.

28. Y. Wada, K. Era, and M. Yamashita. Luminescence lifetimes in halogen bridged mixed valence metal complexes. *Solid State Commun.*, 67:953–956, 1988.

29. Y. Wada, U. Lemmer, E. O. Gobel, and M. Yamashita. Time-resolved luminescence Study of self-trapped exciton relaxation in ordered and disordered one-dimensional MX-chain systems. *J. Luminescence*, 58:146–148, 1994.

30. H. Ooi, M. Yamashita, and T. Kobayashi. Ultrafast optical response in a quasi-one-dimensional Halogen-bridged mixed-valence complex $[Pt(en)_2]$ $[PtBr_2(en)_2](ClO_4)_4$. *Solid State Commun.*, 86:789–793, 1993.

31. H. Ooi, M. Yoshizawa, M. Yamashita, and T. Kobayashi. Ultrafast optical response in a quasi-1-D halogen-bridged mixed-valence complex [Pd(en)$_2$] [PdCl$_2$(en)$_2$](ClO$_4$)$_4$. *Chem. Phys. Lett.*, 210:384–388, 1993.

32. G. S. Kanner, G. F. Strouse, B. I. Swanson, M. Sinclair, J. P. Jiang, and N. Peyghambarian, Subpicosecond dynamics of excitons and photoexcited intrinsic polarons in the quasi-one-dimensional solid PtCl. *Phys. Rev. B*, 56:2501–2509, 1997.

33. R. J. Donohoe, L. A. Worl, C. A. Arrington, A. Bulou, and B. I. Swanson, Photoinduced defects in the bromide-bridged platinum linear chain [Pt(en)$_2$] [Pt(en)$_2$Br$_2$](ClO$_4$)$_4$ (with en=ethylenediamine). *Phys. Rev. B*, 45:13185–13194, 1992.

34. W. T. Pollard, S.-Y. Lee, and R. A. Mathies. Wave Packet theory of dynamic absorption spectra in femtosecond pump-probe experiments. *J. Chem. Phys.*, 92:4012–4028, 1990.

35. V. Romero-Rochin and J. A. Cina. Aspects of impulsive stimulated scattering in molecular systems. *Phys. Rev. A*, 50:763–778, 1994.

36. L. Dhar, J. A. Rogers, and K. A. Nelson. Time-resolved vibrational spectroscopy in the impulsive limit. *Chem. Rev.*, 94:157–193, 1994.

37. S. L. Dexheimer, A. D. Van Pelt, A. Gross, J. A. Brozik, and B. I. Swanson. Vibrational dynamics of self-trapping in a quasi-one-dimensional system. *Optics and Photonics News Special Issue*, Optical Society of America Annual Meeting, 9(8):PD23-1, 1998.

38. S. L. Dexheimer, A. D. Van Pelt, J. A. Brozik, and B. I. Swanson. Ultrafast vibrational dynamics in a quasi-one-dimensional system: femtosecond impulsive excitation of the PtBr(ethylenediamine) mixed-valence linear chain complex. *J. Phys. Chem. A*, 104:4308–4313, 2000.

39. J. Makhoul. Linear prediction: a tutorial review. *Proc. IEEE*, 63:561–580, 1975.

40. S. M. Kay and J. Marple. Spectrum analysis—a modern perspective. *Proc. IEEE*, 69:1380–1419, 1981.

41. S. L. Dexheimer, A. D. Van Pelt, J. A. Brozik, and B. I. Swanson. Femtosecond vibrational dynamics of self-trapping in a quasi-one-dimensional system. *Phys. Rev. Lett.*, 84:4425, 2000.

42. W. T. Pollard, S. L. Dexheimer, Q. Wang, L. A. Peteanu, C. V. Shank, and R. A. Mathies. Theory of dynamic absorption spectroscopy of nonstationary states. 4. Application to 12-fs resonant impulsive raman spectroscopy of bacteriorhodopsin. *J. Phys. Chem.*, 96:6147–6158, 1992.

43. S. L. Dexheimer, Q. Wang, L. A. Peteanu, W. T. Pollard, R. A. Mathies, and C. V. Shank. Femtosecond impulsive excitation of nonstationary vibrational states in bacteriorhodopsin. *Chem. Phys. Lett.*, 188:61–66, 1992.

44. A. B. Myers and R. A. Mathies. Resonance Raman Intensities: A probe of excited-state structure and dynamics. In *Biological Applications of Raman Spectroscopy*, Vol. 2, T. G. Spiro (Ed.), Wiley—Interscience, New York, 1987. pp. 1–58.

45. T. Yamaguchi. Vibrational overtone dephasing in liquids under the influence of non-Gaussian noise. *J. Chem. Phys.*, 112:8530–8533, 2000.

46. A. Bulou, R. J. Donohoe, and B. I. Swanson. Phenomenological description of the longitudinal vibrations of the quasi-one-dimensional solid PtCl: calculation of the valence defect frequencies. *J. Phys. Cond. Matt.*, 3:1709–1726, 1991.

47. S. Tomimoto, S. Saito, T. Suemoto, K. Sakata, J. Takeda, and S. Kurita. Observation of the wave-packet oscillation during the exciton self-traping process in a quasi-one-dimensional halogen-bridged Pt complex. *Phys. Rev. B*, 60:7961–7965, 1999.

48. S. Tomimoto, S. Saito, T. Suemoto, J. Takeda, and S. Kurita. Ultrafast dynamics of lattice relaxation of excitons in quasi-one-dimensional halogen-bridged platinum complexes. *Phys. Rev. B*, 66:115112, 2002.
49. Y. Kayanuma. Resonant secondary radiation in strongly coupled localized electron–phonon system. *J. Phys. Soc. Jpn.*, 57:292–301, 1988.
50. A. D. Van Pelt and S. L. Dexheimer. *Ultrafast Dynamics of Excitonic Self-Trapping: The Role of the Electron–Phonon Interaction*, Springer Series in Chemical Physics, Vol. 66, *Ultrafast Phenomena XII*, T. Elsaesser et al., Eds., Springer-Verlag, Berlin, 2001, pp. 393–397.
51. F. X. Morrissey and S. L. Dexheimer. Ultrafast lattice dynamics in excitonic self-trapping. OSA Trends in Optics and Photonics Series Vol. 97, International Quantum Electronics Conference, IMJ1-1, 2004.
52. A. Sugita, T. Saito, H. Kano, M. Yamashita, and T. Kobayashi. Wave packet dynamics in a quasi-one-dimensional metal-halogen complex studied by ultrafast time-resolved spectroscopy. *Phys. Rev. Lett.*, 86:2158–2161, 2001.
53. F. X. Morrissey and S. L. Dexheimer. *Studies of the Mechanism for the Formation of the Self-Trapped Exciton in Quasi-One-Dimensional Systems*, OSA Trends in Optics and Photonics Series, Vol. 89, *Quantum Electronics and Laser Science*, QThJ18/1, 2003.
54. E. I. Rashba. Self-trapping of excitons. In: *Excitons*, (E. I. Rashba and M. D. Sturge, Eds.), North-Holland, New York, 1982. pp. 543–602.
55. D. W. Brown, K. Lindenberg, and B. J. West. On the dynamics of polaron formation in a deformable medium. *J. Chem. Phys.*, 84:1574–1582, 1986.
56. D. W. Brown, K. Lindenberg, and B. J. West. Polaron formation in the acoustic chain. *J. Chem. Phys.*, 87:6700–6705, 1987.
57. F. X. Morrissey and S. L. Dexheimer. *Vibrational Spectroscopy of Nonlinear Excitations via Excited-State Resonant Impulsive Raman Spectroscopy*, Springer Series in Chemical Sciences, *Ultrafast Phenomena XV*, P. Corkum et al., Eds., Springer-Verlag, Berlin, 2007. pp. 240–242.
58. D. W. McCamant, P. Kukura, S. Yoon, and R. A. Mathies. Femtosecond broadband stimulated Raman spectroscopy: apparatus and methods. *Rev. Sci. Instrum.*, 75:4971–4980, 2004.

Index